Dielectric Spectroscopy
of Polymeric Materials

Dielectric Spectroscopy of Polymeric Materials

Fundamentals and Applications

James P. Runt
Pennsylvania State University

John J. Fitzgerald
General Electric Company

American Chemical Society, Washington, DC

Library of Congress Cataloging-in-Publication Data

Dielectric spectroscopy of polymeric materials : fundamentals and
    applications / James P. Runt, John J. Fitzgerald, [editors].
        p.   cm.—(Professional reference book)
    Includes bibliographical references and index.
    ISBN 0–8412–3335–7
    1. Polymers—Spectra.   2. Dielectric measurements.   I. Runt,
James P. (James Patrick), 1953–   .   II. Fitzgerald, John J.
III. Series: ACS professional reference book.

QC463.P5D54   1997
537′.24—dc21   97–7622
                                                              CIP

Advisory Board

About the Editors

JAMES P. RUNT has been a member of the faculty of the Department of Materials Science and Engineering (in the Polymer Science and Engineering Program) at The Pennsylvania State University since 1980 and is currently Professor of Polymer Science. His Ph.D. degree is in Solid State Science, also from Penn State. He has taught a number of courses at Penn State at the undergraduate and graduate levels including (presently): multicomponent polymer materials (polymer blends, block copolymers and IPNs); mechanical and electrical properties of polymers and composites, and introduction to polymer science. He has also been involved in instruction at short courses and workshops.

He has published more than 75 refereed papers and several book chapters, in the areas of crystalline polymers, polymer blends and conductor and ceramic–polymer composites. His current research interests include: crystallization, microstructure and phase behavior in semicrystalline polymers and polymer blends, including the application of dielectric spectroscopy to these problems. He has also been interested in various aspects of biomedical polymers related to the development of artificial organs (LVAD and the Penn State Total Artificial Heart). This work is being conducted in collaboration with Penn State's Milton S. Hershey Medical Center. Finally, he has a continuing interest in the electrical properties of conductor–polymer composites, particularly polymeric PTC materials.

He is a member of the American Chemical and Physical Societies, among others. He has recently co-organized two symposia for the ACS: "Dielectric Spectroscopy of Polymers" (1994), co-sponsored by the Divisions of Polymeric Materials Science and Engineering and Polymer Chemistry and "Advances in Crystalline Polymers" (1995), sponsored by the Division of Polymer Chemistry.

JOHN J. FITZGERALD is currently the Leader of the Heat Cured Elastomer Group at the General Electric Company. Previous to this John was a Research Associate at Eastman Kodak for eight years. His Ph.D. is in Materials Science From the University of Connecticut and he has a B.S. in Chemical Engineering from Columbia University.

He has published 45 papers and a book chapter and has been awarded more than 20 patents.

His research interests have been in ion-containing polymers, polymer blends, photoresists, electron transport properties of photoconductors, dielectric properties of polymers, and most recently in the cycle fatigue of silicone elastomers.

He has organized two international symposiums at the national meetings of the American Chemical Society, Division of Polymeric Materials.

Table of Contents

Modeling and Techniques

Application of Dielectric Spectroscopy to Polymer Systems

Contributors

Keiichiro Adachi 261
Faculty of Science
Department of Macromolecular
  Science
Osaka University
Toyonaka, Osaka 560, Japan

M. Arndt 67
Fakultät für Physik
Universität Leipzig
04103 Leipzig, Germany

Peter Avakian 379
Central Research and
  Development
DuPont Company
Wilmington, DE 19880-0014

A. Bernes 227
Polymer Physics Laboratory
Paul Sabatier University
31062 Toulouse Cedex, France

Richard H. Boyd 107
Departments of Materials
  Science & Engineering and
  Chemical & Fuels Engineering
University of Utah
Salt Lake City, UT 84112

Ricardo Diaz-Calleja 139
Escuela Tecnicac Superior de
  Ingenieros Industriales
Universidad Politcnica de
  Valencia
46071 Valencia, Spain

John J. Fontanella 379
Physics Department
U.S. Naval Academy
Annapolis, MD 21404-5026

C. Y. Stacey Fu 395
Purdue University
School of Chemical Engineering
West Lafayette, IN 47907–1283

S. Havriliak, Jr. 175
Rohm and Haas Research
Bristol Research Park
Bristol, PA 19007

S. J. Havriliak 175
Havriliak Software Development
  Company
Huntingdon Valley, PA 19006

David E. Kranbuehl 303
Chemistry and Applied Science
  Department
College of William and Mary
Williamsburg, VA 23187–8795

F. Kremer 67, 423
Institut für Physik
Universität Leipzig
Linnéstrasse 5
04103 Leipzig, Germany

C. Lacabanne 227
Polymer Physics Laboratory
Paul Sabatier University
31062 Toulouse Cedex, France

Hilary S. Lackritz 395
Purdue University
School of Chemical Engineering
West Lafayette, IN 47907–1283

Fufuo Liu 107
Departments of Materials Science & Engineering and Chemical & Fuels Engineering
University of Utah
Salt Lake City, UT 84112

Satoru Mashimo 201
School of Science
Department of Physics
Tokai University
Hiratsuka-shi, Kanagawa 259–12, Japan

S. Mezghani 227
Polymer Physics Laboratory
Paul Sabatier University
31062 Toulouse Cedex, France

Mark H. Ostrowski 395
Purdue University
School of Chemical Engineering
West Lafayette, IN 47907–1283

Evaristo Riande 139
Instituto de Ciencia y Tecnologia de Polimeros (CSIC)
Madrid, Spain

James P. Runt 283
Department of Materials Science and Engineering
The Pennsylvania State University
University Park, PA 16802

A. Schönhals 81
Institut für Angewandte Chemie
Berlin-Adlershof e. V.
Rudower Chaussee 5
D-12484 Berlin, Germany

George P. Simon 329
Materials Engineering
Monash University
Clayton, Victoria 3168, Australia

Howard W. Starkweather, Jr. 379
Central Research and Development
DuPont Company
Wilmington, DE 19880–0014

G. Teyssedre 227
Polymer Physics Laboratory
Paul Sabatier University
31062 Toulouse Cedex, France

Mary C. Wintersgill 379
Physics Department
U.S. Naval Academy
Annapolis, MD 21404–5026

Graham Williams 3
Department of Chemistry
University of Wales
Swansea
Singleton Park, Swansea SA2 8PP
United Kingdom

Preface

Interest in dielectric properties and spectroscopy of polymeric materials flourished in the 1950s and 1960s as our fundamental understanding of polymers was developing. A comprehensive review of this early work is contained in the authoritative "Anelastic and Dielectric Effects in Polymer Solids" by N. G. McCrum, B. E. Read, and G. Williams, published in 1967. The area seems to have gone out of fashion for a time but has reemerged in the past decade. This is at least partly the result of the power that dielectric spectroscopy brings to the understanding of dynamics of complex solid polymer systems such as liquid crystals and blends, of polymer solutions, and of polymerization and curing reactions. Advances in instrumentation, as well as in theory and modeling, have also helped to stimulate renewed interest. Another driving force has been the relatively recent introduction of commercial dielectric spectrometers, which in a number of cases are being purchased as modules for existing microprocessor-based thermal analysis systems. As a result, there are many in the polymer thermal analysis community, and in analytical facilities in general, that are being exposed to and using dielectric "thermal analysis" for the characterization of coatings, cured systems, thermoplastic solids, and so on.

A major hurdle however, particularly for the beginning but also for the more frequent practitioner, is the lack of an up-to-date monograph giving broad coverage to the discipline. Responding to this need, the editors conceived the idea for this text while organizing the symposium "Dielectric Spectroscopy of Polymers," which was held at the 1994 spring meeting of the American Chemical Society in San Diego, California. Because the topic is so broad, we felt the best approach was to enlist experts in the various subject areas as contributors. The downside to this is the possibility that the book may become disjointed: Some of this is inevitable but we have taken care to avoid this where possible.

The book is divided into three sections. The first section presents the basics and is the recommended starting point for all beginners. The second section covers modeling and some specialized techniques. The final section is devoted to the application of dielectric methods to a variety of polymer systems.

The editors express their appreciation to all authors of this volume for the time and energy that they devoted to their contributions. We also thank them for their patience and understanding throughout this rather lengthy project.

James P. Runt
Polymer Science and Engineering Program
Department of Materials Science and Engineering
The Pennsylvania State University
319 Steidle Building
University Park, PA 16802

John J. Fitzgerald
Research & Development
General Electric Company
260 Hudson River Road, MS 12–11
Waterford, NY 12188

Fundamentals of Dielectric Materials

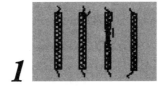

1

Theory of Dielectric Properties

Graham Williams

Phenomenological and molecular theories of dielectric permittivity and dielectric relaxation of polymers are reviewed. Special emphasis is given to amorphous and liquid crystalline (LC) polymer systems. It is shown how the static permittivity and frequency-dependent complex permittivity are related to chemical structure, chain conformation, group dipole moments, and time correlation functions for the reorientational motions of dipoles. Multiple dielectric relaxation processes are shown to arise from partial and total reorientational motions of dipole groups (for amorphous polymers) or from different modes of motion of the components of dipole groups (for LC polymers). Mechanisms for motion in amorphous polymers are described and it is shown that recent experimental evidence suggests that the characteristic broad line shape for the primary dielectric process arises from an intrinsic dynamic heterogeneity of the amorphous solid that produces a broad distribution of relaxation times. The limitations of models based on a generalization of the Flory rotation–isomeric state model to the dynamics of polymer chains are discussed. In addition recent real-time dielectric studies of systems undergoing bulk polymerization are discussed briefly.

Experimental studies of the dielectric properties of materials have been made over wide ranges of frequency and temperature for well over 100 years. Such studies, embracing both conduction and polarization processes and extending from millihertz through gigahertz to terahertz frequencies, received their first systematic interpretations in terms of molecular processes through the work of Debye (*1*)—for which he was awarded the Nobel Prize in Chemistry in 1936. The reorientational motions of dipolar molecules were shown to give rise to dielectric dispersion behavior with accompanying dielectric absorption. The connection between dielectric permittivity (dielectric constant) and molecular dipole moments provided a means of determining molecular structure. Following the pioneering work of Debye, many studies of the dipole moments of molecules, including geometric, positional, and stereoisomers, also were made, especially by Smyth (2) in the United States and by Sutton (3) and Smith (4) in the United Kingdom. Such a traditional method for aiding the determination of molecular structure has long been superseded by modern spectroscopic methods, especially multinuclear NMR spectroscopy, but nevertheless the use of molecular dipole moments retains an important place in the history of the determination of molecular structure.

Dielectric relaxation studies, which date back to the remarkable microwave measurements made by Drude at the turn of the last century, have made slow progress toward providing a well-established method for studying molecular dynamics processes. The frequency range is awesome, from 10^{-6} to 10^{11} Hz, and requires diverse equipment, each band-limited, to obtain comprehensive data for a given material. Although the principles of measurement, which are lumped-circuit elements at low frequencies (up to ~10^7 Hz) and distributed circuits at high frequencies (from 10^7 Hz to 10^{11} Hz), have been well-established for over 50 years, researches into the dielectric relaxation properties of liquids and solids have been largely carried out by using custom-made equipment or adaptations of commercial equipment that were manufactured originally for purposes other than those for dielectric research. These limitations hampered the development of dielectric relaxation spectroscopy and only in the past decade have commercial dielectric spectrometers become available for general use (5).

In practice, dielectric relaxation spectroscopy broadly breaks down into studies below and above ~10^7 Hz. The dielectric dispersion and absorption features for molecular liquids, liquid crystals, and for solutes (e.g., polymers) in solution and for some rotator-phase solids occur in the microwave region (10^8–10^{11} Hz). These features for most rotator-phase solids (e.g., the different polymorphs of ice), for amorphous, crystalline, and liquid-crystalline (LC) polymers, and for glass-forming liquids commonly occur at frequencies less than ~10^7 Hz. The usual microwave measurements are tedious and require manual point-by-point measurements at selected frequencies. The associated practical difficulties meant that few laboratories were established for this work. The laboratory of Smyth at Princeton University during and after the Second World War made extensive studies of simple and associated molecular liquids and rotator-phase solids, and these works together with many other studies are summarized in Smyth's classic book (2) and in the research texts of Hill, Vaughan, Price, and Davies (6) and Böttcher and Bordewijk (7). Such studies yield three essential items of information for each material at a fixed temperature and pressure:

- the *magnitude* of each of the dielectric relaxation processes $\Delta\varepsilon_\alpha$, $\Delta\varepsilon_\beta$, etc.
- their *frequency locations*, and temperature (T) and pressure (P) dependencies, as defined by average relaxation (correlation) times $\tau_\alpha(T,P)$, $\tau_\beta(T,P)$, etc., where $\tau_i = (2\pi f_{mi})^{-1}$ and f_{mi} is the frequency of maximum loss for each process i
- *band-shape* or dielectric absorption contour for each dielectric relaxation process

Given only limited dielectric data in the microwave region for molecular liquids (sometimes only 5–10 frequency points), it has been difficult to ensure that the dielectric processes have been documented adequately. Nevertheless, a considerable body of microwave dielectric data for liquids, liquid crystals, and rotator-phase solids was accumulated and analyzed in terms of molecular-rotation dynamics (*6, 7*). For the low frequency range $<10^7$ Hz a vast body of accurate data has now been accumulated that describes the dielectric dispersion and absorption behavior of amorphous, crystalline and LC polymers, glass-forming liquids, and of flexible and rod-like polymers in solution, including lyotropic phases for rod-like molecules in solution at high concentrations. These data have been reviewed and analyzed in the texts of Smyth (*2*); Hill, Vaughan, Price, and Davies (*6*); and Böttcher and Bordewijk (*7*), whereas that of McCrum, Read, and Williams (*8*) is wholly concerned with amorphous and crystalline solid polymers. In addition, a number of specialized reviews of dielectric properties by Williams focus on amorphous and crystalline polymers (*9–13*), glass-forming liquids (*14*), thermotropic liquid crystals (*15*), and LC polymers (*16, 17*). In all these systems single or multiple dielectric relaxations are observed that give information on partial and total dipole reorientational processes involving chain segments or whole molecules, how these processes vary with temperature and applied pressure, and how they reveal the nature of the phase such as in lyotropic (*18–20*) or thermotropic–nematic or thermotropic–chiral–smectic LC phases (*16, 17*).

Apart from the fundamental studies of the dielectric properties of liquids, solids, and liquid crystals there are many applied dielectric studies, covering the entire frequency range, that are concerned with

- the electric insulation characteristics of polymers such as poly(vinyl chloride), polyethylene, nylon, polyimides, polyesters, and cured polymers and proprietary products that are used in cable insulation or component insulation in electrical machinery or as insulation (e.g., polyimides) in electronic assemblies
- microwave absorption characteristics of insulators and semiconductors that are required in relation to microwave transmission and reflection through or from materials (as in radar applications)

- microwave heating characteristics of materials that are used in microwave cookers, because dielectric absorption is the mechanism whereby food and a variety of organic and inorganic materials are heated in microwave units

The applications of dielectric properties of materials and the means by which they may be measured over the entire frequency range were the major concerns of the Dielectrics Research Laboratory at Massachusetts Institute of Technology during the Second World War. Through the director, Arthur Von Hippel, and his staff, the dielectric properties of many materials were documented by using experimental methods developed in that time, and this work led to the well-known texts by Von Hippel (*21, 22*). Similarly, Willis Jackson and his group at Imperial College, London, developed new microwave cavity techniques (*23*) and studied the dielectric properties of many materials as a part of the war effort.

Because much experimental dielectric data became available from the 1930s onward, parallel developments were made in dielectric theory, especially

- phenomenological theory of the static dielectric permittivity, ε_0, and the *frequency-dependent complex permittivity*, $\varepsilon(\omega)$, where $\omega = 2\pi f/$ Hz

- molecular theory of ε_0 and $\varepsilon(\omega)$ in relation to the dipole moments, molecular structure, and the molecular dynamics of molecules of different chemical structure in the liquid, solid, and LC phases

The texts of Debye (*1*), Smyth (*2*), Fröhlich (*24*), Bordewijk and Böttcher (*7*), and Hill, Vaughan, Price, and Davies (*6*) cover many of the developments in phenomenological theory and molecular theory for liquids and solids composed of small molecules, whereas the texts of McCrum, Read, and Williams (*8*), Hedvig (*25*), and Riande and Saiz (*26*) are concerned with polymeric materials or polymer solutions.

This introductory chapter will be concerned with aspects of the theory of the dielectric properties of polymer systems. It is not possible to include all the theoretical developments here. Extensive accounts are given in the texts and in the specialized reviews cited previously and in the further sources to be referenced subsequently. Rather, we shall give a personal view of some of the fundamental working theories that can be used as a framework for analyzing dielectric properties of different polymer systems as they are studied at this time. The materials to which the theories are applied will be indicated briefly giving some of the essential conclusions, but no attempt will be made to give an in-depth analysis of the application of theory to individual polymer systems.

Equilibrium Theory

Isotropic Amorphous Polymer Systems

The permittivity of a dielectric material is measured relative to that of a vacuum for which the permittivity has the value $\varepsilon_{vac} = 8.85 \times 10^{-12} \, Fm^{-1}$. The static (frequency-independent) relative dielectric permittivity for an isotropic ensemble of N equivalent molecules each having a dipole moment, μ, and contained in a spherical volume, V, may be written according to the relation (8–13, 27–29)

$$\varepsilon_0 - \varepsilon_\infty = \left\{ \frac{4\pi}{3kT} \frac{3\varepsilon_0(2\varepsilon_0 + \varepsilon_\infty)}{(2\varepsilon_0 + 1)^2} \right\} \frac{\langle \mathbf{M}(0) \cdot \mathbf{M}(0) \rangle}{V} \tag{1}$$

where ε_0 and ε_∞ are the limiting low and high frequency permittivities with respect to the entire dielectric dispersion region that arises from the reorientational motions of the dipolar molecules, k is the Boltzmann constant, T is temperature, and V is volume. $\mathbf{M}(0)$ is the instantaneous dipole moment of the macroscopic sphere at the arbitrary time $t = 0$. Because of molecular motions, $\mathbf{M}(t)$ fluctuates in time but the equilibrium average of the scalar product $\langle \mathbf{M}(0) \cdot \mathbf{M}(0) \rangle$ taken over all complexions of the ensemble will be independent of time for a stationary thermodynamic system (i.e., a system that is not undergoing chemical or physical changes with time so it has stationary thermodynamic properties with respect to time).

Equation 1 is the essential starting equation for understanding the static dielectric (relative) permittivity for an isotropic system of dipolar molecules. The original theories of the static permittivity due to Debye (*1*), Onsager (*30*), Kirkwood (*31*), and Fröhlich (*24*) and their subsequent modifications and improvements (*6, 7*) are derived from eq 1 for special cases of the local field factor that relates the applied electric field to the local field experienced by the dipolar molecules. The full details are given in the standard texts (*1, 2, 6–8, 24–26*) and specialized reviews (*27, 28*).

Consider now a system of N polymer chains in the volume V. Following Cook, Watts, and Williams (*29*) we write $\mathbf{P}_i^L(0)$ as the instantaneous *liquid* dipole moment of chain i so

$$\langle \mathbf{M}(0) \cdot \mathbf{M}(0) \rangle = \sum_{i=1}^{N} \sum_{i'=1}^{N} \langle \mathbf{P}_i^L(0) \cdot \mathbf{P}_{i'}^L(0) \rangle \tag{2}$$

where i and i' refer to different chains. $\langle \mathbf{P}_i^L(0) \cdot \mathbf{P}_{i'}^L(0) \rangle$ accommodates autocorrelation terms for a given chain (i.e., i) as $\langle \mathbf{P}_i^L(0) \cdot \mathbf{P}_i^L(0) \rangle$ and cross-correlation terms $\langle \mathbf{P}_i^L(0) \cdot \mathbf{P}_{i'}^L(0) \rangle$, which accommodate angular correlations between the dipole moments on chains i and i'. Equation 2 is particularly appropriate to rod-like polymer chains such as the polypeptides or poly(n-alkyl isocyanates) (18–20) where the elementary dipole moments of the repeat unit add to yield a large cumulative dipole moment for the chain. For dilute solutions of rod-like polymers the cross-correlations may be negligible, but as concentration is raised, angular correlations between molecular axes lead to terms $\langle \mathbf{P}_i^L(0) \cdot \mathbf{P}_{i'}^L(0) \rangle$ between chains i and i' becoming significant in comparison with the autocorrelation terms $\langle \mathbf{P}_i^L(0) \cdot \mathbf{P}_i^L(0) \rangle$. For isotropic solutions of rod-like molecules we may write from eqs 1 and 2

$$\varepsilon_0 - \varepsilon_\infty = \frac{4\pi}{3kT} \left(\frac{3\varepsilon_0}{2\varepsilon_0 + \varepsilon_\infty} \right) \left(\frac{\varepsilon_\infty + 2}{3} \right)^2 \sum_{i=1}^{N} \sum_{i'=1}^{N} \langle \mathbf{P}_i(0) \cdot \mathbf{P}_{i'}(0) \rangle \quad (3)$$

where the *liquid* dipole moment $\mathbf{P}_i^L(0)$ is related to the *vacuum* dipole moment $\mathbf{P}_i(0)$ by the Fröhlich relation (24, 29).

$$\mathbf{P}_i^L(0) = \mathbf{P}_i(0) \left(\frac{\varepsilon_\infty + 2}{3} \right) \left(\frac{2\varepsilon_0 + 1}{2\varepsilon_0 + \varepsilon_\infty} \right) \quad (4)$$

Equation 3 accommodates distributions of molecular species and mixtures of different species (e.g., mixtures of rod-like polymers of different chemical structure) in the *isotropic* state. Equation 3 is also general to include rod-like polymers or flexible polymers. It is not appropriate to lyotropic–nematic polymer solutions where locally there are nematic structures in which molecular axes are strongly correlated. In such systems the domain structure prevents any reference rod-like chain, such as i, from being able to reorientate into 4π solid angle. Therefore, the reorientations of the molecule are limited to motions in a virtual cone defined by neighboring rod-like molecules. As a result the static permittivity ε_0 for solutions of polypeptides or poly(alkyl isocyanates) falls dramatically on going from the isotropic to nematic states on raising polymer–solute concentration (18–20).

For flexible polymer chains it is convenient to write $\mathbf{P}_i^L(0)$ as the sum of the elementary dipole moments $\boldsymbol{\mu}_{ji}^L(0)$ contained in chain i, so that for the special case where orientational correlations between chains is absent we have from eq 1

$$\langle \mathbf{M}(0) \cdot \mathbf{M}(0) \rangle = \sum_{i=1}^{N} \left\{ \sum_{j=1}^{n_{r_i}} \sum_{j'=1}^{n_{r_i}} \langle \boldsymbol{\mu}_{dji}^L(0) \cdot \boldsymbol{\mu}_{j'i}^L(0) \rangle \right\} \quad (5)$$

where n_{r_i} is the number of dipole repeat units of chain i.

If consideration is restricted to chains containing only one type of dipolar group then $\mu_{ji}^L(0) = \mu_{j'i}^L(0) = \mu_{ji'}^L(0) = \mu_{j'i'}^L(0) = \mu^L$ and eq 5 becomes

$$\langle \mathbf{M}(0) \cdot \mathbf{M}(0) \rangle = \sum_{i=1}^{N} \left\{ n_{r_i}(\mu^L)^2 + 2 \sum_{j=2}^{n_{r_i}} \sum_{j'=1}^{j-1} \langle \mu_{ji}^L(0) \cdot \mu_{j'i}^L(0) \rangle \right\} \quad (6)$$

where $\mu_{ji}^L(0)$ is the dipole moment of the jth dipolar group in polymer chain i. Thus $\langle \mathbf{M}(0) \cdot \mathbf{M}(0) \rangle$ is made up of autocorrelation terms involving $(\mu^L)^2$ and cross-correlation terms $\langle \mu_{ji}^L(0) \cdot \mu_{j'i}^L(0) \rangle$ between groups j and j' along a polymer chain. For flexible chains these cross-correlation (CC) terms decrease in magnitude with increasing separation of the dipole units along a chain if there is no persistent cumulative dipole moment along the chain backbone. For the special case of flexible chains of high molecular weight conforming with this criterion, eq 6 becomes

$$\frac{\langle \mathbf{M}(0) \cdot \mathbf{M}(0) \rangle}{V} = c_r \left\{ (\mu^L)^2 + \sum_{\substack{k' \\ k' \neq k}} \langle \mu_k^L(0) \cdot \mu_{k'}^L(0) \rangle \right\} \quad (7)$$

where c_r is the number of dipole groups per unit volume, and k and k' refer to dipoles k and k' along the same polymer chain.

Hence from eqs 1 and 7 and using the relation

$$\mu_k^L(0) = \mu_k(0) \left(\frac{\varepsilon_\infty + 2}{3} \right) \left(\frac{2\varepsilon_0 + 1}{2\varepsilon_0 + \varepsilon_\infty} \right) \quad (8)$$

we may write

$$\varepsilon_0 - \varepsilon_\infty = \frac{4\pi}{3kT} c_r \left(\frac{3\varepsilon_0}{2\varepsilon_0 + \varepsilon_\infty} \right) \left(\frac{\varepsilon_\infty + 2}{3} \right)^2 g(0) \mu^2 \quad (9)$$

where

$$g(0) = \left\{ \mu^2 + \sum_{\substack{k' \\ k' \neq k}} \langle \mu_k(0) \cdot \mu_{k'}(0) \rangle \right\} \Big/ \mu^2 \quad (10)$$

Equation 9 bears a strong resemblance to the Kirkwood–Fröhlich equation (6, 24) derived for associated liquids such as the alcohols and water, but the criteria leading to this equation for polymer chains require

- high molecular weight species so that any reference dipole k is equivalent to any other dipole k'

- flexible chains without a persistent cumulative dipole moment
- chains that contain only one kind of dipole having dipole moment μ

Flexible polymer chains in solution or in the bulk amorphous state above the glass-transition temperature (T_g) conform in the main with these criteria so eq 9 may be used to relate dielectric relaxation strength $\Delta\varepsilon = \varepsilon_0 - \varepsilon_\infty$ to dipole concentration (c_r); the squared dipole moment of a repeat unit, μ^2; and the conformational properties of the chain through the terms $\langle \mu_k(0)\cdot\mu'_k(0) \rangle$. For more complicated systems, such as for copolymers containing two kinds of dipoles or for mixtures of homopolymers or copolymers, it is necessary to start from eq 5 and to express $\langle M(0)\cdot M(0) \rangle$ in terms of the various component dipoles and their individual concentrations.

Earlier derivations of the static dielectric permittivity in terms of constituent dipole moments for polymer solutions and amorphous polymers above T_g were given by Kirkwood and Fuoss (31–33). The first description of the extension of the Fröhlich–Kirkwood theory of static permittivity to amorphous polymers was given by McCrum, Read, and Williams (8). The present account follows closely to that described by Cook, Watts, and Williams (29).

A feature of eqs 9 and 10 is that $\Delta\varepsilon$ for flexible chains in the amorphous state does not depend on the molar mass of the polymer or the distribution of molar mass. The value of $\Delta\varepsilon$ is proportional to the concentration of repeat (dipole) units of the chain, to μ^2, and to $g(0)$. For no angular correlations between dipoles along a chain, then, $\langle \mu_k(0)\cdot\mu'_k(0) \rangle = 0$, and $g(0) = 1$. Such a result would occur for chains containing only a few dipoles—as one can have in lightly oxidized polyethylene studied in its *amorphous* state above the melting temperature T_m. However, for conventional polymers such as poly(vinyl chloride), poly(ethylene terephthalate), and poly(methyl methacrylate) each repeat unit along a chain has a dipolar group so there will be strong angular correlations between dipoles. Because parallel or antiparallel alignments of dipoles along a flexible chain are possible, the CC terms $\langle \mu_k(0)\cdot\mu'_k(0) \rangle$ may be positive or negative in eq 10.

As one example of the importance of CC terms in the determination of $g(0)$, we may quote from the calculations of Cook, Watts, and Williams (29) for model polyether chains based on a model theory of their dipole moments given first by Read (34a) (*see also* Appendix 2 by Read and Williams in reference 34b and McCrum, Read, and Williams, reference 8, p 97). For the polyether series $-[-(CH_2)_{p-1}-O-]-_n$ where $p = 2$, $3 \ldots$, we take a model for the chain (34a, 34b) in which

- correlations of internal rotation states were taken to be absent
- all three-bond systems had, on average, the same internal rotation characteristics

- all valence angles were the same
- the internal rotation potential is symmetric about the trans-conformation

Therefore, Read's eq 3 may be written as (29)

$$\langle \mathbf{P}(0)\cdot\mathbf{P}(0)\rangle = n\mu^2 \left\{ 1 + \left(\frac{n-1}{n}\right) \frac{[2\overline{\mathbf{A}}^p - \overline{\mathbf{A}}^{p+1} - \overline{\mathbf{A}}^{p-1}]_{33}}{(1-\cos\alpha)} \right.$$

$$+ \left(\frac{n-2}{n}\right) \frac{[2\overline{\mathbf{A}}^{2p} - \overline{\mathbf{A}}^{2p+1} - \overline{\mathbf{A}}^{2p-1}]_{33}}{(1-\cos\alpha)} + \cdots$$

$$\left. + \frac{1}{n} \frac{[2\overline{\mathbf{A}}^{(n-1)p} - \overline{\mathbf{A}}^{(n-1)p+1} - \overline{\mathbf{A}}^{(n-1)p-1}]_{33}}{(1-\cos\alpha)} \right\} \quad (11)$$

where n is the number of units in the chain, μ is the ether dipole moment, and $\overline{\mathbf{A}}$ is the average coordinate transformation matrix as defined by Read (34a, 34b)

$$\overline{\mathbf{A}} = \begin{bmatrix} -\eta\cos\alpha & 0 & -\eta\sin\alpha \\ 0 & -\eta & 0 \\ \sin\alpha & 0 & \cos\alpha \end{bmatrix} \quad (12)$$

where $\eta = \langle\cos\phi\rangle$, ϕ is the internal-rotation angle ($\phi = 0$ for trans conformation), $\alpha = \pi -$ valence angle, and $\overline{\mathbf{A}}_{33}$ indicates the 33 element of the matrix $\overline{\mathbf{A}}$. Table I shows the values of the coefficients of $(n-m)/n$ for $p = 2, 3, 4,$ and 5 for tetrahedral valence angles (i.e., $\cos\alpha = 1/3$) and for chosen values of η.

Consider first the results for poly(oxymethylene) ($p = 2$). For $\eta = 0$ the first cross-correlation function terms contribute significantly to $g(0)$ to the extent of $-2 \times 0.111 = -0.222$. Higher order CC terms are very small. For trans-favored conformations ($\eta > 0$) the importance of the CC terms increases dramatically and all terms are positive. This result is consistent with increasing parallel alignment of dipoles because the all-trans form of the chain is shown in **1**.

1

For gauche-favored conformations ($\eta < 0$) the first nearest neighbors contribute negative terms and the second nearest neighbors contribute positive terms to $g(0)$. Therefore, for chains with strong preferences to trans or

Table I. Coefficients of $(n - m)/n$ for $p = $ 2, 3, 4, and 5 for $\cos\alpha = \frac{1}{3}$ and for Chosen Values of $\eta = \langle \cos\phi \rangle$

p Value	η	$C_{(n-1)/n}$	$C_{(n-2)/n}$	$C_{(n-3)/n}$	$C_{(n-4)/n}$
2	−0.50	−1.00	+0.250	+0.062	−0.109
	−0.25	−0.639	+0.084	+0.013	−0.009
	0	−0.222	−0.025	−0.003	0.000
	+0.25	+0.250	+0.016	−0.007	+0.005
	+0.50	+0.778	+0.300	+0.113	+0.040
	+0.75	+1.361	+0.926	+0.630	+0.428
3	−0.50	−0.50	+0.062	+0.024	
	−0.25	−0.183	+0.013	0.000	
	0	−0.074	−0.003	0.000	
	+0.25	−0.187	−0.007	−0.004	
	+0.50	−0.537	+0.113	−0.038	
	+0.75	−1.137	+0.630	−0.361	
4	−0.50	0.250	−0.109		
	−0.25	0.084	−0.009		
	0	−0.025	0.000		
	+0.25	+0.016	+0.005		
	+0.50	+0.300	+0.042		
	+0.75	+0.926	+0.428		
5	−0.50	+0.380	+0.066		
	−0.25	+0.081	+0.001		
	0	−0.008	0.000		
	+0.25	−0.043	−0.002		
	+0.50	−0.219	+0.014		
	+0.75	−0.775	+0.291		

gauche conformations the CC terms make significant contributions to $g(0)$. The magnitude of the CC terms decreases with increasing separation of dipoles along a chain but may persist over long separations (e.g., $\eta = 0.75$, $p = 2$).

Similar results are observed for the other polyethers in Table I (e.g., polyethylene oxide, $p = 3$). As the dipole spacing increases the CC terms decrease in importance, so for $p = 5$ only nearest neighbors to the reference dipole contribute significantly even for highly structured chains.

Thus, the static dielectric permittivity ε_0 and relaxation strength $\Delta\varepsilon$ may be understood for flexible chains in solution or in the amorphous rubbery (elastomeric) state in terms of eqs 9 and 10, which involve dipole concentration, dipole moment strength, and vector correlations of dipole units along a chain. The temperature and pressure dependencies of $\Delta\varepsilon$ may also be analyzed using this theory, but apart from kT, the summed CC terms may have a T dependence and P dependence that will be difficult to interpret in practice.

Equation 3 or 9 may be used to determine the dipole moments of polymer molecules. This subject has not been an active one for many years because most dielectric studies of polymer systems have been concerned with relaxation phenomena rather than the static dielectric properties. It is apparent from the books by Volkenstein (*35*) and Flory (*36*) that $\langle P^2 \rangle$ for chain molecules of different structures and tacticities can be predicted using the rotation–isomeric model for chain conformation but, again, this subject has not been studied actively for several years.

Crystalline Solid Polymers

We have considered above the dielectric permittivity for isotropic amorphous polymer systems. Dielectric studies are widely conducted for crystalline and LC polymer systems, and the directional properties of the static permittivity have to be considered in such cases. The dielectric permittivity for a polycrystalline solid polymer has been considered by Boyd and co-workers (*37*, *38*). In such cases amorphous and crystalline components, each with their own permittivities, coexist so the permittivity of the bulk sample is related in a complicated way to those of its components. Upper and lower bounds for the measured permittivity have been given by Boyd as a function of the volume fraction of the amorphous component and the permittivities ε_x and ε_s for the crystalline and amorphous phases. By using this approach Boyd and co-workers (*38*) were able to estimate dipole moments and g-factors for particular processes in partially crystalline polymers such as polyethylene terephthalate. An extensive consideration of the dielectric properties of crystalline polymers will be given subsequently.

Thermotropic LC Polymers

Thermotropic LC polymers were widely studied in recent years, especially their dielectric relaxation properties (*16*, *17*, *39–43*). For a nematic LC material the ordering of the mesogenic (LC-forming) groups is defined by a local director, **n**, and a local-order parameter, *S*. For smectic LC materials the mesogenic groups form layers, and the relative orientation of the molecular axes from one layer to another defines the type of smectic ordering (S_A, S_B, S_C, S_E, etc.). The local ordering in individual layers is again defined in terms of **n** and *S*. LC polymer materials may be aligned macroscopically in directing electric (*E*) or magnetic (*B*) fields to yield homeotropically aligned (director **n** $\parallel$ **E**$_{meas}$), planarly aligned (director **n** $\perp$ **E**$_{meas}$), and intermediately aligned samples. The static permittivity of such an LC polymer is a tensorial quantity that for an axially symmetric LC phase takes the simple form

$$\varepsilon_0 = diag\ (\varepsilon_{0\parallel}, \varepsilon_{0\perp}, \varepsilon_{0\perp}) \tag{13}$$

where $\varepsilon_{0\parallel}$ is measured for **n** $\parallel$ **E**$_{meas}$ and $\varepsilon_{0\perp}$ is measured for **n** $\perp$ **E**$_{meas}$.

The measured dielectric permittivity for an LC material having axial symmetry with respect to the measuring field direction (Z) varies systematically with the macroscopic order of the sample. Writing a director order parameter S_d as

$$S_d = \langle 3 \cos^2 \theta_{nE} - 1 \rangle / 2 \tag{14}$$

where θ_{nE} is the polar angle between the local director n and the measuring field E and $\langle \dots \rangle$ indicates a statistical average taken over the sample. Assuming the electric field is uniform through the specimen, we have, as one bound

$$\varepsilon_{0Z} = \frac{(1 + 2S_d)}{3} \varepsilon_{0\parallel} + \frac{2}{3} (1 - S_d) \varepsilon_{0\perp} \tag{15}$$

where ε_{0Z}, $\varepsilon_{0\parallel}$, and $\varepsilon_{0\perp}$ are the measured permittivities along the principal axis of the axially symmetric sample and those for homeotropic and planar samples, respectively. Equation 15 was derived by Attard, Araki, and Williams (40) and gave an excellent representation of experimental dielectric data for siloxane-chain LC side-chain polymers (16, 39–43). If it is assumed that the dielectric displacement is uniform for a sample then the second bound is obtained as (40)

$$\frac{1}{\varepsilon_{0Z}} = \frac{(1 + 2S_d)}{3} \frac{1}{\varepsilon_{0\parallel}} + \frac{2}{3} (1 - S_d) \frac{1}{\varepsilon_{0\perp}} \tag{16}$$

Thus, upper and lower bounds for $\varepsilon_{0Z} = f(\varepsilon_{0\parallel}, \varepsilon_{0\perp})$ may be determined from eqs 15 and 16.

For a homeotropic sample $S_d = 1$ and for unaligned and planarly aligned samples we have $S_d = 0$ and -0.5, respectively. For the unaligned sample $S_d = 0$, we obtain from eq 15

$$(\varepsilon_{0Z})_u = (\varepsilon_{0\parallel} + 2\varepsilon_{0\perp})/3 \tag{17}$$

with an analogous result from the other bound, eq 16. Equations 15 and 17 are the normal working equations to be applied to practical data (16, 17, 39–43).

The principal static permittivities for an LC polymer may be written as a function of the longitudinal component μ_ℓ and the transverse component μ_t of the dipole moment of a mesogenic (LC-forming) group. For a nematic monodomain containing anisometric molecules Maier and Meier wrote (44)

$$\varepsilon_{0\parallel} = \varepsilon_{\infty\parallel} + \frac{G}{3kT}[\mu_\ell^2(1 + 2S) + \mu_t^2(1 - S)] \tag{18a}$$

$$\varepsilon_{0\perp} = \varepsilon_{\infty\perp} + \frac{G}{3kT}[\mu_\ell^2(1 - S) + \mu_t^2(1 + S/2)] \tag{18b}$$

where G is a term involving the concentration of dipolar mesogenic groups. The value of μ_ℓ is taken to be parallel to the principal (z) axis of this group and μ_t is perpendicular to that axis. The value S is the local-order parameter

$$S = \langle 3 \cos^2 \theta_{zZ} - 1 \rangle / 2 \tag{19}$$

where θ_{zZ} is the angle between the z axis in the molecular frame and the Z axis of the laboratory frame. For thermotropic LC polymers well below the clearing temperature S lies in the range 0.4–0.6 and decreases sharply to 0 as the material is heated from the LC state to the isotropic liquid state. The values of $\varepsilon_{\infty\parallel}$ and $\varepsilon_{\infty\perp}$ are the limiting high-frequency permittivities measured for $\mathbf{n} \parallel \mathbf{E}$ and $\mathbf{n} \perp \mathbf{E}$, respectively. In practice $(\varepsilon_{0\parallel}, \varepsilon_{\infty\parallel})$ are obtained for the homeotropic materials ($S_d = 1$), whereas $(\varepsilon_{0\perp}, \varepsilon_{\infty\perp})$ are obtained for the planar material (or a homogeneous monodomain) for which $S_d = -0.5$ (*see* reference 40 for further details).

Equations 18a and 18b do not include angular correlations between dipolar mesogenic groups between chains or along chains. Clearly this consideration is an important one. Bordewijk (*45*) generalized the theory of Maier and Meier to include angular correlations between dipoles and made an improvement of the local field to include the anisotropic environment of the molecules. His equation may be expressed as (*15, 45*)

$$(\varepsilon_{0\gamma} - \varepsilon_{\infty\gamma})[\varepsilon_{0\gamma} + A_\gamma(\varepsilon_{\infty\gamma} - \varepsilon_{0\gamma})]/\varepsilon_{0\gamma} = \left(\frac{4\pi N}{kT}\right) g_\gamma \langle \mu_\gamma^2 \rangle \tag{20}$$

where N is dipole concentration and where $\gamma = \parallel$ or $\perp$, so g_γ is $g_\parallel$ or $g_\perp$ and corresponds to longitudinal and transverse Kirkwood g factors (as introduced for isotropic systems; *see* eqs 9 and 10). In addition

$$\langle \mu_\parallel^2 \rangle = \mu_\ell^2[(1 + 2S)/3]f_\ell + \mu_t^2[(1 - S)/3]f_t \tag{21a}$$

$$\langle \mu_\perp^2 \rangle = \mu_\ell^2[(1 - S)/3]f_\ell + \mu_t^2[(1 + S/2)/3]f_t \tag{21b}$$

where f_ℓ and f_t are local field factors, and $A_\parallel$ and $A_\perp$ are functions of molecular shape. Bordewijk (*7*) and Dunmur and Miller (*46*) have applied eqs 20 and 21 to experimental data for low-molar-mass liquid crystals of nematic and smectic type. These equations are equally applicable to LC side-chain polymers but this process has not been applied to experimental data up to this time.

To summarize, we see that the static dielectric permittivity of an LC polymer will be characterized, in the simplest case of an axially symmetric phase, by the principal static permittivities $\varepsilon_{0\parallel}$ and $\varepsilon_{0\perp}$. These permittivities are functions of μ_ℓ, μ_t; the local order parameter, S; local field factors, $f_\parallel$ and $f_\perp$; shape factors $A_\parallel$ and $A_\perp$; and anisotropic Kirkwood g factors $g_\parallel$ and $g_\perp$ that involve angular correlations between (μ_ℓ, μ_t) on one chain unit with (μ_ℓ, μ_t) on adjacent units, both along a representative chain and between chains. Thus, $\varepsilon_{0\parallel}$ and $\varepsilon_{0\perp}$ not only reflect the dipole moments of the mesogenic groups but also the structure–ordering of the LC phase as expressed in S and in the g factors that are sensitive to chain conformation. Note that for low-molar-mass alkylcyanobiphenyls there is a tendency toward antiparallel alignment of the longitudinal component μ_ℓ of the different dipoles, which leads to a lowering of $\varepsilon_{0\parallel}$ compared with the case $g_\parallel = 1$.

A link between the isotropic and LC state dielectric permittivities for an LC-forming material can be established as follows: for a randomly orientated LC phase, $S_d = 0$ in eq 15 leading to $(\varepsilon_{0Z})_u$ given by eq 17. (Note, if $\varepsilon_{0\parallel}$ and $\varepsilon_{0\perp}$ are known experimentally, then $(\varepsilon_{0Z})_u$ can be calculated from eq 17, thus avoiding the uncertainty of the measured permittivity for samples that are nominally unaligned but that may be aligned to some extent through polymer flow into the measuring cell or surface-induced alignment in the cell.) By using eq 18 we readily show that

$$(\varepsilon_{0Z})_u = \bar{\varepsilon}_\infty + \frac{G}{3kT}\mu^2 \tag{22}$$

where $\bar{\varepsilon}_\infty = (\varepsilon_{\parallel\infty} + 2\varepsilon_{\perp\infty})/3$, which within the approximations involved is the same result as that for the static permittivity for the *isotropic phase*. Thus, the mean permittivity for the LC phase, defined by eq 17, is equal to the permittivity of the isotropic phase when data for the LC phase and isotropic phase are extrapolated to a common temperature (i.e., the clearing temperature). This result is true, approximately, for experimental data for LC polymers (*39–43*) and low-molar-mass liquid crystals (*15, 46*), but where deviations are observed information is obtained on the preference for parallel or antiparallel ordering of dipoles in the LC phase.

Dynamic Behavior: Phenomenological Theory

Time- and Frequency-Dependent Dielectric Properties

The phenomenological theory of dielectric relaxation in polymer systems and solids and liquids generally is summarized in the well-known standard texts (*1, 2, 6–8, 24–26*) and in specialized reviews (*9–14, 27, 28*). In general we write the total complex relative permittivity $\varepsilon_t(\omega)$ as

$$\varepsilon_t(\omega) = \varepsilon'(\omega) - i\varepsilon''(\omega) + i\frac{\sigma}{\omega}\varepsilon_{vac} \tag{23}$$

where $\omega = 2\pi f/\mathrm{Hz}$ is the angular frequency of the measuring electric field; ε' and ε'' are the real permittivity and dielectric loss factor, respectively, measured at ω; and σ is the frequency-independent dc conductivity of the sample that arises from the motion of charge carriers (ionic or electronic). Phenomenological theory requires (*6–8*) that the relaxation part of the complex permittivity (i.e., $\varepsilon(\omega) = \varepsilon'(\omega) - i\varepsilon''(\omega)$) is given, for an isotropic dielectric, by

$$\frac{\varepsilon(\omega) - \varepsilon_\infty}{\varepsilon_0 - \varepsilon_\infty} = 1 - i\omega\Im[\Phi(t)] \tag{24}$$

where $\varepsilon_0 - \varepsilon_\infty$ is the total relaxation strength, $\Im$ indicates a one-sided Fourier transform (a pure imaginary Laplace transform), and $\Phi(t)$ is a macroscopic relaxation function that can be measured as the transient charge decay function following the step withdrawal of a steady applied electric field from a sample, or from the transient charge rise function $(1 - \Phi(t))$ when a step electric field is applied at $t = 0$ to an unperturbed sample. For the single relaxation time model

$$\Phi(t) = \exp - t/\tau \tag{25}$$

where τ is a single macroscopic relaxation time. On insertion of eq 25 into eq 24 the familiar single relaxation time equations are obtained:

$$\varepsilon'(\omega) = \varepsilon_\infty + \frac{\varepsilon_0 - \varepsilon_\infty}{1 + \omega^2\tau^2} \tag{26a}$$

$$\varepsilon''(\omega) = (\varepsilon_0 - \varepsilon_\infty)\frac{\omega\tau}{1 + \omega^2\tau^2} \tag{26b}$$

so $\varepsilon'(\omega)$ exhibits dispersion in the frequency domain, falling from ε_0 to ε_∞ with increasing frequency, and $\varepsilon''(\omega)$ exhibits a bell-shaped absorption curve whose maximum value occurs at $\omega_m\tau = 1$, hence τ is determined.

   For polymer materials multiple dielectric relaxation regions are typically observed (*6–14, 24–28*) where each relaxation is considerably broader than a single relaxation time process. We may write for a discrete distribution of relaxation processes

$$\frac{\varepsilon(\omega) - \varepsilon_\infty}{\varepsilon_0 - \varepsilon_\infty} = R(\omega) = \sum_j a_j R_j(i\omega) = \sum_j a_j\{1 - i\omega\Im[\Phi_j(t)]\} \quad (27)$$

where a_j is the fraction of the total relaxation magnitude that is relaxed by process i and $R_j(\omega) = (1 + i\omega\tau_j)^{-1}$ is the relaxation function (a complex quantity) for process j, expressed in the f-domain, and $\Phi_j(t)$ is the corresponding relaxation function (a real quantity) for process j expressed in the t-domain, $\Sigma a_j = 1$. For a continuous distribution of relaxation times characterized by a distribution function $H(\tau)$ we may write

$$\frac{\varepsilon(\omega) - \varepsilon_\infty}{\varepsilon_0 - \varepsilon_\infty} = \int \frac{H(\tau)}{1 + i\omega\tau} \, d\tau \quad (28)$$

Several phenomenological approaches to the dielectric properties of polymers, in bulk or solution, may be used for f-domain data and t-domain data.

Phenomenological Distribution Functions

For a continuous distribution the relaxation function may be written as

$$\Phi(t) = \int H(\tau) \exp - (t/\tau) \, d\tau \quad (29)$$

Thus, by choice of a suitable distribution function, data obtained in the f domain or t domain may be fitted numerically (*see* Ferry, reference 47). Note that if $\Phi(t)$ or $R(\omega)$ is known, then $H(\tau)$ may be determined numerically: for example, data for the dielectric α-relaxation in poly(methylacrylate) (*48*).

Empirical Expressions for R(ω)

Cole and Cole (*49*), Davidson and Cole (*50*), and Havriliak and Negami (*51*) modified the single relaxation time function

$$R(\omega) = \frac{1}{1 + i\omega\tau} \quad (30)$$

to give broadened dispersion and absorption curves that gave an improved representation of experimental data for solid polymers and glass-forming liquids. All three expressions are contained in the Havriliak–Negami (HN) function

$$R(\omega) = \frac{1}{[1 + (i\omega\tau)^a]^b} \quad (31)$$

where a and b are empirical constants $0 < a \leq 1$, $0 < b \leq 1$. For $b = 1$ the Cole–Cole expression emerges, whereas for $a = 1$ the Davidson–Cole equation emerges. At the present time the HN function is one of the most widely used functions for fitting dielectric relaxation data for polymers. It gives loss curves that are skewed to high frequencies but to a lesser extent than those for the Davidson–Cole function for comparable widths of the plots of ε'' versus log frequency (52, 53).

Note that the transient-current relaxation function $\phi(t) = -d\Phi(t)/dt$ may be expressed as follows:

1. Davidson–Cole (50, 54)

$$\phi(t) = \frac{1}{\tau\Gamma(b)} \left(\frac{t}{\tau}\right)^{-(1-b)} \exp-(t/\tau) \tag{32}$$

2. Cole–Cole (55)

$$\left.\begin{aligned}\phi(t) &= \frac{a}{\tau\Gamma(1+a)} \left(\frac{t}{\tau}\right)^{-(1-a)} \quad \text{for} \quad \frac{t}{\tau} \ll 1 \\[2mm] &= \frac{a}{\tau\Gamma(1-a)} \left(\frac{t}{\tau}\right)^{-(1+a)} \quad \text{for} \quad \frac{t}{\tau} \gg 1\end{aligned}\right\} \tag{33}$$

where Γ indicates the gamma-function.

Empirical Relaxation Function in the t-Domain

One empirical relaxation function, the Köhlrausch–Williams–Watts (KWW) function, has received considerable use not only for the dielectric relaxation behavior of polymers but also for other polymer relaxation phenomena including volume and enthalpy relaxations, specific heat relaxation, NMR and dynamic mechanical relaxations, quasielastic light scattering, quasielastic neutron scattering, transient optical relaxation, and t-dependent fluorescence depolarization. It was first used by Köhlrausch ($56a$, $56b$) to fit the transient decay function for electric charge in the Leyden jar. It was introduced independently by Williams and Watts (52) to fit dielectric relaxation data for amorphous solid polymers in the t domain and the f domain. The KWW function may be written as

$$\Phi(t) = \exp-(t/\tau)^{\bar{\beta}} \tag{34}$$

For the special case $\bar{\beta} = 0.5$ insertion of eq 34 into eq 24 yields the result (52)

$$\frac{\varepsilon(\omega) - \varepsilon_\infty}{\varepsilon_0 - \varepsilon_\infty} = (\pi)^{1/2} \frac{(1 - i)}{\rho} w(z) \tag{35}$$

where $\rho = (8\omega\tau)^{1/2}$, $z = (1 + i)/\rho$, $i = \sqrt{-1}$, and $w(z) = (\exp - z^2)\text{erfc}(-iz)$. The values of $w(z)$ have been tabulated (57), hence Williams and Watts were able to show that plots of $\varepsilon'(\omega)$ and $\varepsilon''(\omega)$ versus log ω were skewed in the Davidson–Cole (DC) sense but were not as asymmetrical for a given half width of the loss curve. A comparison of the DC and KWW functions in the time-domain may be made if we write from eq 34

$$\phi(t) = \frac{\overline{\beta}}{\tau} \left(\frac{t}{\tau}\right)^{-(1-\overline{\beta})} \exp - (t/\tau)^{\overline{\beta}} \tag{36}$$

which may be compared with the DC function eq 32.

The transformation of the KWW function into the frequency domain for general values of $\overline{\beta}$ has been described by Williams, Watts, Dev, and North (53) who wrote the two infinite series

$$R(\omega) = \frac{\varepsilon(\omega) - \varepsilon_\infty}{\varepsilon_0 - \varepsilon_\infty} = \sum_{n=1}^{\infty} (-1)^{\overline{\beta}-1} \frac{1}{(\omega\tau)^{n\overline{\beta}}} \frac{\Gamma(n\beta + 1)}{\Gamma(n + 1)}$$
$$\times \left[\cos \frac{n\overline{\beta}\pi}{2} - i \sin \frac{n\overline{\beta}\pi}{2}\right] \tag{37}$$

$$R(\omega) = \frac{\varepsilon(\omega) - \varepsilon_\infty}{\varepsilon_0 - \varepsilon_\infty} = \sum_{n=1}^{\infty} (-1)^{n-1} \frac{(\omega\tau)^{n-1}}{\Gamma(n)} \frac{\Gamma(n + \overline{\beta} - 1)}{\beta}$$
$$\times \left[\cos \frac{(n - 1)\pi}{2} - i \sin \frac{(n - 1)\pi}{2}\right] \tag{38}$$

Equations 37 and 38 were used to determine $R(\omega) = R'(\omega) - iR''(\omega)$ over the range $-4 \le \log \omega\tau \le 4$ (53). Moynihan, Boesch, and Laberge (58) carried out the transform of the KWW function from $\Phi(t) \to \varepsilon(\omega)$ by first fitting the KWW function to a weighted sum of single exponential decay functions and then transforming each term in the series analytically. Koizumi and Kita (59) tabulated $R(\omega)$ over large ranges of log $\omega\tau$ and β and provided tables of the transform by use of eqs 37 and 38.

Williams (60) reviewed the properties of the KWW function including a discussion of

- the analytic $t \to \omega$ transformations for $\overline{\beta}$ equal to 2/3 (Whittaker function), 0.50 (Fresnel function), and general $\overline{\beta}$ (Wintner series expansion)

- the empirical distribution of relaxation times involved in the KWW function (*61*)
- physical mechanisms for molecular motion that give this function
- the applications of the function to different polymer relaxation phenomena

Note that the HN function and KWW functions give very similar behavior if (*a*, *b*) values are chosen in certain ranges for a given value of β. The nature and form of the distribution of relaxation times for the KWW function have been analyzed (*61*).

In addition to the empirical CC, DC, HN, and KWW functions, other empirical functions may be used; for example, that involving a box distribution of relaxation times as described by Fröhlich (*24*) and used by Shears and Williams (*62*) for dielectric relaxation in glass-forming liquids. That function due to Fuoss and Kirkwood (*32*), eq 39, continues to be particularly useful for symmetrical loss peaks in plots of $\varepsilon''(\omega)$ versus log ω.

$$\varepsilon''(\omega) = \varepsilon''_{max} \operatorname{sech}(m\ell n\omega\tau) = (\varepsilon_0 - \varepsilon_\infty) m \frac{(\omega\tau)^m}{1 + (\omega\tau)^m} \tag{39}$$

$0 < m \le 1$

Thus a number of empirical functions may be used to express dielectric data for polymers in the t domain or f domain. At the present time the HN and KWW functions are used extensively for multiple dielectric relaxations (α,β,γ, etc.) in amorphous crystalline and LC polymers.

Note that eq 24 may be inverted giving (*27–29*) the following relations

$$\Phi(t) = \frac{2}{\pi} \int_0^\infty \left[\frac{\varepsilon_0 - \varepsilon'(\omega)}{\varepsilon_0 - \varepsilon_\infty} \right] \sin \omega t \, d\ell n\omega$$

$$\Phi(t) = \frac{2}{\pi} \int_0^\infty \left[\frac{\varepsilon''(\omega)}{\varepsilon_0 - \varepsilon_\infty} \right] \cos \omega t \, d\ell n\omega \tag{40}$$

Equation 40 provides a convenient means of determining the relaxation function numerically from data for dielectric dispersion of $\varepsilon'(\omega)$ or from the dielectric absorption $\varepsilon''(\omega)$ as was demonstrated some years ago by Cook, Watts, and Williams (*29*).

Temperature- and Pressure-Dependent Dielectric Properties

Even though the strengths of individual relaxation processes $\Delta\varepsilon_i = (\varepsilon_{0i} - \varepsilon_{\infty i})$ depend on T and P through the density dependence of dipole concentration (*see* eqs 9 and 10), such dependencies are small compared with changes in average relaxation time $\langle \tau_i \rangle = \langle \tau_i(T,P) \rangle$ with T and P, for a given process i. Two common dependencies on T at fixed P are observed experimentally:

1. Vogel-Fulcher behavior

$$\langle \tau(T) \rangle = A \exp B/(T - T_\infty) \tag{41}$$

   where A, B, and T_∞ are fitting parameters.

2. Arrhenius behavior

$$\langle \tau(T) \rangle = C \exp Q/RT \tag{42}$$

   where C is a fitting parameter, and Q is an apparent activation energy (J/mol).

The dielectric α-relaxations in amorphous systems (polymeric and non-polymeric) are well fitted by the Vogel–Fulcher equation (which can be rearranged to give the well-known Williams–Landel–Ferry equation (*8, 47*)) and has been applied and discussed extensively in the literature (*7–14, 47*). Secondary relaxations in amorphous and crystalline polymers often obey eq 42, which is surprising given the fact that such relaxations are unusually broad in plots of $\varepsilon''(\omega)$ versus log ω.

The effect of pressure on $\langle \tau \rangle$ for individual dielectric processes in polymers has received less attention. The pressure dependence of the α- and β-relaxations in amorphous polymers was studied 30 years ago (*63–69*). $(\partial \ell n \langle \tau \rangle / \partial P)_T$ was in the range 1–4 kb^{-1} for α-processes but was considerably smaller for β-processes. Sayre et al. (*70*) studied the dielectric α_c- and γ-relaxations in polyethylenes and observed large shifts in the loss peak for the α_c-process as pressure was increased. For amorphous materials it is usual to assume that $\langle \tau_\alpha \rangle$ is a function of a free volume, V_f, and that an increase in pressure acts so as to decrease V_f and hence increase $<\tau_\alpha>$. However, the free volume theories (*47, 63, 71*) have some difficulties in reconciling the experimental data of Williams for the α-relaxation in poly(methyl acrylate) (*63, 71*) and poly(propylene oxide) (*64*). The difficulties concern the constant volume activation energy $Q_v(T,V)$, as was first pointed out by Hoffman, Williams, and Passaglia (*72*) and emphasized by Williams (*11*). We may write (*63*)

$$Q_v(T,V) = -RT^2(\partial \ell n \langle \tau_\alpha \rangle / \partial T)_v \tag{43}$$

$$Q_P(T,P) = -RT^2(\partial \ell n \langle \tau_\alpha \rangle / \partial T)_p \tag{44}$$

$$Q_v(T,V) = Q_p(T,P) - RT^2 \left(\frac{\partial P}{\partial T}\right)_v \left(\frac{\partial \ell n \langle \tau_\alpha \rangle}{\partial P}\right)_T \tag{45}$$

where $(\partial P/\partial T)_v$ is the thermal pressure coefficient that is equal to the ratio of the thermal expansion coefficient to the isothermal compressibility of the polymer. The constant volume activation energy $Q_v(T,V)$ is numerically less than the constant pressure activation energy $Q_p(T,P)$, according to eq 45. The experimental data for the dielectric α-relaxation in amorphous polymers show (63, 64, 71) that $Q_v(T,V)$ is only about 20% smaller than $Q_p(T,P)$. For the simplest free volume theory the volume $V = V_0 + V_f$, where V_0 is bound volume and V_f is free volume.

The free volume theory gives (64)

$$\langle \tau \rangle = D \exp E/V_f \qquad (46)$$

where D and E are fitting parameters. If $V_0 = $ constant (hard sphere model) then eq 46 predicts that $\langle \tau \rangle$ will not change if temperature is raised under constant volume conditions: that is, $Q_v = 0(!)$, which is not in accord with the experimental data. To bring the free volume theory into line with experimental data a strongly negative value for $(\partial V_0/\partial T)_V$ is required, which is physically unrealistic, as has been emphasized earlier (11, 72). Thus, while variations of $\langle \tau \rangle$ with T at constant P or $\langle \tau \rangle$ with P at constant T appear to be reconciled by using a simple free volume theory (eq 46), the constant volume experiments reveal considerable difficulties with this approach. Dielectric relaxation processes arise from molecular motions that are driven by temporal fluctuations in thermal energy for a local ensemble of molecules, so theories of relaxation need to be based on dynamic models, not a time-averaged quantity such as free volume V_f.

Dynamic Behavior: Molecular Theory

Isotropic Amorphous Polymers

The original theory of Debye (1) for dielectric relaxation of dipolar molecules in the liquid and solid states involved the elucidation of the macroscopic dipole moment of an ensemble of dipoles in the presence of an ac or dc perturbing electric field. The motions of the molecules could be modelled by equations for rotational diffusion in the presence of a field or as motion in barrier systems in the presence of a field. The mathematical complexities of such theories are apparent and it was difficult to obtain a physical insight into the true meaning of the relaxation times that were deduced. McCrum, Read, and Williams (8) reviewed the theories of Kirkwood and Fuoss (33), of Yamafuji and Ishida (73) for rotational diffusion of chains, and those of Hoffman (*see* reference 72) for motion in barrier systems.

In 1960 Glarum (74) adapted the linear response theory of Kubo (75) to the situation of an ensemble of dipolar molecules in the amorphous state and found that the complex permittivity $\varepsilon(\omega)$ could be related to the time-autocorrelation function $\Phi_\mu(t)$ for the reorientational motions of the dipole vector μ of a molecule in space and time in the absence of a perturbing electric field:

$$\Phi_\mu(t) = \langle \mu(0) \cdot \mu(t) \rangle / \mu^2 = \langle \cos \theta(t) \rangle \tag{47}$$

where $\theta(t)$ is the angle between the dipole vector at $t = 0$ and $t = t$. The average is taken over the equilibrium ensemble. Subsequent developments of the theory by Cole (76) and several others (see reference 29 and reference therein) are mainly concerned with important details of the relationships between applied macroscopic E fields and the local E fields experienced by the individual molecules. Developments in the time-correlation function approach to dielectric relaxation up to 1972 were reviewed by Williams (28). The questions of local field corrections were effectively resolved by Titulaer and Deutch (77) and by Sullivan and Deutch (78) in the mid 1970s, so well-established practical working relations exist between $\varepsilon(\omega)$ and the time-correlation functions for the motions of dipoles. Such theories do not specify the mechanism for dipole motion, so they are general—within the approximations that relate local and macroscopic electric fields.

One common difficulty with the use of the time-correlation function approach to dielectric relaxation is that the principles and practice of time-dependent statistical mechanics are not commonly encountered in taught courses in chemistry, physics, materials science, or polymer science at universities and are not given much attention in the standard texts of physical chemistry, chemical physics, polymer science, or materials science. The *language* of time-correlation functions is, therefore, unfamiliar to many polymer scientists, who may be familiar with advanced aspects of equilibrium and nonequilibrium thermodynamics and quantum mechanics (for spectroscopy). Therefore, a brief summary is appropriate and instructive on how $\varepsilon(\omega)$ may be related to time-correlation functions by using linear-response theory, as described earlier by the author in a review of time-correlation functions and molecular motion (28). The derivation follows closely that of Glarum (74) and Cole (76).

The phase-space distribution function $f(p,q)$ for N equivalent dipolar molecules contained in a volume V, where $f(p,q) \, dp \, dq$ is the probability of obtaining the ensemble in a configuration p with associated momenta q ($p = f(p_i, p_j \ldots)$), ($q = F(q_i, q_j \ldots)$) in the region $dp \, dq$ around p,q, is given by the Liouville equation of motion.

$$\frac{df}{dt} = \frac{\partial f}{\partial t} + \mathscr{L}f = \frac{\partial f}{\partial t} + \sum_i^N \left[\frac{\partial H}{\partial p_i} \frac{\partial f}{\partial q_i} - \frac{\partial H}{\partial q_i} \frac{\partial f}{\partial p_i} \right] \tag{48}$$

where $\mathscr{L}$ is the Liouville operator, $\mathscr{L}f = -\{f,H\} = \{H,f\}$, H is the Hamiltonian of the system, and the brackets are Poisson brackets. If a uniform electric field $E_Z(t)$ is applied to the system

$$H(p,q:t) = H_0(p,q) - M_Z(q)E_Z(t) \tag{49}$$

where $M_z(q) \cdot E_i(t) = -\sum_i m_i(q)E_z(t)$ is the sum of the energy of interactions between the individual dipole moments and the applied field. Linear response theory proceeds by writing $\mathscr{L} = \mathscr{L}_0 + \mathscr{L}_1$ and noting that $M_Z(q)$ is independent of moments p_i. We have

$$\mathscr{L}_0 = \sum_i^N \left[\frac{\partial H_0}{\partial p_i} \frac{\partial}{\partial q_i} - \frac{\partial H_0}{\partial q_i} \frac{\partial}{\partial p_i} \right] \tag{50a}$$

$$\mathscr{L}_1 = E_Z(t) \sum_i^N \left[\frac{\partial M_Z}{\partial q_i} \frac{\partial}{\partial p_i} \right] \tag{50b}$$

Writing $f = f_0 + f_1$ then from eqs 48–50 we obtain the paired equations

$$\frac{\partial f_0}{\partial t} = -\mathscr{L}_0 f_0 \tag{51a}$$

$$\frac{\partial f_1}{\partial t} = -[\mathscr{L}_0 f_1 + \mathscr{L}_1 f_0] \tag{51b}$$

$\mathscr{L}_1 f_1$ is omitted in eqs 51a and 51b in order that $f_1 = O(E_Z)$. This value is the linear response condition that the change from f_0 to $(f_0 + f_1)$ is linear in the applied field. Solving eqs 51a and 51b we obtain (28)

$$f_0 = A \exp(-\beta H_0) \tag{52a}$$

$$f_1(t) = -\int_{-\infty}^t \{\exp[-(t - t')\mathscr{L}_0]\}\mathscr{L}_1 f_0 \, dt' \tag{52b}$$

where $\beta^{-1} = kT$.

The average dipole moment in the field direction is

$$\langle \mathbf{M}_Z(t) \rangle = \iint \mathbf{M}_Z(q) f(p,q:t) \, dp \, dq \tag{53}$$

Combination of eqs 52a and 52b with eq 53 gives, after considerable manipulation, the result

$$\langle \mathbf{M}_Z(t) \rangle = - \frac{\langle \mathbf{M}(0) \cdot \mathbf{M}(0) \rangle}{3kT} \int_{-\infty}^{t} E_Z(t') \dot{\Phi}_\mu(t - t') \, dt' \qquad (54)$$

where $\mathbf{M}(0) = \sum_i \mathbf{\mu}_i(0)$ and $\Phi_\mu(t)$ is the dipole-moment time-correlation function for the ensemble that is defined as

$$\Phi_\mu(t) = \frac{\langle \mathbf{M}(0) \cdot \mathbf{M}(t) \rangle}{\langle \mathbf{M}(0) \cdot \mathbf{M}(0) \rangle} \qquad (55)$$

Thus, the macroscopic dipole moment $\langle M_Z(t) \rangle$ is proportional to the equilibrium quantity $\langle \mathbf{M}(0) \cdot \mathbf{M}(0) \rangle$, which is related to ε_0 (see eq 1) and to a superposition integral in eq 54, involving field history, $E(t')$, and the intrinsic molecular quantity $\Phi_\mu(t)$, which is a time-correlation function (eq 55) that is deduced in the absence of the field. Note the formal similarity of eq 54 to the superposition equation used in phenomenological theory of dynamic mechanical relaxation and dielectric relaxation (see reference 8, p 104, eq 4.5a).

Equation 54 may be solved for three cases of interest (28). Writing, for convenience, $\langle \mathbf{M}(0) \cdot \mathbf{M}(0) \rangle / 3kT = p_\mu$ and writing $\Im$ to indicate a one-sided Fourier transform we have

1. E_0 applied as a step at $t' = 0$

$$\langle M_Z(t) \rangle = p_\mu E_0 [1 - \Phi_\mu(t)] \qquad (56a)$$

2. E_0 applied at $t = -\infty$ is removed as a step at $t' = 0$

$$\langle M_Z(t) \rangle = p_\mu E_0 \Phi_\mu(t) \qquad (56b)$$

3. steady-state ac field $E(t') = E_0 \exp(i\omega t')$

$$\langle M_Z(t) \rangle = - p_\mu E_0 \exp(i\omega t') \Im[\dot{\Phi}(t)]$$
$$= p_\mu E_0 \exp(i\omega t') \{1 - i\omega \Im[\Phi_\mu(t)]\} \qquad (57)$$

These equations connect the transient experiments (eqs 56a and 56b) to the steady-state ac experiments (eq 57). Equations 56a and 56b show that the time functions involved in step-on and step-off experiments are equivalent (a requirement for linear systems) and that Fourier transformation of the transient function $\Phi_\mu(t)$ yields the steady-state ac behavior, eq 57. Because $\langle M_Z(t) \rangle = (\varepsilon(\omega) - \varepsilon_\infty) E_Z(t) K(t) \varepsilon_0$, where $K(t)$ is an internal field factor connecting applied and local fields we have from eq 57.

$$\frac{\varepsilon(\omega - \varepsilon_\infty)}{\varepsilon_0 - \varepsilon_\infty} p(i\omega) = 1 - i\omega \Im[\Phi_\mu(t)] \tag{58}$$

where $K(\omega)/K(0) = p(i\omega)$. Equation 58 is the important result of this linear response theory: $\varepsilon(\omega)$ is related to a molecular time-correlation function $\Phi_\mu(t)$. There are complications associated with the internal field factor. For example Klug, Kranbuehl, and Vaughan (79) wrote

$$p(i\omega) = \frac{\varepsilon_0}{\varepsilon(\omega)}\left[\frac{2\varepsilon(\omega) + \varepsilon_\infty}{2\varepsilon_0 + \varepsilon_\infty}\right] \tag{59}$$

Also the relationship between liquid dipole moment and vacuum dipole moments needs to be considered both at equilibrium ($t = 0$) and with respect to time (t) so a close examination of $\Phi_\mu(t)$ will reveal further internal field factors. These internal field complications were discussed by Cole (76), Deutch and co-workers (77, 78), Brot (80), Cook, Watts, and Williams (29), and Williams (27, 28).

There is little doubt that for applications of linear response theory to dielectric data for amorphous polymer systems, eq 58 setting $p(i\omega) = 1$ gives a good approximation for the determination of $\Phi_\mu(t)$ and hence its functional form. By using this approach Cook, Watts, and Williams (29) deduced $\Phi_\mu(t)$ for the α-relaxation in poly(ethyl acrylate). This result was achieved by using the inversion of the imaginary part of eq 58 (*see* eq 40) so that

$$\Phi(t/\langle\tau\rangle) = \frac{\int_{-\infty}^{\infty}\left(\frac{\varepsilon''(\omega)}{\varepsilon_m''}\right)\cos\left(\frac{\omega}{\omega_m}\frac{t}{\langle\tau\rangle}\,d\ell n(\omega/\omega_m)\right)}{\int_{-\infty}^{\infty}\left(\frac{\varepsilon''(\omega)}{\varepsilon_m''}\right)d\ell n(\omega/\omega_m)} \tag{60}$$

where $\omega_m\langle\tau\rangle = 1$. The plot of $\varepsilon''/\varepsilon_m''$ versus $\log f/f_m$ (a master loss curve) is obtained from the experimental data where ε_m'' is the maximum loss factor observed at $\omega = \omega_m = 2\pi f_m$. The cosine transform of the loss data yields a relaxation function $\Phi(t)$ that is approximately equal to the dipole moment correlation function $\Phi_\mu(t)$ (*compare* eqs 24 and 58).

To summarize, linear response theory yields the working relation eq 58, which connects $\varepsilon(\omega)$ to $\Phi_\mu(t)$ for an amorphous polymer system, independent of any mechanism for dipole motion. The value of $\Phi_\mu(t)$ may be deduced for any chosen model for motion in the absence of a perturbing E-field. Following Cook, Watts, and Williams (29), $\Phi_\mu(t)$ may be expressed in terms of time-correlation functions for the reorientational motions of molecular dipoles, so the generalization of eq 2 to the dynamic situation gives

$$\Phi_\mu(t) = \frac{\sum\limits_{i}^{N} \sum\limits_{i'}^{N} \langle \mathbf{P}_i^{\mathrm{L}}(0) \cdot \mathbf{P}_{i'}^{\mathrm{L}}(t) \rangle}{\sum\limits_{i}^{N} \sum\limits_{i'}^{N} \langle \mathbf{P}_i^{\mathrm{L}}(0) \cdot \mathbf{P}_{i'}^{\mathrm{L}}(0) \rangle} \tag{61}$$

where $\mathbf{P}_i^{\mathrm{L}}(t)$ is the instantaneous dipole moment of molecule i. Expanding eq 61 gives autocorrelation terms $\langle \mathbf{P}_i^{\mathrm{L}}(0) \cdot \mathbf{P}_i^{\mathrm{L}}(t) \rangle$ for the motions of molecule i and cross-correlation terms $\langle \mathbf{P}_i^{\mathrm{L}}(0) \cdot \mathbf{P}_{i'}^{\mathrm{L}}(t) \rangle$ between molecules i and i'. For rod-like chains having persistent dipole moments along the long axis and no transverse dipole moments

$$\langle \mathbf{P}_i^{\mathrm{L}}(0) \cdot \mathbf{P}_i^{\mathrm{L}}(t) \rangle = \langle \mathbf{P}_i^{\mathrm{L,2}} \rangle \langle \cos \theta_{ii}(t) \rangle \tag{62}$$

where $\theta_{ii}(t)$ is the angle projected by dipole vector i at $t = t$ back on its direction at $t = 0$. Cross-correlation terms involve equilibrium angular correlations between dipoles i and i' because we may write

$$\langle \mathbf{P}_i^{\mathrm{L}}(0) \cdot \mathbf{P}_{i'}^{\mathrm{L}}(t) \rangle = \langle \mathbf{P}_i^{\mathrm{L}}(0) \cdot \mathbf{P}_{i'}^{\mathrm{L}}(0) \rangle \, \lambda_{ii'}(t) \tag{63}$$

where $\lambda_{ii'}(t)$ is a normalized cross-correlation function.

For the special case of an ensemble of flexible chains each containing only one type of dipolar group and with no orientation correlations between chains, then, for species of high molecular weight (29)

$$\Phi_\mu(t) = \frac{\langle \boldsymbol{\mu}_k^{\mathrm{L}}(0) \cdot \boldsymbol{\mu}_k^{\mathrm{L}}(t) \rangle + \sum\limits_{k' \neq k} \langle \boldsymbol{\mu}_k^{\mathrm{L}}(0) \cdot \boldsymbol{\mu}_{k'}^{\mathrm{L}}(t) \rangle}{\mu_k^2 + \sum\limits_{k' \neq k} \langle \boldsymbol{\mu}_k^{\mathrm{L}}(0) \cdot \boldsymbol{\mu}_{k'}^{\mathrm{L}}(0) \rangle} \tag{64}$$

Terms in the denominator of the right-hand side of eq 64 represent equilibrium angular correlations between dipoles along a chain (see eq 7). Autocorrelation terms $\langle \boldsymbol{\mu}_k^{\mathrm{L}}(0) \cdot \boldsymbol{\mu}_k^{\mathrm{L}}(t) \rangle$ refer to the motion of a reference dipole k along a chain, whereas cross-correlation terms may be written as

$$\langle \boldsymbol{\mu}_k^{\mathrm{L}}(0) \cdot \boldsymbol{\mu}_{k'}^{\mathrm{L}}(t) \rangle = \langle \boldsymbol{\mu}_k^{\mathrm{L}}(0) \cdot \boldsymbol{\mu}_{k'}^{\mathrm{L}}(0) \rangle \, \lambda_{kk'}(t) \tag{65}$$

As we have seen, for flexible chains, terms in $\langle \boldsymbol{\mu}_k^{\mathrm{L}}(0) \cdot \boldsymbol{\mu}_{k'}^{\mathrm{L}}(0) \rangle$ decay in magnitude as $|k - k'|$ increases. Therefore, we have the important result that the dielectric relaxation behavior of an amorphous polymer material composed of flexible chains is expressed through eqs 58 and 64. In physical terms the most important contributions to $\varepsilon(\omega)$ arise from the autocorrelation terms, that is the contribution from the motion of any reference dipole k. However, cross-correlation terms make important contributions through their magnitudes ($\langle \boldsymbol{\mu}_k^{\mathrm{L}}(0) \cdot \boldsymbol{\mu}_{k'}^{\mathrm{L}}(0) \rangle$) and their time-dependencies $\lambda_{kk'}(t)$, which may, in principle, be different from that of the autocorrelation term $\lambda_{kk}(t)$. However, it has been reasoned (*81*) that under certain conditions the $\lambda_{kk'}(t)$ will have exactly the time dependence of $\lambda_{kk}(t)$. This result means that although cross-correlation terms contribute to the relaxation strength $\Delta\varepsilon$, the time dependence of $\Phi_\mu(t)$ may be understood simply in terms of the motions of any reference dipole moment in a chain.

Certain flexible polymers exhibit a persistent cumulative dipole moment along the chain contour together with a transverse dipole moment component for each repeat unit that may be relaxed by local chain motions. First observed by Stockmayer and Baur for polypropylene oxides (*82, 83*) and subsequently by Adachi, Kotaka, and Imanishi for polyisoprene in bulk and in solution (e.g., references 84–87 and references therein) it has been amply demonstrated that such polymers exhibit two dielectric relaxation regions:

- a high frequency process due to segmental motions of the transverse component $\mu_\perp$ of the dipole moment of chain segments

- a low frequency process due to long-range motions of the cumulative dipole moment $\mathbf{P}$ along a chain

This process in the bulk polymer exhibits different dependencies on polymer molecular weight below and above a critical molecular weight, M_c, for the formation of chain entanglements. For $M < M_c$, for $\langle \tau \rangle$ the low frequency process varies as M^1 (*84–87*), but above M_c it varies according to $M^{3.7}$ (*84–87*), which is similar to the dependence observed for mechanical relaxation in an entangled polymer system. For $M > M_c$, a number of models for motion may be considered to apply, prominent among which is the reptation model of de Gennes for a chain moving in a virtual tube defined by neighboring chains in the polymer melt. The dipole moment correlation function for an entire chain i is written as (*87*)

$$\Phi_{\mu i}(t) = \frac{\mu^2 \langle \mathbf{r}_i(0) \cdot \mathbf{r}_i(t) \rangle + \sum_j \sum_{j'} \langle \boldsymbol{\mu}_{ji}^\perp(0) \cdot \boldsymbol{\mu}_{j'i}^\perp(t) \rangle}{\mu^2 \langle \mathbf{r}_i(0) \cdot \mathbf{r}_i(0) \rangle + \sum_j \sum_{j'} \langle \boldsymbol{\mu}_{ji}^\perp(0) \cdot \boldsymbol{\mu}_{j'i}^\perp(0) \rangle} \tag{66}$$

where μ is the parallel dipole moment component of a chain unit, $\mathbf{r}_i(t)$ is the vector sum of bond vectors along the chain so $\langle \mathbf{r}_i(0) \cdot \mathbf{r}_i(0) \rangle$ is the mean-square end-to-end distance, and $\langle \mathbf{r}_i(0) \cdot \mathbf{r}_i(t) \rangle$ is the time-correlation function for end-to-end vector motion. The terms $\langle \boldsymbol{\mu}_{ji}^\perp(0) \cdot \boldsymbol{\mu}_{j'i}^\perp(t) \rangle$ are time-correlation functions for the motions of the perpendicular (transverse) components of the dipole moments per repeat unit and $\langle \boldsymbol{\mu}_{ji}^\perp(0) \cdot \boldsymbol{\mu}_{j'i}^\perp(0) \rangle$ are equilibrium autocorrelation ($j = j'$) and cross-correlation ($j \neq j'$) terms for the transverse dipole moments.

If a chain has a cumulative dipole moment along its contour (as for polypropylene oxides and polyisoprenes) then the long-range motions may be observed dielectrically through $\langle \mathbf{r}_i(0) \cdot \mathbf{r}_i(t) \rangle$. This result was demonstrated by Stockmayer and Baur (82, 83), Adachi, Kotaka, and Imanishi (84–87), and by Boese et al. (88) who studied multiarmed star polymers of bulk cis-1,4-polyisoprene. In all these studies a low frequency process due to long-range motions was observed whose frequency location was strongly dependent on molecular weight and was interpreted as being a normal-mode process (Zimm–Rouse) for bulk materials below M_c and as a reptation-like process for $M > M_c$, where M_c is the critical molecular weight for the onset of entanglements. In addition a high-frequency process whose magnitude and frequency location were essentially independent of molecular weight was observed and was due to the local motions of the transverse dipole moments along a chain. Thus, dielectric theory provides a very satisfactory account of the multiple relaxations seen in bulk amorphous polypropylene oxides and polyisoprenes.

We have shown how the complex dielectric permittivity is related to certain dipole-moment time-correlation functions (auto- and cross-) for a polymer chain in an isotropic (amorphous) material. This linear-response theory may be generalized to include anisotropic motions of individual dipole groups along a chain. Rosato and Williams (89) expressed the dipole moment of a dipolar unit as follows:

$$\mu'^{(m)} = \sum_{n=-1}^{n=1} D_{mn}^{1*}(\Omega)\mu^{(n)} \tag{67}$$

where primed and unprimed notations denote laboratory and molecular coordinates, respectively, and the D_{mn}^1 elements are of a Wigner rotation matrix of order 1 in Euler space (α,β,γ) (90). By using linear response theory, these equations show that the dipole moment autocorrelation function for a representative dipole in an ensemble of dipoles is given by

$$\Phi_\mu(t) = \sum_{m=-1}^{1} (-1)^m \mu^{(-m)} \mu^{(m)} a_{10} \langle D_{0m}^1[\Omega(t)] D_{0m}^{1*}[\Omega(0)] \rangle \tag{68}$$

and the $\mu^{(m)}$ are the magnitudes of the dipole moment components along the Cartesian axes chosen for the dipole group.

The value of a_{10} is a known coefficient and the time-correlation functions (TCF) $\langle D_{0m}^1 \Omega(t) D_{0m}^{1*} \Omega(0) \rangle$ are associated with the motions of the defined components of the dipole moments. Thus, anisotropic motions of the individual dipole groups along a chain can lead to a weighted sum of dielectric relaxation processes. Because polymer chains are highly anisotropic it seems appropriate to include the anisotropy of the motions of chain units when the nature of α-, β-, and $\alpha\beta$-relaxations is considered for amorphous polymers but this has not been done as far as the author is aware. The treatment of Rosato and Williams (*89*) may, in principle, be generalized to include cross-correlation functions along a chain, but this has not been described in the literature. Rosato and Williams (*89, 91*) have also given a molecular theory for the dynamic electrooptic Kerr effect by using nonlinear response theory and have considered the application of theories of dielectric and Kerr-effect relaxations to low frequency relaxations in liquids and amorphous polymers.

Anisotropic Polymers: Thermotropic Liquid Crystals

We have seen that the static dielectric permittivity of thermotropic liquid crystals involves longitudinal and transverse dipole moments, μ_ℓ and μ_t, of the mesogenic group, anisotropic internal field factors and the Kirkwood g factors $g_\parallel$ and $g_\perp$. The generalization of this theory to the dynamic situation is extremely complicated but has been achieved by Bordewijk (*7, 92*), Luckhurst and Zannoni (*93*), and by Edwards and Madden (*94*). The complex permittivities $\varepsilon_\parallel(\omega)$ and $\varepsilon_\perp(\omega)$ reflect the anisotropy of the motions of the LC-forming (mesogenic) groups in the anisotropic potential of the LC phase, and eq 20 is generalized to the dynamic situation giving (*see* reference 7, p. 470)

$$\frac{\varepsilon_\gamma(\omega) - \varepsilon_{\infty\gamma}]}{[\varepsilon_{0\gamma} - \varepsilon_{\infty\gamma}]} \frac{\{\varepsilon_\gamma(\omega) - A_\gamma^*(\omega)[\varepsilon_\gamma(\omega) - \varepsilon_{\infty\gamma}]\}\varepsilon_{0\gamma}}{\{\varepsilon_{0\gamma} - A_\gamma^*(\omega)[\varepsilon_{0\gamma} - \varepsilon_{\infty\gamma}]\}\varepsilon_\gamma(\omega)} = 1 - i\omega\Im[\Phi_{\mu\gamma}(t)]$$

(69)

Here, $\gamma = \parallel$ or $\perp$, the A_γ^* values are theoretical factors analogous to the A_γ in eq 20, and the TCF are given by

$$\Phi_{\mu\gamma}(t) = \frac{\sum_k \sum_{k'} \langle \boldsymbol{\mu}_{k\gamma}(0)\cdot\boldsymbol{\mu}_{k'\gamma}(t)\rangle}{\sum_k \sum_{k'} \langle \boldsymbol{\mu}_{k\gamma}(0)\cdot\boldsymbol{\mu}_{k'\gamma}(0)\rangle}$$

(70)

where the superscripts ℓ on the μ terms were omitted for brevity. Equation 70 contains both auto- and cross-correlation functions for longitudinal and transverse dipole components of mesogenic groups. Thus, eqs 69 and 70 relate the principal components $\varepsilon_\parallel(\omega)$ and $\varepsilon_\perp(\omega)$ of the complex permittivity to the component dipole moments $\mu_{k\parallel}$ and $\mu_{k\perp}$ of individual groups k, and to equilibrium correlation functions $\langle \boldsymbol{\mu}_{k\gamma}(0)\cdot\boldsymbol{\mu}_{k'\gamma}(0)\rangle$ for different groups, along a chain and between chains, and to the TCF for the motions of groups that include both auto- and cross-TCF for dipole motion.

For the special case where the internal field factors are taken to be negligible, eq 69 reduces to

$$\frac{\varepsilon_\gamma(\omega) - \varepsilon_{\infty\gamma}}{\varepsilon_{0\gamma} - \varepsilon_{\infty\gamma}} = 1 - i\omega\Im[\Phi_{\mu\gamma}(t)] \tag{71}$$

$$\Phi_{\mu\gamma}(t) = \frac{\langle \mu_{k\gamma}(0)\cdot\mu_{k\gamma}(t)\rangle}{\langle \mu_{k\gamma}^2\rangle} \tag{72}$$

Because $\mu_{k\gamma}$ can be resolved into longitudinal and parallel components of molecular dipole moments in the mesogenic group, $\Phi_{\mu\parallel}(t)$ and $\Phi_{\mu\perp}(t)$ will generally contain more than one time correlation function. Nordio and Segre, Maier and Meier, and Bordewijk (*see* reference 7 for a recent account) have shown how these TCF may be represented for the special case of anisotropic rotational diffusion of mesogenic groups in the LC phase. More recently, Araki et al. (*41*) generalized the earlier treatments to obtain general relations between $\varepsilon_\parallel(\omega)$ and $\varepsilon_\perp(\omega)$ and the time correlation functions for dipole motion. The orientational distribution function of molecular axes of the mesogenic groups in the absence of an applied electric field is written as (*41*)

$$f^0(\Omega_0) = \sum_{J=0}^{\infty} \left(\frac{2J+1}{8\pi^2}\right) \overline{D}_{00}^J D_{00}^J(\Omega_0) \tag{73}$$

where $D_{mn}^J(\Omega_0)$ values are Wigner rotation matrix elements expressed in the laboratory frame $\Omega_0 \equiv \Omega(\alpha,\beta,\gamma)$, where α, β, and γ are the Euler angles. Here

$$\overline{D}_{00}^J = \langle D_{00}^J\rangle = \int_{\Omega_0} f^0(\Omega_0)D_{00}^{*J}(\Omega_0)\ d\Omega_0 \tag{74}$$

The field-perturbed distribution $f^E(\Omega_0)$ in the presence of a steady electric field **E** is written as (*41*)

$$f^E(\Omega_0) = \left(1 + \frac{\boldsymbol{\mu}\cdot\mathbf{E}}{kT} + \cdots\right)f^0(\Omega_0) \tag{75}$$

whereas the conditional probability function $f(\Omega,t/\Omega_0,0)$ of obtaining the molecule in the orientation around Ω at time t given it was around Ω_0 at $t = 0$ is written as

$$f(\Omega,t/\Omega_0,0) = \sum_{J'mP'} D_{P'm}^{J'}(\Omega_0)D_{P'm}^{*J'}(\Omega)\Phi_{P'm}^{J'}(t) \tag{76}$$

If the steady field is withdrawn at $t = 0$ then the dipole moment of the sample decays according to the relation

$$\langle \mu_{\text{lab}}^{(1,P)}(t) \rangle = \int_\Omega \int_{\Omega_0} f^E(\Omega_0) f(\Omega, t/\Omega_0, 0) \mu_{\text{lab}}^{(1,P)} \, d\Omega \, d\Omega_0 \qquad (77)$$

where we use the spherical representation of the vector so that $\mu_{\text{mol}}^{(1,0)} = \mu_z$, $\mu_{\text{mol}}^{(1 \pm 1)} = (1/\sqrt{2})(\mu_x \pm i\mu_y)$ and subscripts lab and mol refer to the laboratory (X, Y, Z) and molecular (x, y, z) frames, respectively.

Combining eqs 73–77, $\mu_z^{\text{lab}}(t)$ and $\mu_x^{\text{lab}}(t)$ may be deduced (*41*) and lead to the following expressions for the principal complex permittivities

$$\varepsilon_\parallel(\omega) = \varepsilon_{\infty\parallel} + \frac{G}{3kT} [(1 + 2S)\mu_\ell^2 F_\parallel^\ell(\omega) + (1 - S)\mu_t^2 F_\parallel^t(\omega)] \qquad (78a)$$

$$\varepsilon_\perp(\omega) = \varepsilon_{\infty\perp} + \frac{G}{3kT} [(1 - S)\mu_\ell^2 F_\perp^\ell(\omega) + (1 + S/2)\mu_t^2 F_\perp^t(\omega)] \qquad (78b)$$

Equation 78 is seen to be the generalization of eq 18 to the dynamic situation. The terms $F_j^i(\omega)$ are given by the Fourier-transform relation

$$F_j^i(\omega) = 1 - i\omega \Im[F_j^i(t)] \qquad (79)$$

where the $F_j^i(t)$ are real time-correlation functions composed of linear combinations of certain complex time correlation functions, as given by Araki et al. (*41*)

$$F_\parallel^\ell(t) = \Phi_{00}^1(t); \quad F_\parallel^t(t) = \Phi_{01}^1(t) + \Phi_{0-1}^1(t)$$
$$F_\perp^\ell(t) = \Phi_{-10}^1(t) + \Phi_{10}^1(t) \qquad (80)$$
$$F_\perp^\ell(t) = \Phi_{-1-1}^1(t) + \Phi_{-11}^1(t) + \Phi_{1-1}^1(t) + \Phi_{11}^1(t)$$

where

$$\Phi_{Pm}^J(t) = \langle D_{Pm}^{J*}(\Omega_0) D_{Pm}^J(\Omega) \rangle \qquad (81)$$

Equation 78 shows that there are four relaxation modes, two for $\varepsilon_\parallel(\omega)$ and two for $\varepsilon_\perp$. The lowest frequency mode $\Phi_{00}^1(t)$ is due to the motion of μ_ℓ with respect to the local director **n** and involves only the polar angle β. All four relaxation modes, involving μ_ℓ and μ_t in the molecular frame, have been shown pictorially by several authors (*95, 96*). An alternative theoretical treatment using linear response theory that again leads to eqs 78–81 was given by Kozak and Moscicki (*97*).

The anisotropic dielectric properties of thermotropic LC polymers have been extensively investigated during the past 10 years (*39–43, 98–116*, and references therein) and these and related studies will be considered in this volume (*see* Chapter 12 by G. P. Simon). Samples are prepared in different states of macroscopic alignment (by using directing electric or magnetic fields). Attard, Araki, and I (*40*) showed that the complex permittivity for an LC material having axial symmetry with respect to the measuring field direction (*z*-direction) is given by

$$\varepsilon_Z(\omega) = \left(\frac{1 + 2S_d}{3}\right)\varepsilon_\parallel(\omega) + \frac{2}{3}(1 - S_d)\varepsilon_\perp(\omega) \tag{82}$$

Equation 82 is seen to be the generalization of eq 15 to the dynamic situation. Separating eq 82 into real and imaginary parts we have the dual relations

$$\varepsilon'_Z(\omega) = \left(\frac{1 + 2S_d}{3}\right)\varepsilon'_\parallel(\omega) + \frac{2}{3}(1 - S_d)\varepsilon'_\perp(\omega) \tag{83}$$

$$\varepsilon''_Z(\omega) = \left(\frac{1 + 2S_d}{3}\right)\varepsilon''_\parallel(\omega) + \frac{2}{3}(1 - S_d)\varepsilon''_\perp(\omega) \tag{84}$$

If $\varepsilon'_z(\omega)$ is measured for a sample of intermediate alignment between homeotropic ($\mathbf{n}\|\mathbf{E}$) and homogeneous or planarly aligned ($\mathbf{n}\perp\mathbf{E}$) samples, and $\varepsilon'_\parallel(\omega)$ and $\varepsilon'_\perp(\omega)$ are known (from measurements on *H*-aligned and *P*-aligned samples, respectively) then S_d is determined from eq 83. The calculation can be repeated at each frequency at which $\varepsilon'_Z(\omega)$, $\varepsilon'_\parallel(\omega)$, and $\varepsilon'_\perp(\omega)$ are measured and yield a table of S_d values whose consistency provides a quantitative test of the method. Similarly, S_d values may be determined from measurements of the dielectric loss factors $\varepsilon''_Z(\omega)$, $\varepsilon''_\parallel(\omega)$, and $\varepsilon''_\perp(\omega)$. Attard, Araki, and I (*40*) have shown that consistent S_d values may be obtained from permittivity and from loss-factor measurements in this way by using eqs 83 and 84 for siloxane LC side-chain (LCSC) polymers. In addition, the changes with time of the macroscopic ordering of LCSC polymers in the presence or absence of applied fields can be monitored by measurements of $\varepsilon'_z(\omega,t)$ or $\varepsilon''_z(\omega,t)$ and using eqs 83 and 84 (*109–112*).

Note that in plots of $\varepsilon'_\parallel(\omega)$ versus $\log f$ and $\varepsilon'_\perp(\omega)$ versus $\log f$ that a common value may occur at a crossover frequency $f_c = \omega_c/2\pi$ at which $\varepsilon'_\parallel(\omega_c) = \varepsilon'_\perp(\omega_c)$. If $\Delta\varepsilon'(\omega) = \varepsilon'_\parallel(\omega) - \varepsilon'_\perp(\omega)$ is positive then application of a directing E field having $f > f_c$ or $f < f_c$ drives the LC material to the planar ($\mathbf{n}\perp\mathbf{E}$) or homeotropic ($\mathbf{n}\|\mathbf{E}$) orientations, respectively, and therefore a method is provided for obtaining *P*-aligned or *H*-aligned or intermediately aligned ($-0.5 < S_d < 1$) samples (*16, 17, 98, 104–107*). For a field applied at $f = f_c$ no directed alignment will be obtained. The alignment mechanism is a dielectrically driven process that involves the orientation of the LC director $\mathbf{n}$ in line with the field E accompanied by a flow of the phase. The theory of Martins et al. (*117, 118*) is a development of the continuum theories for liquid crystals as originally described by Leslie and Ericksen and by de Gennes. The equations of motion for the director are written as

$$\sigma \frac{\partial^2 \theta}{\partial t^2} + \gamma_1 \frac{\partial \theta}{\partial t} - \frac{\Delta \varepsilon}{8\pi^2} E^2 \sin 2\theta + (\alpha_2 - \gamma_2 \sin^2 \theta) v_{x,z} - K(\theta) = 0$$

(85)

$$\rho \frac{\partial v_{x,z}}{\partial t} = \frac{\partial^2}{\partial z^2} \left[j(\theta) \frac{\partial \theta}{\partial t} + g(\theta) v_{x,z} \right]$$

(86)

where the viscosities γ_1, α_2, and γ_2, the terms σ and ρ, and the functions $j(\theta)$ and $g(\theta)$ and the elastic constant term $K(\theta)$ are defined by Martins et al. (*117, 118*). The value $\theta = \theta(t)$ is the orientation of the director **n** with respect to the field direction (z); $v_{x,z}$ is the flow velocity of the aligning region. By using eqs 85 and 86 the alignment of an LC polymer with time in a directing field may be predicted for comparison with experimental data. This process was done for siloxane-LCSC polymers by Williams and co-workers (*17, 109, 119*). Equations 85 and 86 show *inter alia* that

1. Viscosity and elastic terms are involved in the behavior of $\theta(t)$. In the absence of flow ($v_{z,x}(t) = 0$), alignment will not take place, as was observed experimentally (*113*) for a lightly cross-linked LCSC polymer.

2. For $\partial \theta / \partial t \to 0$ and $\partial^2 \theta / \partial t^2 \to 0$ then two terms remain in eq 85 that balance the dielectric term $\Delta \varepsilon E^2 \sin 2\theta / 8\pi^2$ with the elastic term $K(\theta)$. Thus, for moderate E fields the alignment level will tend to plateau below that for full alignment, as is observed experimentally (*109, 119*).

3. The time required to align LCSC materials depends on flow viscosities and elastic constants whose values depend strongly on temperature. Therefore, the kinetic rate of the alignment process will decrease markedly as temperature is reduced below the clearing temperature T_c. The dominant factor for LCSC polymers is the marked increase in viscosity as temperature is lowered, which retards the flow process required for the orientation of the LC director **n**. Reviews of the alignment behavior of LCSC polymers in directing E fields were given recently (*17, 120*).

Note also that f_c is an isosbestic frequency so plots of $\varepsilon'_Z(\omega)$ versus $\log f$ for samples of different S_d values all pass through the common point $\varepsilon'_Z(\omega_c)$ = $\varepsilon'_{\parallel}(\omega_c)$ = $\varepsilon'_{\perp}(\omega_c)$. Similarly for loss data, all curves pass through the common point $\varepsilon''_Z(\omega'_c) = \varepsilon''_{\parallel}(\omega'_c) = \varepsilon''_{\perp}(\omega'_c)$ at the isosbestic frequency $f'_c = \omega'_c/2\pi$.

In summary, the theory of the dielectric relaxation of thermotropic LC polymers is rather complicated and shows that the principal complex permittivities $\varepsilon_{\parallel}(\omega)$ and $\varepsilon_{\perp}(\omega)$ are related to the time-correlation functions for the anisotropic motions of the components μ_{ℓ} and μ_t of the mesogenic dipolar groups. Both auto- and cross-TCF are involved (eqs 69 and 70). Consideration of the auto-TCF gives, for this special case, the result that $\varepsilon_{\parallel}(\omega)$ and $\varepsilon_{\perp}(\omega)$ each contain two relaxation modes (eq 78) whose strength factors are functions of μ_{ℓ}^2, μ_t^2 and the local order parameter, S, and whose frequency dependencies are given by Fourier transforms of relaxation functions that are given by linear combinations of TCF for the reorientational motions of μ_{ℓ} and μ_t in Euler space (eqs 80 and 81). Dielectric relaxation spectroscopy provides a direct and convenient nonoptical method for determining the director order parameter S_d (eq 14) for LCSC samples having a macroscopic alignment lying between homeotropic ($S_d = 1$) and planar or homogeneous ($S_d = -0.5$) through the application of eqs 83 and 84 to experimental data.

Dielectric Relaxation and Molecular Reorientations

We have shown above that the macroscopic static permittivity ε_0 and macroscopic complex permittivity $\varepsilon(\omega)$ may be related to the molecular quantities–properties of dipole moments and TCF for molecular reorientation. The theoretical relations thus obtained required knowledge of the nature of the phase (liquid, crystal, or liquid crystal) but did not specify the mechanism for dipole reorientational motions. Thus for an isotropic liquid, $\varepsilon(\omega)$ is related to the auto-TCF $\langle \mu(0) \cdot \mu(t) \rangle$ for dipole motions but the decay of this function in time could occur by a variety of, as yet, unspecified processes such as rotational diffusion, collision-interrupted rotation, hopping in a local barrier system, rotation via defect diffusion, rotation by thermal activation, cooperative reorientational processes, or by some other means. The question arises of what specific information on the mechanism of the reorientational process is obtained from a knowledge of the dipole moment auto-TCFs for dielectric relaxations in polymers (amorphous, crystalline, or liquid-crystalline).

Consider first the simple case of an isotropic liquid composed of small rigid dipolar molecules having no orientational (angular) correlations between them so $\varepsilon(\omega)$ gives information on the autocorrelation function.

$$\lambda_{ii}(t) = \frac{\langle \mu_i(0) \cdot \mu_i(t) \rangle}{\mu^2} = \iint_{\Omega,\Omega_0} f(\Omega, t/\Omega_0, 0) \cos \theta \, d\Omega \, d\Omega_0$$

$$= \langle \cos \theta(t) \rangle = \langle P_1(\cos \theta(t)) \rangle \tag{87}$$

where θ is the angle between the dipole vector $\mu(0)$ at $t = 0$ and $\mu(t)$ at $t = t$. The value of $f(\Omega,t/\Omega_0,0)d\Omega \, d\Omega_0$ is the conditional probability of obtaining the vector μ around Ω at t given it was around Ω_0 at $t = 0$ (*see* references 27, 28, and 121 for a description of $f(\Omega,t/\Omega_0,0)$ and its expansion in terms of TCFs). For this simple system we may write

$$f(\Omega,t/\Omega_0,0) = \sum_n a_n P_n (\cos \theta) \Psi_n(t) \tag{88}$$

where a_n is a known coefficient and P_n indicates the nth Legendre polynomical. The $\Psi_n(t)$ values are time-correlation functions for the isotropic reorientation of the dipole vector and are obtained from eq 88, as

$$\langle P_n (\cos \theta(t)) \rangle = \Psi_n(t) = \iint_{\Omega,\Omega_0} f(\Omega,t/\Omega_0,0) P_n (\cos \theta) \, d\Omega \, d\Omega_0 \tag{89}$$

Comparisons of eqs 87 and 89 show that

$$\lambda_{ii}(t) = \langle P_1(\cos \theta(t)) \rangle = \Psi_1(t) \tag{90}$$

Thus, the dielectric experiment gives information on only *one* moment of the time-dependent orientation distribution function $f(\Omega,t/\Omega_0,0)$, and therefore insufficient information is obtained to specify the mechanism for motion, which would need complete information on $f(\Omega,t/\Omega_0,0)$ itself. Thus, no matter how accurately $\Psi_1(t)$ is determined over its range $0 < \Psi_1(t) \leq 1$ from dielectric experiments, the mechanism for the process cannot be determined from the results of this experiment alone. Further information on higher moments, $\Psi_2(t)$, $\Psi_3(t)$, etc., is needed to describe $f(\Omega,t/\Omega_0,0)$ and hence allow certain mechanisms to be ruled out and others favored. This fundamental problem with relaxation experiments was recognized many years ago by the author in connection with dielectric relaxation in glass-forming liquids (*122*). As a result Clarkson and Williams (*122*) devised an information-theory method that allowed $f(\Omega,t/\Omega_0,0)$ to be reconstructed from a knowledge of $\Psi_1(t)$ or $\Psi_1(t)$ and $\Psi_2(t)$ by using a development of earlier work by Berne and co-workers (*123–125*) who had shown how $\Psi_2(t)$, $\Psi_3(t)$, etc., could be determined from a knowledge of $\Psi_1(t)$. Berne et al. showed that if $\Psi_1(t)$ was known then (*123–125*)

$$f(\Omega,t/\Omega_0,0) = \frac{\exp \beta(t) \cos \theta}{4\pi B_{1/2}(\beta(t))} \tag{91}$$

where $\beta(t)$ is a Lagrange undetermined multiplier, and B indicates a modified Bessel Function. They showed that

$$\Psi_n(t) = B_{n+1/2}(\beta(t))/B_{1/2}(\beta(t)) \qquad (92)$$

Hence, if $\Psi_1(t)$ is known then $\beta(t)$ is determined from eq 92 that, on insertion into eq 91, yields $f(\Omega,t/\Omega_0,0) = f(\cos\theta(t))$ as was first demonstrated by Clarkson and Williams (122). The method can be improved by adding further information that both $\Psi_1(t)$ and $\Psi_2(t)$ are known. In this case it was shown that (122)

$$f(\Omega,t/\Omega_0,0) = \exp(\alpha + \beta\cos\theta + \gamma\cos^2\theta) \qquad (93)$$

where α is a normalization constant.

$$\Psi_1(t) = \frac{\sum_n (\gamma^n/n!)I(\cos^{2n+1}\theta)}{\sum_n (\gamma^n/n!)I(\cos^{2n}\theta)} \qquad (94)$$

$$\Psi_2(t) = \frac{\left(\frac{3}{2}\right)\left[\sum_n (\gamma^n/n!)I(\cos^{2n+2}\theta)\right] - \left(\frac{1}{2}\right)\left[\sum_n (\gamma^n/n!)I(\cos^{2n}\theta)\right]}{\sum_n (\gamma^n/n!)I(\cos^{2n}\theta)} \qquad (95)$$

where β and γ are undetermined multipliers, and the integrals $I(\cos^m\theta)$ are evaluated in terms of $B_{m+1/2}(\beta)$ given by Berne (see reference 124, eqs 6.74 and 6.79). Thus, $(\Psi_1(t), \Psi_2(t))$ paired values at each time t yield $\beta(t)$ and $\gamma(t)$, which are inserted into eq 93 to yield values of $f(\Omega,t/\Omega_0,0)$ over the range of time investigated. Values of $\Psi_1(t)$ may be obtained from dielectric experiments, whereas $\Psi_2(t)$ can be determined for liquids, in suitably constructed apparatus, by Kerr-effect relaxation of the polarizable dipolar molecules. Such studies were made for glass-forming molecular liquids, such as solute:o-terphenyl solutions, di-n-butyl phthalate (121, 126–129).

Therefore, it is possible to reconstruct $f(\Omega,t/\Omega_0,0)$ from a knowledge of $\Psi_1(t)$ from dielectric experiments or more accurately from $\Psi_1(t)$ (dielectric) and $\Psi_2(t)$ (Kerr-effect relaxation). This result appears to be the maximum information obtainable by using dielectric results alone or in combination with complementary relaxation data before one begins to compare results (for $\Psi_1(t)$, $f(\Omega,t/\Omega_0,0)$) with models for relaxation. Beevers and I (130) gave an extensive account of time-dependent orientational distribution functions and their pictorial representations for motions of small molecules in the gaseous and simple-liquid states.

All of the previous discussions are concerned with the form of $f(\Omega,t/\Omega_0,0)$ for the motions of simple dipolar molecules in the nonassociated liquid state: that is, the discussion is concerned with the auto-TCF $\lambda_{ii}(t) = \langle P_1(\cos\theta(t))\rangle$ for representative molecules in the liquid state. For polymer chains in the amorphous isotropic state above T_g, where dipole group vectors are able to access 4π solid angle of Euler space in time due to micro-Brownian motions, then the same considerations apply to the auto-TCF $\lambda_{ii}(t)$ for representative dipoles along a chain if the motions are isotropic. Thus, $\lambda_{ii}(t)$ may be determined in principle from measurements of $\varepsilon(\omega)$ for an amorphous system in which the motions of dipoles are isotropic and the contributions to the relaxation from cross-correlation terms are negligible. In the more general case for an amorphous polymer where the cross-correlation terms are negligible but the motions are anisotropic, then $\varepsilon(\omega)$ measurements may yield, through eqs 60 and 68, a linear combination of defined auto-TCF $\langle D_{0m}^1(\Omega(t))D_{0m}'(\Omega(0))\rangle$ that cannot be further separated into each component without assuming a model for the anisotropic reorientational process.

For a polymer chain in the amorphous state where cross-correlation terms are significant we have to consider the meaning of the cross-TCF $\langle \mu_i(0)\cdot\mu_j(t)\rangle$. In this case the following conditional distribution functions arise

$$f(\Omega_i,0;\Omega_j,0) = \frac{8\pi^2}{Np}\int \rho^{(2)}(\mathbf{R}_i,\mathbf{R}_j)\ dr_{ij} \tag{96}$$

where $f(\Omega_i,0;\Omega_j,0)d\Omega_i\,d\Omega_j$ is the conditional probability of obtaining dipole i around the orientation Ω_i at $t = 0$ given that the dipole j was around the orientation Ω_j at $t = 0$. Here $\rho^{(2)}(\mathbf{R}_i,\mathbf{R}_j)$ is the pair distribution function for the particles i and j, r_{ij} is their relative distance, and ρ is sample density. The value of $\rho^{(2)}(\mathbf{R}_i,\mathbf{R}_j)$ may be expanded in terms of rotation matrices introducing further equilibrium correlation factors $g_{K_1M_1-\kappa_1\kappa_2}^{J_1J_2}$ as has been described by Pecora (*131*). The second distribution function is time dependent (*131*).

$$f(\Omega_j,t/\Omega_i,0) = \sum_{JKMM'} D_{KM}^{J*}(\Omega_i)D_{KM'}^{J}(\Omega_j)C_{KMM'}^{J}(t) \tag{97}$$

where $f(\Omega_i,t/\Omega_j,0)d\Omega_i\,d\Omega_j$ is the probability of obtaining dipole j in the orientation Ω_j at time t given dipole i had the orientation Ω_i at $t = 0$. The expansion is made in terms of the D functions of each dipole, and the $C_{KMM'}^{J}(t)$ values are cross-correlation functions for the motions of dipoles j and i. Hence

$$\langle \mu_i(0)\cdot\mu_j(t)\rangle = \mu_i\mu_j\int f(\Omega_i,0;\Omega_j,0)f(\Omega_j,t/\Omega_i,0)\cos(\Omega_j-\Omega_i)\ d\Omega_i\,d\Omega_j \tag{98}$$

where $(\Omega_j - \Omega_i)$ is the angle between the dipole vector i at $t = 0$ and the dipole vector j at $t = t$. Thus, the cross-correlation function is the averaged projection of the vector j at time t with respect to vector i at $t = 0$ further averaged over all allowed initial relative orientations of dipoles i and j at $t = 0$ and of the trajectories of dipole j in time from each initial relative orientation of these dipoles.

As we have seen in eqs 55, 58, and 59, $\varepsilon(\omega)$ is a function of autocorrelation terms $\lambda_{ii}(t)$ and $\lambda_{ij}(t) = \langle \mathbf{\mu}_i(0) \cdot \mathbf{\mu}_j(t) \rangle / \langle \mathbf{\mu}_i(0) \cdot \mathbf{\mu}_j(0) \rangle$ and, given the complexity of these relations for the cross-correlation functions, it might be thought that the problem of relating $\varepsilon(\omega)$ to the motions of dipoles in a polymer chain in the amorphous state would be intractable because $\varepsilon(\omega)$ will be given by a weighted sum of terms involving auto- and cross-TCF. However, in certain dipolar copolymers ($10, 81$) the form of the relaxation function, $\Phi(\tau)(\approx\Phi_\mu(t))$ is independent of composition thus indicating that the time dependence of the auto-TCF $\lambda_{ii}(t)$ and cross-TCF $\lambda_{ij}(t)$ is approximately the same for the α-process (as can be demonstrated by a simple geometrical argument for isotropic motions of chain molecules, as has been discussed in some detail ($10, 81$)). Thus, despite the complexity of eqs 96–98, those amorphous polymer systems that exhibit a constant shape for the plots of $\varepsilon_\alpha''(\omega)$ versus log f with changing dipole concentration along a chain (e.g., styrene–chlorostyrene copolymers) demonstrate that although cross-correlation terms are present, and may have appreciable magnitude ($\langle \mathbf{\mu}_i(0) \cdot \mathbf{\mu}_j(0) \rangle$) relative to the autocorrelation terms ($\langle \mathbf{\mu}_i(0) \cdot \mathbf{\mu}_j(0) \rangle = \langle \mathbf{\mu}_i^2 \rangle$), the $\lambda_{ij}(t)$ have approximately the same time dependence as the ACF $\lambda_{ii}(t)$ for the α-process (micro-Brownian motions of chains).

Multiple Relaxations in Amorphous Polymers

As we have seen, the reorientation of dipole vectors gives rise to dielectric relaxation in a system composed of polymer chains in the amorphous state (*see* eqs 55–60, 64). Multiple relaxations generally occur ($8, 13, 29, 48, 60, 63, 64–69, 72, 81$) and are as follows:

α'-Process

The long-range motions of the end-to-end dipole moment vector along a chain as is observed for polypropylene oxides and polyisobutylene by Stockmayer and Baur ($82, 83$) and Adachi and Kotaka ($84–87$ and references therein) and by Boese et al. (88). The theory of such motions has been introduced and is based on the Rouse–Zimm model for normal-mode motions of chains that also occur in viscoelasticity theory (47), for $M < M_c$ and is based on reptation theory for $M > M_c$.

α-Process

The micro-Brownian motions of chain segments that give rise to the dynamic glass-transition process ($8, 47$) also give rise to the primary (α) relaxation process in glass-forming liquids (*14, 132–135*). Its functional form is well approximated by the KWW function eq 34, and $\bar{\beta}$ is in the range 0.35–0.70, depending on the polymer (*132, 133*). The application of the KWW function to the α-relaxation in amorphous polymers with and without rotatable dipolar side groups is extensive (*9, 13, 27, 52, 53, 81*) and will not be discussed here in great detail.

β-Process

The local motions of chain segments involving partial reorientation of dipoles rigidly or flexibly attached to chains (as for polymethacrylates) give rise to a broad dielectric β-process (half-width of ε'' versus $\log f$ plots ~ 3–6 decades) (*8–13*).

A unified theory of α- and β-processes that leads, at high temperatures, to the merged process, the αβ-process, was given many years ago without prescribing specific mechanisms for the individual α-, β-, and αβ-process (*132*). Partial reorientation of a dipole vector in a temporary local environment, r, is assumed to make a contribution to the overall β-process and is followed at longer times by fluctuations that lead to complete reorientation of the vector. This simple model, which does not involve chain connectivity and is applicable to both amorphous polymers and glass-forming molecular liquids (*14*), gives the following dipole-moment autocorrelation function for the reference dipole (omitting subscripts *ii* for brevity) (*132*).

$$\lambda(t) = \frac{\langle \boldsymbol{\mu}(0) \cdot \boldsymbol{\mu}(t) \rangle}{\langle \mu^2 \rangle} = \varphi_\alpha(t) \left[\sum_r {}^0p_r q_r + \sum {}^0p_r(1 - q_r)\varphi_{\beta r}(t) \right]$$

$$= \varphi_\alpha(t) \left[A_\alpha + (1 - A_\alpha)\Psi_\beta(t) \right] \tag{99}$$

where ${}^0p_r$ is the equilibrium probability of obtaining environment r, $\varphi_\alpha(t)$ and $\varphi_{\beta r}(t)$ are normalized relaxation functions for the α- and β_r-processes that decay from 1 to 0 with time, $q_r = \langle \mu_r \rangle^2 / \langle \mu^2 \rangle$ where $\langle \mu_r \rangle$ is the mean moment residing in environment r when the β_r-process is completed. $A_\alpha = \sum_r {}^0p_r q_r$ and $\Psi_\beta(t)$ is the overall decay function for the β-process.

$$\Psi_\beta(t) = \frac{\sum {}^0p_r(1 - q_r)\varphi_{\beta r}(t)}{(1 - \sum_r {}^0p_r q_r)} \tag{100}$$

Equation 99 predicts that $\lambda(t)$ decays generally in two stages. The faster process (β-process) is due to partial reorientational motions of dipoles in a range of local environments, and the slower process (α-process) is due to the gross micro-Brownian motions of dipoles that relax the remainder of $\langle\mu^2\rangle$ that persists after the β-process is completed.

Equation 99 predicts the following behavior irrespective of detailed mechanisms for the α- and β-processes:

1. The total relaxation strength $\Delta\varepsilon$ is partitioned between the α- and β-processes, $\Delta\varepsilon = \Delta\varepsilon_\alpha + \Delta\varepsilon_\beta$, so any increase or decrease in the strength of one process with an external variable (T,P) will result in the decrease or increase in the strength of the other process. Thus, the dielectric data for poly(n-butylmethacrylate) (65) and poly(ethylmethacrylate) (67), where pressure led to a significant decrease in $\Delta\varepsilon_\beta$ accompanied by a corresponding increase in $\Delta\varepsilon_\alpha$, were rationalized in terms of eq 99 (see reference 10 for a discussion).

2. The α- and β-processes coexist in a certain range of temperature above T_g and show that the β-process is not unique to the glassy state.

3. For $T < T_g$, $\varphi_\alpha(t)$ becomes so slow that only the β-process can be observed in the accessible frequency range. Thus, the A_α contribution to $\varepsilon(\omega)$ is not observable and only the partial reorientations of dipoles are detected as the β-process (see eq 99). Because the chain-backbone motions are suppressed when the α-process is suppressed, it follows that the r environments do not permute in time between different chain segments so a physical picture results in which individual groups partially relax in fixed local environments. This model resembles the model of *islands of mobility* described by Johari and co-workers (135).

4. For $T > T_g$ as T is increased, the α- and β-relaxations coalesce to form the combined αβ-process. According to eq 99 when $\varphi_\alpha(t)$ relaxes far faster than $\Psi_\beta(t)$, the relaxation is dominated by the α-process. Thus, in a relaxation map of $\log f_m$ versus $1/T$ for α- and β-processes, the β-process will join the α-process forming the αβ($\equiv\alpha$)-process after the coalescence. This model was proposed by the author many years ago (10, 132, 133), and appears to give a good representation of the α-, β-, and αβ-processes in poly(alkyl-methacrylates) (10, 67–69) for which $\Delta\varepsilon_\beta > \Delta\varepsilon_\alpha$. However, in certain polymers such as poly(ethylene terephthalate) (8) and for simple glass-forming liquids (134, 135) the dielectric β-process approaches the α-process tangentially in the relaxation map. Johari (134, 135) suggested that in such cases the β-process extrapolated to high temperatures is the dominant process at high temperatures. However, the merged high temperature process in amorphous polymers and glass-forming liquids will ultimately approach

that of a single relaxation time (SRT) process, whereas the β-process observed for $T < T_g$ and in a limited range above T_g has a half-width in the ε'' versus $\log f$ plot many times greater than that (1.14) for a SRT process.

A full discussion of the nature and occurrence of α-, β-, and αβ-dielectric relaxations in amorphous polymers was given by Williams and co-workers (9–13). Simple glass-forming molecular liquids also exhibit a similar pattern of α-, β-, and αβ-dielectric processes and these have been reviewed and discussed especially in relation to polymer relaxations (14, 136, 137). The simple theory outlined previously leading to eq 99 allowed a prediction that the primary dielectric relaxation observed in higher polyalkylmethacrylates and poly(vinyl chloride) (PVC) for $T > T_g$ was predicted to be αβ-processes rather than an α-process, as was demonstrated through experiments at high pressures where the single process transformed into α- and β-processes (*see* reference 138 for the data for poly(*n*-lauryl and *n*-nonyl)-methacrylates and reference 68 for PVC).

Even though eq 99 gives a good representation of α-, β-, and αβ-relaxations in all glass-forming materials (polymeric and nonpolymeric), the α- and β-processes are probably not completely independent in the coalescence region and at higher temperatures. In this region the backbone α-motions occur in a time scale comparable with the local motions so the backbone motions are facilitated by the local motions leading to an increased rate for the backbone motions. This result has not been observed experimentally for amorphous polymers, but simulations of the dipole-moment correlation function for a dense assembly of chains by Pakula have shown that such an increase in the rate for the α-process is obtained when α- and β-processes coalesce at high temperatures (139).

Equation 99 concerns the auto-TCF for the motion of a reference chain unit. Its generalization to include cross-correlation terms has been given by Williams (10). All qualitative conclusions of the simple model are retained: that is, the β-process is due to the weighted sum of contributions of local motions in a range of local environments and the α-process is due to the gross micro-Brownian motions of chain segments leading again to the conditions 1–4 for α-, β-, and αβ-relaxations. However, additional strength factors due to equilibrium cross-correlations between dipoles and cross-TCF for dipole motion, $\lambda_{ij}(t)$, appear in the general expression for $\Phi_\mu(t)$ (*see* eqs 21–26 of reference 10).

Mechanisms for α-Process in Amorphous Polymers

Interest is continuing to see if the dielectric α-process (which exhibits KWW behavior, approximately, in all amorphous polymers and most glass-forming liquids) can be described in terms of particular mechanisms for dipole motion. The similarity in behavior between both classes of material suggests that mechanisms based on chain connectivity and chain dynamics are not required to explain the polymer behavior (Occam's razor, as pointed out originally by Johari and Goldstein). By a comparison of $\Psi_1(t)$ obtained from dielectric relaxation and $\Psi_2(t)$ obtained from Kerr-effect relaxation for selected glass-forming liquids it was shown that small-step rotational diffusion could be ruled out as the mechanism for motion and a strong-collision model could be favored (121, 126–129). For small-step rotational diffusion (30)

$$\Psi_n(t) = \exp - n(n + 1)D_r t \tag{101}$$

where D_r is the rotational diffusion coefficient for the dipole group.

Thus, the correlation functions $\Psi_1(t)$ (dielectric) and $\Psi_2(t)$ (Kerr-effect) would both be exponential in time and the correlation (or relaxation) times would be in the ratio $\tau(\text{Kerr})/\tau(\text{dielectric}) = 1/3$. This result was not found experimentally for the α-process in solute/o-terphenyl solutions (129) or for tritolyl phosphate (121, 126), but KWW-relaxation was observed in all cases by both techniques and $\langle\tau(\text{Kerr})\rangle/\langle\tau(\text{dielectric})\rangle \sim 1$. This observation favored a *strong-collision* model for which (121)

$$f(\Omega, t/\Omega_0 0) = \zeta(t)\delta(\Omega - \Omega_0) + \frac{[1 - \zeta(t)]}{4\pi} \tag{102}$$

when δ indicates a delta-function, and $\zeta(t)$ is a normalized decay function.

In physical terms, molecules having dipole moments $\mu(0)$ at $t = 0$ would randomize not by a continuous reorientational motion but by a disappearance of dipole orientation caused by strong collisions that randomize the dipoles instantly and completely. From equation 102 we obtain

$$\Psi_n(t) = \zeta(t) \tag{103}$$

for all values of n, so this strong-collision model rationalizes the dielectric and Kerr-effect data for the glass-forming liquids studied by Williams and co-authors (121, 126, 129). The form of $\zeta(t)$ corresponds in the cases studied to the KWW function with $\bar{\beta} \approx 0.5$–0.7 depending on the liquid. Even though small-step rotational diffusion was ruled out as a mechanism for the α-process in these systems and the strong-collision model was favored, the nature of the dynamics of fluctuation that lead to $\zeta(t)$ of the form observed for these liquids was not established even by such comparative experiments.

Beevers et al. (140) studied multiple samples of polypropylene glycols by using the dielectric (D) and Kerr-effect (K) techniques. All materials were of low molecular weight being 1000–4000 and below M_c; their data suggest that the small-step rotational diffusion model is applicable to the observed α-process. In addition they observed the normal-mode process discussed previously, as observed originally by Stockmayer and Baur (*82, 83*), by both techniques. In later studies the α-process in a poly(phenyl-methyl siloxane) was studied simultaneously by using D and K techniques and it was found (*141*) that $\langle\tau(K)\rangle/\langle\tau(D)\rangle \sim 1$ and that the D-functions and K-functions were both of KWW-type with $\bar\beta_D \simeq 0.38$ and $\bar\beta_K \simeq 0.6$. Although there were difficulties in analyzing these data (*141*) it appears that to a first approximation the strong-collision model, found applicable to the α-process in glass-forming liquids, is also applicable to segmental motions that lead to the α-relaxation in poly(phenylmethyl siloxane).

Little doubt exists that the origin of KWW-relaxation behavior for the α-process in amorphous polymer systems and glass-forming molecular liquids has been one of the great challenges in polymer science and condensed-matter physics during the past decade. The interest stems from the fact that the α-process may be observed by dielectric, mechanical, NMR, electron spin resonance, fluorescence, volume, enthalpy, and specific heat techniques, and being the dynamic glass-transition, the α-process has a determining effect on nonequilibrium phenomena associated with glass formation such as volume relaxation and physical aging in the T_g range. The KWW function and similar functions (e.g., the HN-function) provide the phenomenological description for steady state and nonequilibrium processes in all such glass-forming materials irrespective of their chemical structure (*see* Matsuoka (*142*) for a comprehensive account of mechanical and thermodynamic properties of polymers in the amorphous state and Hodge (*143*) for a definitive account of enthalpy relaxation in such systems).

Williams reviewed the applicability of the KWW function to relaxations in amorphous polymers (*10, 11, 60*) and the different models that have been proposed that lead to such behavior. These models include defect-diffusion models in one, two, and three dimensions; motions of chains on a tetrahedral lattice (*see also* reference 144); motions in Ising chains; and modified rotational diffusion models. Prominent among the various models is that according to Shlesinger and Montroll (*145*), who consider a lattice (suitably defined) whose sites are occupied by molecules (or chain units) and lattice vacancies. Diffusion of the lattice vacancies according to controlled random time walk (CRTW) processes that involve a *pausing-time* distribution leads to relaxation of the target dipole unit on the lattice. Assuming a pausing-time function that is a power law in time, then the dipole moment TCF take the KWW form and $\bar\beta$ is the stretched-exponential parameter whose origin lies in this power law. This approach focuses on the conditional diffusion of defects as the means to relax the dipole so dispersive diffusion, which is brought about by the pausing time function, becomes the means whereby $\Psi(t)$ takes the KWW form.

The dielectric α-relaxation in most amorphous polymers is well represented by the KWW function, as has been amply demonstrated by Ngai and co-workers (146–148) who take as the starting point for their analyses the first time derivative of $\Phi(t)$ (i.e., $\dot{\Phi}^Y(t)$ in eq 36). Also, the α-process, as observed by different relaxation and scattering techniques, is well represented by the KWW function, eq 34, in the time domain or its transform in the frequency domain (see references 149 and 150 for many examples and for further theoretical models for KWW behavior).

Two fundamental propositions may be posed concerning the mechanisms for the α-relaxation in such systems.

1. The process has a natural nonexponential dependence so there is an equation of motion for $f(\Omega, t/\Omega_0, 0)$ or for its moments $\Psi_n(t)$, which gives this result based on physical mechanisms for reorientational motions of chain segments or whole molecules.

2. The process corresponds to a weighted sum of elementary processes each having a correlation function that *may* be exponential in time, giving an overall correlation function of KWW form. This result may be equivalent to a distribution of relaxation times (see eq 28).

Models for defect diffusion (see reference 60 for a review) and strong collision models, eqs 102 and 103, where $\zeta(t)$ can be modelled in different ways, fall into proposition 1. Another approach that leads to nonexponential relaxation functions is that involving first-order or higher order memory functions, which incorporate mode–mode coupling theories, that are presently attracting considerable interest. As co-workers and I emphasized recently (151) the memory function approach makes the fundamental assumption that the memory functions derived from experimental data have a physical significance but this is not necessarily true for the α-process in polymers and other glass-forming materials.

It is a general property of time-correlation functions for a homogeneous ergodic system that they obey a set of coupled memory-function (Volterra) equations (124)

$$\frac{dK_n}{dt} = -\int_0^t K_{n+1}(t - \tau) K_n(\tau) \, d\tau \tag{104}$$

where $K_n(t)$ is the nth-order memory function of the time-correlation function $C(t) = K_0(t)$. Fourier transformation of eq 104 gives the Mori-continued-fraction representation of $\tilde{C}(\omega) = \Im(C(t))$ (124)

$$\tilde{C}(\omega) = \cfrac{K_0}{i\omega + \cfrac{K_1(0)}{i\omega + K_2(0)\ldots\ldots}} \tag{105}$$

If a suitable form for $K_n(\omega)$ is chosen empirically or is derived from an appropriate model for motion, then $\tilde{C}(\omega)$ may be obtained from eq 105, which with eq 58 yields a functional form for $\varepsilon(\omega)$ (*124*). A special case of eq 104 involves the correlation function $C(t)$ and the first-order memory function $K_1(t)$

$$\frac{dC(t)}{dt} = -\int_0^t K_1(t - \tau) C(\tau)\, d\tau \tag{106}$$

Douglas and Hubbard pointed out (*152*) that if

$$K_1(t,\tau) = \frac{\delta(t - \tau)}{\tau^m} \tag{107}$$

where $0 < m \leq 1$, then eqs 106 and 107 give $C(t)$ as the KWW function, eq 34. However, this memory function appears to be unphysical and it is not clear what realistic models for motion could lead to eq 107. Douglas and Hubbard proposed a general empirical memory function involving two spread parameters $0 < \bar{a} \leq 1, 0 < \bar{b} \leq 1$, that gives the dielectric permittivity in the frequency domain, as the HN, CC, or KWW functions for particular $(\bar{a}, \bar{b})$ combinations (*152*).

Thus, this memory function approach changes the focus of attention from the distribution function $f(\Omega, t/\Omega_0, 0)$ and the time-correlation functions $C_n(t)$ for the motions of a particular molecular probe to a consideration of the physical meaning of the corresponding memory functions $K_1(t)$, $K_2(t)$, etc., depending on the level taken in the hierarchy of such functions. If specific models for motion can be generated that lead to memory functions that give KWW behavior for the correlation functions then progress will be made but the author is unaware of such developments.

Götze and co-workers (*see* references 153–159 and references therein) developed a mode–mode coupling approach for the α-relaxation in glass-forming materials. Their method is based essentially on a Mori-continued fraction for $\tilde{C}(\omega)$ terminated at the second-order memory function so that

$$\tilde{C}(\omega) = \cfrac{K_0}{i\omega + \cfrac{K_1(0)}{i\omega + K_2(\omega)}} \tag{108}$$

The next step makes the assumption that $K_2(\omega)$ may be written as a function of $\tilde{C}(\omega)$

$$K_2(\omega) = F(\tilde{C}(\omega)) \tag{109}$$

thus, closing the problem in eq 108 and allowing $\tilde{C}(\omega)$ to be determined analytically from eq 108. Proceeding in this way Götze and co-workers have calculated $\tilde{C}(\omega)$ and hence the susceptibility $\chi(\omega) = \chi'(\omega) - i\chi''(\omega)$ (for dielectric relaxation $\chi(\omega) \equiv \varepsilon(\omega)$). Through choices of the assumed relationships between $K_2(\omega)$ and $\tilde{C}(\omega)$ they have generated plots of $\chi'(\omega)$ and $\chi''(\omega)$ versus $\log f$ for comparison with experimental data. They have fitted data for dielectric relaxation ($\varepsilon(\omega)$), electrical conductivity relaxation (dielectric modulus, $M(\omega) = 1/\varepsilon(\omega)$), and wave-vector ($q$)-dependent scattering functions $S(q, t)$, $S(q, \omega)$ for quasi-elastic scattering of light and neutron beams. Good fits have been obtained to experimental data for the α-process in amorphous polymers and glass-forming liquids; the high frequency behavior of $\tilde{C}(\omega)$ has received attention and questions of bifurcation in relaxation behavior above T_g involving ergodicity have emerged as a result of these interesting works (153–159).

Equation 108 corresponds to a master equation of motion for $C(t)$ in the time domain of the following form (155, 156)

$$\ddot{C}(t) + \nu\dot{C}(t) + D \int_0^t m(t - t')C(t')\,\mathrm{d}t' = 0 \tag{110}$$

where ν and D are constants and $m(t)$ is a memory function that is linearly related to $K_2(t)$. Thus, the choice of the form of $m(t - t')$ allows $C(t)$ to be determined from eq 110. By suitable choice of the relationship between $K_2(\omega)$ and $\tilde{C}(\omega)$, with its attendant parameters, the form of $\tilde{C}(\omega)$ may be varied to give a wide range of results and these, together with the assumption that the memory functions are physically meaningful, are the main features of their mode–mode coupling approach that leads to nonexponential relaxation for the α-process, in line with proposition 1 described previously. The mode–mode coupling theory in this case focuses on the second-order memory function of the correlation function rather than on $f(\Omega, t/\Omega_0, 0)$ or the correlation functions themselves.

For proposition 2 if each elementary process in a distribution is characterized by a single exponential decay function in time with its associated correlation time the overall relaxation function will be broad and could be of KWW form, because it is a weighted sum of the parallel processes. Because a single-exponential decay function, $\exp{-t/\tau}$, has no proper memory function (a δ-function only) there will be no meaningful memory function $K_1(t)$ for an overall relaxation function containing a distribution of relaxation times even though apparent (or virtual) memory functions $K_1(t)$, $K_2(t)$ can be determined numerically from data for $\tilde{C}(\omega)$ or $C(t)$ by using eqs 104–108.

The question arises as to whether either of propositions 1 and 2 can be resolved through experiments. As was emphasized previously, dielectric experiments give only the first moment (the dipole moment correlation function, $\Phi_1(t)$) in the expansion of $f(\Omega, t/\Omega_0, 0)$ and thus are unable, on their own, to distinguish between 1 and 2. Combined dielectric and Kerr-effect experiments allow distinctions between different relaxation mechanisms to be made, as was described previously, but again these experiments are unable to distinguish between propositions 1 and 2. Recently considerable progress in this direction was made by Schmidt–Rohr and Spiess and co-workers (*160–162*) who have developed a multinuclear, multidimensional NMR technique that is able to obtain information on $f(\Omega, t/\Omega_0, 0)$ directly for the α-relaxation of selected groups in poly(vinylacetate), polystyrene, and other amorphous polymers. Selecting a subensemble of groups in poly-(vinyl acetate) (*160*), they found that some appeared to be highly mobile whereas some appeared to be fairly immobile at the arbitrary time $t = 0$.

The system appeared to be spatially heterogeneous suggesting that a distribution of relaxation processes was present. In time the groups that appeared immobile at $t = 0$ became mobile, whereas those that were mobile at $t = 0$ became immobile, as further sampling of subensembles of the original subensembles revealed by using their technique (*160*). Thus, the heterogeneity is dynamic and all relaxors, being chemically equivalent, are shown to be dynamically equivalent as is required for an ergodic system. The system conforms in the time scale of the α-relaxation to a weighted sum of parallel processes whose distribution is extremely broad and may be fitted with log-normal or KWW distribution functions (*61*) of relaxation times.

Thus, this NMR evidence (*160*) and further work (*161, 162*) show that the α-relaxation function in such polymers may be considered to arise from a broad distribution of relaxation processes (in line with proposition 2). Because each elementary process in such a distribution may have no proper memory (because $K_1(t, \tau) = \delta(t - \tau)$ for each single relaxation-time process), it follows that the correlation function for the broad α-process in the *f*-domain or *t*-domain may have no associated memory function, although virtual memory functions may be calculated from experimental data. The question of virtual memory functions obtained from experimental dielectric data and from analytical relaxation functions (e.g., the KWW function) was discussed further by Fournier and Williams (*151*). Schmidt–Rohr, Spiess, and co-workers (*see* reference 161 and references therein) show that $f(\Omega, t/\Omega_0, 0)$ may be determined for the motion of particular groups in amorphous (and crystalline and LC) polymers by using their modern NMR techniques, so their new results, taken together with comparable dielectric, dynamic mechanical, optical (e.g., fluorescence), and dynamic scattering data, are beginning to provide a basis for a new understanding of cooperative motions in amorphous glass-forming polymers.

We have not to this point described models for motion that relate specifically to the dynamics of polymer chains. The long-range motions of chains are well described by the theories of Rouse (*163*) and Zimm (*164*), whereas chain-dynamic models incorporating molecular structural parameters were developed by Kranbuehl and Verdier (*165–167*) (*see* especially reference 167 for references to recent work). Geny, Monnerie, and co-workers (*144, 168a, 168b*) described chain motions on cubic and tetrahedral lattices that result in broad correlation functions for segmental motions. These and further theoretical models were discussed by Williams (*60*) and by Beevers and Williams (*169*) including the dynamic Ising models of Work and Fujita, Anderson and Isbister, and McQuarrie.

The rotational isomeric state model for polymer chains as developed by Flory and co-workers was very successful in describing the equilibrium statistics of polymer chains (*36*) and is probably the most satisfactory method available for the treatment of the equilibrium and dynamic properties of chains because details of chemical structure and the conformational properties of chains are included specifically. Jernigan (*170*) extended the Flory model to the dynamic situation by writing the time-dependent equations governing the occupational probabilities of the available conformations in compact matrix form as

$$\frac{d\mathbf{P}(t)}{dt} = \mathbf{A}\mathbf{P}(t) \tag{111}$$

where $\mathbf{P}(t)$ is the column vector of occupation probabilities of the available conformational states of the entire chain and the matrix $\mathbf{A}$ is given by

$$\begin{aligned}
\mathbf{A} = &\ \mathbf{A}_2 \otimes \mathbf{I}_3 \otimes \mathbf{I}_3 \otimes \cdots \otimes \mathbf{I}_3 + \\
&\ \mathbf{I}_3 \otimes \mathbf{A}_3 \otimes \mathbf{I}_3 \otimes \cdots \otimes \mathbf{I}_3 + \\
&\ \vdots \qquad \vdots \qquad \vdots \qquad\qquad \vdots \\
&\ \mathbf{I}_3 \otimes \mathbf{I}_3 \quad\ \mathbf{I}_3 \otimes \cdots \otimes \mathbf{A}_{n-1}
\end{aligned} \tag{112}$$

where $\mathbf{I}_3$ is a 3×3 identity matrix, $\otimes$ indicates the direct product of two matrices, and $\mathbf{A}_n$ is a matrix for bond n involving the elementary transition probabilities k_1, k_2, and k_3 for local conformational transitions between gauche $\rightarrow$ trans, trans $\rightarrow$ gauche, and gauche$^+$ $\rightarrow$ gauche$^-$, respectively.

Equation 111 is the master equation of motion governing the changes in conformation of the chain with time whose general solution is written as (*169, 170*)

$$\mathbf{P}(t) = \mathbf{B} \exp \mathbf{Q}\, t\, \mathbf{B}^{-1}\mathbf{P}(t = 0) \tag{113}$$

where $\mathbf{Q} = \mathbf{B}^{-1}\mathbf{A}\,\mathbf{B}$ and $\mathbf{B}$ is the matrix that diagonalizes $\mathbf{A}$. The eigen values of the diagonalized matrix are relaxation rates (inverse relaxation times), and hence $\mathbf{P}(t)$ in eq 113 becomes expressible in terms of weighted sums of individual relaxation functions that each have a single exponential dependence on time with relaxation times that are simple linear functions of k_1, k_2, and k_3. *See* references 28 and 171 for the theory of relaxation in simple barrier models for which $\mathbf{P}(t)$ and $\langle \mathbf{\mu}(0) \cdot \mathbf{\mu}(t) \rangle$ are deduced in a similar way.

For the case of dielectric relaxation the dipole moment correlation function for the entire chain $\Phi_\mu(t)$ (*see* eq 64) averaged over all available states at $t = 0$ can be expressed (see eq 19 in reference 169) by specifying the placements of dipoles along the chain and specifying a reference coordinate system. Dielectric relaxation involves vector–vector correlations of dipole orientations (including auto- and cross-correlation terms, as has been detailed previously) so a reference frame within the molecule must be chosen to deduce $\mathbf{M}(0)$ and $\mathbf{M}(t)$. For a chain starting in a particular conformation at $t = 0$, $\mathbf{M}(0)$ is expressed with respect to this internal reference frame to the chain. At a later time t the new occupation probabilities of obtaining the different conformations, given the chain had the specified conformation at $t = 0$, are known from eq 113. Therefore, $\mathbf{M}(t)$ averaged over all these states can be determined with respect to the internal reference coordinate frame of the chain.

Hence $\Phi_\mu^c(t)$ may be determined in suitable cases. Jernigan (*170*) carried out calculations of $\Phi_\mu^c(t)$ for model chains $Br—(CH_2)_{N-1}—Br$ and multiplied this function by a further decay function $\phi_{0v}(t)$ to allow for overall rotational motions. Thus, $\Phi(t)$ was considered to decay as a combination of internal and overall motions of chains. Internal motions led to relaxation that was a weighted sum of elementary processes (*170*). Beevers and Williams (*169*) calculated $\Phi_\mu^c(t)$ for α,ω-dibromoalkanes by using the Jernigan method for chains having $N - 4$ and 5. They found that $\Phi_\mu^c(t)$ depended on the choice of the internal reference frame and did not decay to 0, which is unacceptable for isotropic systems. The reason is clear: the dipole moments of the groups in the chain have a net projection on the internal reference frame that is not relaxed by the conformational changes of the chain, and this net projection will depend on the choice of this frame.

Thus, a fundamental difficulty exists with this approach that cannot be overcome by multiplying $\Phi_\mu^c(t)$ by $\Phi_{0v}(t)$. The normalized functions $\Phi'(t)$ $= [\Phi_\mu^c(t) - \Phi^c(\infty)]/[\Phi_\mu^c(0) - \Phi^c(\infty)]$ calculated for $N = 4$ and 5 with realistic conformational energy barrier systems and for different chosen reference frames decay as a weighted sum of closely spaced decay functions (KWW parameter $\bar{\beta}$ in the range 0.88 to 0.98) (*169*). Therefore, internal relaxation of the chain gives dipole relaxation functions far removed from the KWW behavior observed for glass-forming polymers and liquids, where $\bar{\beta}$ is in the range 0.35–0.70. The apparent failure of these generalizations of the Flory rotation-isomeric model to predict dielectric relaxation behavior where vector–vector correlations are involved (*169, 170*) resides in the difficulty in formulating an acceptable reference coordinate system. The rate equations for chain dynamics, eq 111, are scalar equations but the dielectric relaxation function involves the directions of group dipoles.

Applications of eq 111 to predict NMR, fluorescence, and quasielastic light-scattering behavior where directional quantities are again involved will also encounter this fundamental difficulty, which is not resolved at this time. For real systems the random forces may reorient chains as a whole and cause internal conformational changes simultaneously so the problem of the reference coordinate frame is not easily resolved. Such a difficulty disappears for a chain that is tethered to a surface because in this case the directions of all chain vectors can be referenced to the X, Y, and Z coordinates of that surface.

Rosato and Williams proposed a multistate barrier model for multiple dielectric relaxation processes in amorphous polymers. It was assumed that movement of dipoles between localized states within a general potential well led to a β-process, whereas movement from one well to another leads to an α-process. The formalism follows closely that developed by Williams and Cook for dipole motion in barrier systems (*171*) and that previously described for internal relaxation of chains (*169, 170*). The result of these calculations (*172*) revealed a cluster of fast dielectric processes (β-process) and a single slow relaxation mode (α-process), which corresponds to jumps from one well to another. Thus, despite the detail and complexity of this model (*172*) the slow process thus derived does not have a stretched-exponential form so the model does not presently provide a useful approach to the α- (or β-) relaxations in amorphous polymers.

Mechanisms for Dielectric Relaxation in Thermotropic LC Polymers

Even though much attention has been given in the literature to the dielectric relaxations of amorphous polymers, as was described previously, considerations of the mechanisms for dipole motion in LC polymers are far less extensive. The variations in loss spectra as the macroscopic alignment of an LC polymer is varied from homeotropic to planar are striking and provide a means for resolving the component processes, in particular the dominant low-frequency δ-process (*39–43, 95–116*). Equations 78–81 make it clear that the δ-process arises due to the 00-relaxation mode, whereas the α-process corresponds to a weighted sum of the remaining relaxation modes (*41–44, 95–116*). Pictorial representations of the four modes of motion of μ_ℓ and μ_t have been given (*see* references 95 and 96) and show, for example, that the motions of μ_ℓ with respect to the director **n** that involve only the polar angle β give rise to the δ-process. However, the detailed mechanisms for motion remain undecided.

For polymers having dipolar mesogenic side groups Zentel et al. (*173*) have given pictorial representations of the δ-, α-, and further processes that involve local motions of flexible groups (e.g., OCH_3). The dielectric relaxation behavior of low molar-mass liquid crystals and SCLC polymers has been reviewed (*15, 174, 175*).

In the original theory of Martin, Maier, and Saupe (*175*) it was considered that μ_ℓ moved in the P_2-potential of the LC environment by a small-step rotational diffusion process, yielding two relaxation modes with relaxation times $\tau_\| = \gamma_\| \tau_0$ (homeotropic) and $\tau_\perp = \gamma_\perp \tau_0$ (planar) where τ_0 is the relaxation time for isotropic reorientations and $\gamma_\|$, $\gamma_\perp$ are retardation factors that are determined by the height of the nematic potential. Subsequent developments in the theory by Nordio and Segre and others included a consideration of the motions of μ_ℓ and μ_t by rotational diffusion (*see* reference 15 for a review). However, it seems unlikely that in low-molar-mass liquid crystals, and especially, in LCSC polymers that motions having correlation times typically in the range of 10^{-8} to 10^2 s could be described by small-step rotational diffusion in a weak P_2-potential imposed by a mean field of the surrounding molecules. For polymers the restriction of freedom of the mesogenic groups imposed by chain connectivity makes it extremely difficult to picture small-step rotational diffusion of the mesogenic groups. It seems more likely that a model of cooperative motions involving fluctuations in local energy, as envisaged in strong-collision models, would be more appropriate.

Despite the complexity of LCSC systems it is a remarkable fact that the dielectric δ-process observed in many materials is only slightly broader than that for a single-relaxation-time process (*39–43, 95–115*). The precise mechanisms for the relaxation modes in LCSC polymers remain an unsolved problem. Also, it is hoped that comparative relaxation studies using different techniques or the new NMR techniques of Schmidt–Rohr, Spiess, and their co-workers (*160–162*) will provide new insight into the mechanisms of these processes.

Concluding Remarks

We have seen that the molecular theory of the static permittivity and of the complex permittivity is well established for amorphous, LC, and crystalline polymers, although I have not discussed the crystalline polymers in any detail here (*see* Chapter 4 for an extensive account of the dielectric properties of semicrystalline polymers). These molecular theories give general relationships between $\varepsilon_0(T,P)$ and $\varepsilon(\omega,T,P)$ and the molecular quantities of dipole moments, equilibrium and dynamic angular-correlation terms between dipoles along and between chains, and auto- and cross-time-correlation functions for dipole motion. Considerable interest exists in devising simple physical models to rationalize each dielectric relaxation process in a particular class of polymer. As we have seen a simple phenomenological model for α-, β-, and $\alpha\beta$-relaxations gives a qualitative explanation for the dielectric properties of all amorphous polymers and supercooled molecular liquids, but this model does not specify the particular mechanism for each process (jump-diffusion, fluctuation relaxation, dynamic heterogeneity model, etc.).

We have seen how particular models for relaxation may be introduced, but in common with studies of relaxation phenomena using the complementary techniques of NMR, mechanical, fluorescence, quasielastic light-scattering, and neutron scattering, there continues to be a lively debate as to their applicability in particular materials. The interest and activity in understanding relaxation processes in complex systems including polymers, liquids, and ionic glasses are reflected in the large numbers of recent publications (*see* references 149 and 150 for work presented, respectively, at the 1990 Crete Meeting and 1993 Alicante Meeting or Relaxations in Complex Systems organized by Ngai, Wright, and Riande). In reviewing many of the recent publications on mechanisms for different processes (e.g., the α- and β-processes in amorphous polymers) it is important to remember that particular experiments probe the motions (translational and rotational) of groups and the relationships between experimental observables and molecular properties (e.g., between $\varepsilon(\omega)$ and $\Phi_\mu(t)$ for dielectric relaxation) are well-established from molecular theory, and hence provide the means for establishing molecular mechanisms for motional processes as we have seen for the recent NMR works of Schmidt–Rohr and Spiess. As presented, many of the models proposed for relaxation especially in amorphous systems are phenomenological, as we have seen for multiple relaxations in amorphous polymers (eq 99), for mode–mode coupling theory (eqs 108 and 109), for diffusion by controlled random walks that assume a particular form of pausing distribution (*145*), and for theories involving particle percolation (*176, 177*). Note that in many of these models phenomenological relations for the correlation functions are proposed that take no account of the fact that those correlation functions may refer in each case to particular vector–vector correlations (for dielectric relaxation) or tensor–tensor correlations (for NMR, electron spin resonance, fluorescence, and depolarized quassielastic scattering) in the angular motions of molecular groups or position–position correlations (for quasielastic light scattering).

It is important that molecular models should focus on master equations for $f(\Omega, t/\Omega_0, 0)$ for angular motions and $G(\mathbf{R}, t/\mathbf{R}_0, 0)$ for translational motions (*28, 124*) that will also accommodate experimental data for the different time-correlation functions. The mode–mode coupling theory, as presented and applied (*153–159*), makes no specific reference to angular or translational motions of molecules, thus its origins seem rooted in density fluctuations of a hard-sphere liquid (*157*). This result does not appear to be immediately relevant to the angular motions of bonds, group dipoles, fluorophores, etc., in molecules (*see* discussion comments in reference 149, p 387).

Even though we have not been able to give a detailed outline of many aspects of the theory of dielectric relaxation in this account, several important recent developments are noted. The precise shape of the dielectric α-relaxation in polymers and glass-forming liquids has received much attention with the aim of establishing scaling laws for plots of ε'' vs $\log f$ (*see* Nagel et al. (*178*), Götze and Sjögren (*179*), and subsequent papers by Kremer and co-workers (*180–187*)). Considerable debate has concerned the high frequency behavior of the loss data, but general scaling laws are unlikely to be established because secondary relaxations make a contribution to loss data at high frequencies.

Considerable current interest exists in real-time dielectric studies of amorphous systems undergoing polymerization. The literature is considerable: for example Kranbuel (*188*), Johari and co-workers (*189, 190*), Stephan and co-workers (*191*) and references therein to earlier works concerning epoxide-amine thermosetting systems, and Rolla and co-workers (*192, 193*) for microwave studies of acrylate bulk polymerization. During polymerization the material transforms from a liquid to a glassy solid so the loss peaks move from microwave frequencies through the MHz–kHz range down to sub-Hz frequencies. From a theoretical aspect such systems do not have Fourier transform relationships between $\varepsilon(\omega)$ and relaxation functions or molecular time functions involving dipole reorientation. Linear response theory that relates $\varepsilon(\omega)$ to $\Phi(t)$ or $\Phi_\mu(t)$ is only applicable to stationary systems: that is, systems whose thermodynamic properties are independent of time. For a system that changes with time $\varepsilon(\omega)$ can only be measured at frequencies for which negligible changes in $\varepsilon(\omega)$ occur within the time scale of measurement.

Epoxide-amine reactions are fairly slow (hours to minutes), so measurement of $\varepsilon(\omega)$ is possible for $f > 10^2$ Hz, but at low frequencies ($f \leq 1$ Hz) the dielectric properties change with time during the measurement so $\varepsilon(\omega)$ cannot be measured under these circumstances. Theoretical expressions are needed for $\varepsilon(\omega)$ in terms of molecular quantities in the range where it can be measured for a reacting system. Williams (*151*) considered the Hamon approximation (*8, 194*) that relates a measurement of $(\varepsilon''(\omega), \omega)$ to $(\dot{\Phi}(t), t)$ for a steady-state system, where

$$\varepsilon''(\omega) = \frac{\dot{\Phi}(t)}{\omega}; \qquad \omega t = 0.1 \times 2\pi \qquad (114)$$

Equation 114 is accurate for broad loss curves (*52, 53*), which is the case for thermosetting epoxy systems. Thus, the measurement of $\varepsilon''(\omega)$ at ω gives information on $\dot{\Phi}(t)$ at time t, where $\Phi(t)$ has the meaning that it is the value that would be measured at time t after removal of a step voltage for a hypothetical chemically arrested system at time t_r into the reaction. Clearly $\Phi(t)$ cannot be realized in practice. We await progress with a molecular theory that will relate $\varepsilon(\omega, t_r)$ to molecular time-dependent quantities.

We discussed previously the inherent difficulties concerning a reference coordinate frame when the generalized Flory equations, eqs 111–113, are used to deduce $\varepsilon(\omega)$ for model chains. Molecular dynamics (MD) simulations of the motions of assemblies of model molecules overcome the problem of reference coordinates but are limited to fairly small numbers of particles and have a limited time scale for the simulation ($< 10^{-9}$ s). In addition there is a need to avoid artifacts in the results due to the necessary choice of periodic-boundary conditions. Roe (195) recently made a realistic MD simulation of a model polyethylene chain containing 125 spherical particles, mimicking CH_2 units, moving in a box containing four such molecules. In addition to calculating the mean-square displacement and the scattering functions $F(q,t)$ and $S(q, \omega)$ for the molecule, Roe calculated the angular correlation functions

$$\Psi_n(t) = \langle P_n(\cos \theta(t)) \rangle \tag{115}$$

for $n = 1$ to 5 for the bond vector between two neighboring carbon atoms along a chain.

By using a Cray YMP-864 computer, calculations were made from picoseconds into the nanoseconds range (the calculations required 20 min of real-time computing), and the following results were demonstrated:

- $\Psi_n(t)$ exhibited short and long parts to the decay function where the short-time part was approximately an exponential function with time and the long-time part was approximately a KWW function with time with $\bar{\beta} \simeq 0.56 + 0.04$ for all n values

- τ_{KWW} values were inversely proportional to $n(n + 1)$, which is the scaling law for simple isotropic rotational diffusion (28). Because $\Psi_1(t)$ was dominated by the slow KWW-type process it is evident that the dielectric α-process in amorphous polymers can be understood in terms of the cooperative chain motions that are implicit in the MD simulations of Roe (195).

The new studies of the α-relaxation in amorphous polymers and glass-forming liquids by Götze and co-workers (mode–mode coupling theory) (153–159), by Schmidt–Rohr and Spiess (NMR experiments defining chain-vector motions that imply dynamic heterogeneity of species) (160–162), and by Roe (MD simulations yielding KWW behavior) (195)

have provided a stimulus for further experimental work and have provided new insight into the nature of this process that is important to polymer dynamics (*8–14, 142–150*) and the dynamics of glass-forming liquids (*196*). Further evidence for the nature of the α-process in such systems was provided recently by Ediger and co-workers (*197–202*) by using fluorescence probe techniques. In particular, Cicerone and Ediger (*202*) have shown that a nonequilibrium distribution of fluorescent probe-molecule mobilities can be obtained by preferential bleaching of the most mobile species in a supercooled *o*-terphenyl system. Near T_g the equilibrium distribution is reformed in a time scale ~10^3 longer than the average correlation time for the α-relaxation of the probe molecule. This result suggests that the distribution of τ for the probe molecules (KWW in form) arises from dynamic heterogeneity of species as suggested by Schmidt–Rohr and Spiess (*160–162*) for polymers.

This result also brings into question natural nonexponential behavior for this process and the physical meaning of apparent memory functions determined from experimental correlation functions (*203*) or those constructed to fit dielectric and related relaxation data and scattering data (*152–159*). It would be of interest to see if the MD-simulations of Roe (*195*) contain information on the trajectories in time for a subensemble of units that would allow the components $\boldsymbol{\mu}(0)\cdot\boldsymbol{\mu}(t)$ of the average correlation function $\langle\boldsymbol{\mu}(0)\cdot\boldsymbol{\mu}(t)\rangle$ for bond vectors to be inspected. This inspection would show if the model system exhibits the dynamic heterogeneity suggested from the experiments of Schmidt–Rohr and Spiess (*160–162*) and Cicerone and Ediger (*202*) or if all units behave on average in the same way, implying a natural nonexponential function for the motion and hence implying, possibly, an intrinsic memory for the process.

It has not been possible to describe all the theoretical developments of the dielectric relaxation behavior of polymers in this account. Even though we have described several aspects of the theory for amorphous and LC systems, for detailed accounts of crystalline polymers the reader is referred to Chapter 4 by Boyd and Liu.

Even though this account for amorphous polymers covers phenomenological models, molecular theories of chain dynamics, mode–mode coupling theories, and molecular dynamics simulations, it is not comprehensive. Jonscher (*204*) and Dissado and Hill (*205–215*) have given extensive accounts of the theory of multiple relaxations in amorphous polymers. Jonscher referred to a universal response in such systems (*204*). Dissado and Hill (*205–215*) described phenomenological models, including a cluster model, to account for the low frequency and high frequency power law relations ($\varepsilon'' \sim \omega^m, \omega^{-n}$) obeyed approximately by the dielectric α-relaxations of amorphous polymers. Ngai and co-workers (*216–233*) developed a "coupling model" scheme for application to dielectric relaxation, dynamic mechanical relaxations, quasielastic light-scattering, quasielastic neutron scattering, volume relaxation, and conductivity relaxation in amorphous polymers and glass-forming liquids. This approach considers that at short times a relaxing system will obey a simple relaxation function—the single-relaxation time function in eq 25.

At longer times it is considered that the relaxation function obeys the master equation

$$\frac{d\Phi(t)}{dt} = \left(\frac{t}{\tau}\right)^{\overline{\beta}-1} \exp - (t/\tau)^{\overline{\beta}} \qquad (116)$$

which is seen to be the first-time derivative of the KWW function, eq 34. As a part of their scheme, τ in eq 116 is related to an elementary barrier W and to $\overline{\beta}$ according to an Arrhenius law. The extensive works of Ngai and co-workers cannot be reviewed here in any detail but we note the following. The origin of eq 116 was sought by Ngai and co-workers by using different theoretical approaches (146–149, 216–228), including those involving further master equations of motion, chaotic Hamiltonians, hierar- chical models, constraint dynamics, and molecular dynamics. They have extended the well-known models of relaxation due to Adam and Gibbs and to Hall and Helfand to include the coupling scheme (229–231). In addition to their many other applications of eq 116 to different relaxation and scat- tering phenomena in amorphous polymers and glass-forming liquids, they have given reviews of relaxation that include dielectric behavior (146, 232–236). Further work by them relating to dielectric behavior includes:

- applications of the coupling scheme to poly(vinyl acetate) (237)
- comparison of dielectric, mechanical, and light scattering data for polymers (233)
- effects of diluent on the dielectric relaxation of polymers (238)
- the shape of the α-relaxation in polymer blends (239)
- the nature of the α-relaxation in poly(lauryl methacrylate) (240)
- the effect of spacer-length on relaxation in LC polymers (241)

Finally, we note that although the dielectric relaxation behavior of poly- mer systems discussed arises from the reorientational motions of dipole groups, chemical relaxation effects may be observed in special cases. The theory of the dielectric relaxation behavior of model systems capable of chemical relaxation was given first by Schwarz (242) using a field-perturba- tion approach. He elucidated the dielectric behavior of model systems undergoing chemical exchange as $A \rightleftharpoons B$, $A + B \rightleftharpoons C$, where A, B, and C are, generally, dipolar species in the liquid state. Using Kubo–Glarum linear-response theory and time-correlation functions, the author obtained (243) the same results as Schwarz and extended the model to more complex situations. The dielectric relaxation behavior may have arisen (*242, 243*) from both dipole reorientation and chemical exchange. Recently Müller and co-workers (*244*) made extensive dielectric studies of thermoreversible gels based on polyisoprene in which physical cross-links were formed by polar stickers involving hydrogen-bonding groups. As a part of this work they developed the theory of Schwarz (*242*) and Williams (*243*) to apply to the case where the physical cross-links undergo chemical exchange and hence allow chemical relaxation to contribute, in principle, to the dielectric behavior.

Acknowledgment

The author thanks the EPSRC for contributed support for dielectrics research at Swansea.

References

1. Debye, P. W. *Polar Molecules;* Chemical Catalog Co.: New York, 1927.
2. Smyth, C. P. *Dielectric Behaviour and Structure;* McGraw Hill: New York, 1955.
3. Sutton, L. E. *Bull. Soc. Chim. Fr.* **1949,** D448.
4. Smith, J. W. *Electric Dipole Moments;* Butterworths: London, 1955.
5. For a recent review of commercially available dielectrics measuring equipment *see* Pochan, J. M.; Fitzgerald, J. J.; Williams, G. In *Determination of Electronic and Optical Properties,* 2nd ed.; Rossiter, B. W.; Baetzold, R. C., Eds.; Physical Methods of Chemistry Series; Wiley-Interscience: New York, 1993; Vol. VIII.
6. Hill, N.; Vaughan, W. E.; Price, A. H.; Davies, M. M. *Dielectric Properties and Molecular Behaviour;* Van Nostrand Reinhold: New York, 1969.
7. Böttcher, C. J. F.; Bordewijk, P. *Theory of Electric Polarization,* 2nd ed.; Elsevier: Amsterdam, Netherlands, 1978; Vol. 2.
8. McCrum, N. G.; Read, B. E.; Williams, G. *Anelastic and Dielectric Effects in Polymeric Solids;* Wiley: London, 1967; Dover Ed.; Dover: New York, 1991.
9. Williams, G. In *Dynamic Properties of Solid Polymers;* Pethrick, R.; Richards, R. W., Eds.; NATO ASI, Reidel: Dordrecht, Holland, 1982; pp 213–239.
10. Williams, G. *Adv. Polym. Sci.* **1979,** *33,* 60.
11. Williams, G. *IEEE Trans. Electr. Insul.* **1982,** *EI-17,* 469.
12. Williams, G. In *Comprehensive Polymer Science;* Allen, G.; Bevington, J. C., Eds.; Booth, C.; Price, C., Eds.; Pergamon: Oxford, England, 1989; Vol. 2, Chapter 18, p 601.
13. Williams, G. In *Materials Science and Technology, Vol. 12, The Structure and Properties of Polymers;* Thomas, E. L., Ed.; VCH: Weinheim, Germany, 1992; Chapter 11, pp 471–528.
14. Williams, G. In *Dielectric and Related Molecular Processes;* Davies, M., Ed.; Special Periodical Reports; The Chemical Society: London, 1975; Vol. 2, p 151.

15. Williams, G. In *The Molecular Dynamics of Liquid Crystals;* Luckhurst, G.; Veracini, C. A., Eds.; NATO ASI; Kluwer Acadamic: Dordrecht, Netherlands, 1994; p 431.
16. Attard, G. S.; Araki, K.; Moura-Ramos, J. J.; Williams, G. *Liq. Cryst.* **1988**, *3*, 861.
17. Williams, G. *Polym. Adv. Technol.* **1992**, *3*, 157.
18. Moscicki, J. K.; Aharoni, S. M.; Williams, G. *Polymer* **1981**, *22*, 1361.
19. Moscicki, J. K.; Williams, G.; Aharoni, S. M. *Macromolecules* **1982**, *15*, 642.
20. Moscicki, J. K.; Williams, G. *J. Polym. Sci., Polym. Phys. Ed.* **1983**, *21*, 213.
21. Von Hippel, A. *Dielectrics and Waves;* Wiley: New York, 1959.
22. Von Hippel, A. *Dielectric Materials and Applications;* Wiley: New York, 1954.
23. Horner, F. T.; Taylor, A.; Dunsmur, R.; Lamb, J.; Jackson, W. *J. Inst. Electr. Eng.* **1946**, *93*, 53.
24. Fröhlich, H. *Theory of Dielectrics;* Oxford University: Oxford, England, 1949.
25. Hedvig, P. *Dielectric Properties of Polymers;* Adam Hilger; Bristol, England, 1977.
26. Riande, E.; Saiz, E. *Dipole Moments and Birefringence of Polymers;* Prentice Hall: Englewood Cliffs, NJ, 1992.
27. Williams, G. *Chem. Rev.* **1972**, *72*, 55.
28. Williams, G. *Chem. Soc. Rev.* **1978**, *7*, 89.
29. Cook, M.; Watts, D. C.; Williams, G. *Trans. Faraday Soc.* **1970**, *66*, 2503.
30. Onsager, L.; *J. Am. Chem. Soc.* **1936**, *58*, 1486.
31. Kirkwood, J. G. *J. Chem. Phys.* **1939**, *7*, 911.
32. Fuoss, R. M.; Kirkwood, J. G. *J. Am. Chem. Soc.* **1941**, *63*, 385.
33. Kirkwood, J. G.; Fuoss, R. M. *J. Chem. Phys.* **1941**, *9*, 329.
34. (a) Read, B. E. *Trans. Faraday Soc.* **1965**, *61*, 2140. (b) Williams, G. *Trans. Faraday Soc.* **1963**, *59*, 1397.
35. Volkenstein, M. V. *Configurational Statistics of Polymeric Chains;* Interscience: New York, 1963.
36. Flory, P. J. *Statistical Mechanics of Chain Molecules;* Interscience: New York, 1969.
37. Boyd, R. H. *J. Polym. Sci., Polym. Phys. Ed.* **1983**, *21*, 505.
38. Coburn, J. C.; Boyd, R. H. *Macromolecules* **1986**, *19*, 2238.
39. Attard, G. S.; Williams, G. *Liq. Cryst.* **1986**, *1*, 253.
40. Attard, G. S.; Araki, K.; Williams, G. *Br. Polym. J.* **1987**, *19*, 119.
41. Araki, K.; Attard, G. S.; Kozak, A.; Williams, G.; Gray, G. W.; Lacey, D.; Nestor, G. *J. Chem. Soc. Faraday Trans. II* **1988**, *84*, 1067.
42. Nazemi, A.; Kellar, E.; Williams, G.; Karasz, F. E.; Gray, G. W.; Lacey, D.; Hill, J. S. *Liq. Cryst.* **1991**, *9*, 307.
43. Kellar, E.; Williams, G.; Karasz, F. E.; Gray, G. W.; Lacey, D.; Hill, J. S. *Liq. Cryst.* **1991**, *1*, 331.
44. Maier, W.; Meier, G. *Z. Naturforsch.* **1961**, *162*, 262.
45. Böttcher, C. J. F.; Bordewijk, P. *Theory of Electric Polarization*, 2nd ed.; Elsevier: Amsterdam, Netherlands, 1978; Vol. 2, p 265
46. Dunmur, D. A.; Miller, W. H.; *Mol. Cryst. Liq. Cryst.* **1980**, *60*, 281.
47. Ferry, J. D. *Viscoelastic Properties of Polymers*, 2nd ed.; Wiley: New York, 1970.
48. Williams, G. *Trans. Faraday Soc.* **1964**, *60*, 1556.
49. Cole, K. S.; Cole, R. H. *J. Chem. Phys.* **1941**, *9*, 341.
50. Davidson, D. W.; Cole, R. H. *J. Chem. Phys.* **1950**, *18*, 1417.
51. Havriliak, S.; Negami, S. *J. Polym. Sci.* **1966**, *C14*, 99.
52. Williams, G.; Watts, D. C. *Trans. Faraday Soc.* **1970**, *66*, 80.
53. Williams, G.; Watts, D. C.; Dev, S. B.; North, A. M. *Trans. Faraday Soc.* **1971**, *67*, 1323.
54. Davidson, D. W. *Can. J. Chem.* **1961**, *39*, 571.

55. Cole, R. H.; Cole, K. S. *J. Chem. Phys.* **1942**, *10*, 98.
56. (a) Köhlrausch, R. *Poggendorff's Ann. Phys.* **1854**, *91*, 198. (b) Köhlrausch, R. *Ann. Phys.* **1847**, *12*, 393.
57. Abramowitz, M.; Stegun, I. A. *Handbook of Mathematical Functions;* National Institute of Science and Technology: Gaithersburg, MD, 1964; p1028, eq 118.
58. Moynihan, C. T.; Boesch, L. P.; Laberge, N. L. *Phys. Chem. Glasses* **1973**, *14*, 122.
59. Koizumi, N.; Kita, Y. *Bull. Inst. Chem. Res. Kyoto Univ.* **1978**, *56*, 300.
60. Williams, G. *IEEE Trans. Electr. Insul.* **1985**, *EI-20*, 843.
61. Lindsey, C. P.; Patterson, G. D. *J. Chem. Phys.* **1980**, *73*, 3348.
62. Shears, M. F.; Williams, G. *J. Chem. Soc. Faraday Trans. II* **1973**, *69*, 608.
63. Williams, G. *Trans. Faraday Soc.* **1964**, *60*, 1556.
64. Williams, G. *Trans. Faraday Soc.* **1965**, *61*, 1564.
65. Williams, G.; Edwards, D.A. *Trans. Faraday Soc.* **1966**, *62*, 1329.
66. Williams, G. *Trans. Faraday Soc.* **1966**, *62*, 1321.
67. Williams, G. *Trans. Faraday Soc.* **1966**, *62*, 2091.
68. Sasabe, S.; Saito, S. *J. Polym. Sci. A-2* **1968**, *6*, 1401.
69. Saito, S.; Sasabe, H.; Nakajima, T.; Yada, K. *J. Polym. Sci. Part A-2* **1968**, *6*, 1297.
70. Sayre, J. A.; Swanson, S. R.; Boyd, R. H. *J. Polym. Sci., Polym. Phys. Ed.* **1978**, *16*, 1739.
71. Williams, G. *Trans. Faraday Soc.* **1964**, *60*, 1548.
72. Hoffman, J. D.; Williams, G.; Passaglia, E. *J. Polym. Sci.* **1966**, *C14*, 173.
73. Yamafuji, K.; Ishida, Y. *Koll. Z.* **1962**, *183*, 15.
74. Glarum, S. H. *J. Chem. Phys.* **1960**, *33*, 1371.
75. Kubo, R. *J. Phys. Soc. Jpn.* **1957**, *12*, 570.
76. Cole, R. H. *J. Chem. Phys.* **1965**, *42*, 637.
77. Titulaer, U. M.; Deutch, J. M. *J. Chem. Phys.* **1974**, *60*, 1502.
78. Sullivan, D. E.; Deutch, J. M. **1975**, *62*, 2130.
79. Klug, D. D.; Kranbuehl, D. E.; Vaughan, W. E. *J. Chem. Phys.* **1969**, *50*, 3904.
80. Brot, C. In *Dielectric and Molecular Processes;* Davies, M., Ed.; Special Periodical Reports; The Chemical Society: London, 1972; Vol. 1, p 1.
81. Williams, G.; Cook, M.; Hains, P. J. *J. Chem. Soc. Faraday Trans. II* **1972**, *68*, 1045.
82. Baur, M.; Stockmayer, W. H. *J. Chem. Phys.* **1965**, *43*, 4319.
83. Stockmayer, W. H. *Pure Appl. Chem.* **1967**, *15*, 539.
84. Adachi, K.; Imanishi, Y.; Kotaka, T. *J. Chem. Soc. Faraday Trans. I* **1989**, *85*, 1065.
85. Adachi, K.; Imanishi, Y; Kotaka, T. *J. Chem. Soc. Faraday Trans. I* **1989**, *85*, 1075.
86. Adachi, K.; Imanishi, Y.; Kotaka, T. *J. Chem. Soc. Faraday Trans. I* **1989**, *85*, 1083.
87. Adachi, K.; Kotaka, T. *Macromolecules*, **1988**, *21*, 157.
88. Boese, D.; Kremer, F.; Fetters, L. J. *Macromolecules* **1990**, *23*, 1826.
89. Rosato, V.; Williams, G. *J. Chem. Soc. Faraday Trans.* **1981**, *77*, 1767.
90. Rose, M. E. *Elementary Theory of Angular Momentum;* Wiley: New York, 1957.
91. Rosato, V.; Williams, G. In *Molecular Interactions;* Ratajczak, H.; Orville-Thomas, W. J., Eds.; Wiley: New York, 1982; Vol. 3, p 373.
92. Bordewijk, P. *Physica* **1974**, *75*, 146.
93. Luckhurst, G. R.; Zannoni, C. *Proc. Roy. Soc. A* **1973**, *343*, 380.
94. Edwards, D. M. F.; Madden, P. A. *Mol. Phys.* **1983**, *48*, 471.
95. Attard, G. S. *Mol. Phys.* **1986**, *58*, 1087.
96. Kozak, A.; Moscicki, J. K.; Williams, G. *Mol. Cryst. Liq. Cryst.* **1991**, *201*, 1.
97. Kozak, A.; Moscicki, J. K. *Liq. Cryst.* **1992**, *12*, 377.

98. Attard, G. S. *Mol. Phys.* **1986,** *58,* 1087.
99. Attard, G. S. *Polymer* **1989,** *30,* 438.
100. Attard, G. S.; Araki, K. *Mol. Cryst. Liq. Cryst.* **1986,** *141,* 69.
101. Attard, G. S.; Williams, G. *Polymer* **1986,** *27,* 2.
102. Attard, G. S.; Williams, G. *Polymer* **1986,** *27,* 66.
103. Attard, G. S.; Williams, G.; Gray, G. W.; Lacey, D.; Gemmel, P. A. *Polymer* **1986,** *27,* 185.
104. Attard, G. S.; Williams, G. *J. Mol. Electron.* **1986,** *2,* 107.
105. Attard, G. S.; Williams, G. *Chem. Br.* **1986,** *22,* 919.
106. Attard, G. S.; Moura-Ramos, J. J.; Williams, G. *J. Polym. Sci., Polym. Phys.* **1987,** *25,* 1099.
107. Attard, G. S.; Williams, G. *J. Mol. Electron.* **1987,** *3,* 1.
108. Araki, K.; Attard, G. S.; Williams, G. *Polymer* **1989,** *30,* 432.
109. Kozak, A.; Simon, G. P.; Williams, G. *Polym. Commun.* **1989,** *30,* 102.
110. Kozak, A.; Moura-Ramos, J. J.; Simon, G. P.; Williams, G. *Makromol. Chem.* **1989,** *190,* 2463.
111. Nazemi, A.; Kellar, E.; Williams, G.; Karasz, F. E.; Gray, G. W.; Lacey, D.; Hill, J. S. *Liq. Cryst.* **1991,** *9,* 307.
112. Williams, G.; Nazemi, A.; Karasz, F. E.; Hill, J. S.; Lacey, D.; Gray, G. W. *Macromolecules* **1991,** *24,* 5134.
113. Attard, G. S.; Williams, G.; Fawcett, A. H. *Polymer* **1990,** *31,* 928.
114. Haase, W.; Pranoto, H. In *Polymeric Liquid Crystals;* Blumstein, A., Ed.; Plenum: New York, 1985.
115. Haase, W.; Pranoto, H.; Bormuth, F. J. *Ber. Bunsenger* **1985,** *89,* 1229.
116. Kremer, F.; Vallerien, S. U.; Zentel, R.; Kapitza, H. *Macromolecules* **1989,** *22,* 4040.
117. Martins, A. F.; Esnault, P.; Volino, F. *Phys. Rev. Lett.* **1986,** *57,* 1745.
118. Esnault, P.; Casquilho, J. P.; Volino, F.; Martins, A. F.; Blumstein, A. *Liq. Cryst.* **1950,** *7,* 607.
119. Kozak, A.; Simon, G. P.; Moscicki, J. P.; Williams, G. *Mol. Cryst. Liq.Cryst.* **1990,** *193,* 155.
120. Williams, G. In *Conferencias Plenarias Convidadas, do 1-Encontro de Qumiica-Fisica;* Costa, S. M. B.; Galvao, A. M., Eds.; Edicas da Sociedade Portugesa de Quimica: Lisbon, Portugal, 1994 (ISBN 972–96065–0–1).
121. Beevers, M. S.; Crossley, J.; Garrington, D. C.; Williams, G. *J. Chem. Soc. Faraday Trans II* **1976,** *72,* 1482.
122. Clarkson, T. S.; Williams, G. *J. Chem. Soc. Faraday Trans. II* **1974,** *70,* 1705.
123. Berne, B. J.; Harp, G. D. *Adv. Chem. Phys.* **1970,** *17,* 63.
124. Berne, B. J. In *Physical Chemistry, an Advanced Treatise, Vol. VIIIB, The Liquid State;* Eyring, H.; Henderson, W.; Jost, W., Eds.; Academic: Orlando, FL, 1971; pp 540–713.
125. Berne, B. J.; Pechukas, P.; Harp, G. D. *J. Chem. Phys.* **1968,** *49,* 3125.
126. Beevers, M. S.; Crossley, J.; Garrington, D. C.; Williams, G. *J. Chem. Soc. Faraday Trans. II* **1977,** *73,* 458.
127. Crossley, J.; Elliott, D. A.; Williams, G. *J. Chem. Soc. Faraday Trans. II* **1979,** *75,* 88.
128. Beevers, M. S.; Elliott, D. A.; Williams, G. *J. Chem. Soc. Faraday Trans. II* **1980,** *76,* 112.
129. Beevers, M. S.; Crossley, J.; Garrington, D. C.; Williams, G. *J. Chem. Soc. Faraday Symp.* **1976,** *11,* 38.
130. Beevers, M. S.; Williams, G. *Adv. Mol. Relax. Interact. Processes* **1980,** *16,* 175.
131. Pecora, R. *J. Chem. Phys.* **1968,** *50,* 2650.

132. Williams, G.; Watts, D. C. In *Nuclear Magnetic Resonance, Basic Principles and Progress, Vol. 4, NMR of Polymers;* Springer Verlag: Heidelberg, Germany, 1971, p 271.
133. Williams, G.; Watts, D. C. In *Dielectric Properties of Polymers;* Karasz, F. E., Ed.; Plenum: New York, 1971; pp 17–44.
134. Johari, G. P.; Goldstein, M. *J. Chem. Phys.* **1970,** *53,* 2372.
135. Johari, G. P. In *Molecular Dynamics and Relaxation Phenomena in Glasses;* Dorfmüller, Th.; Williams, G., Eds.; Springer Lecture Notes in Physics; Springer Verlag: Berlin, Germany, 1987; p 90.
136. Williams, G.; Crossley, J. *Ann. Rep. Phys. Chem. A* **1977,** 77–105.
137. Williams, G. In *Dynamics of Molecular Liquids;* Yarwood, J.; Orville-Thomas, W. J., Eds., NATO ASI; Reidel: Dordrecht, Netherlands, 1983; p 239.
138. Williams, G.; Watts, D. C. *Trans. Faraday Soc.* **1971,** *67,* 2793.
139. Pakula, T. Presented at the 2nd International Meeting on Relaxations in Complex Systems, Alicante, Spain, 1993.
140. Beevers, M. S.; Elliott, D. A.; Williams, G. *Polymer* **1980,** *21,* 13.
141. Beevers, M. S.; Elliott, D. S.; Williams, G. *Polymer* **1980,** *21,* 279.
142. Matsuoka, S. *Relaxation Phenomena in Polymers;* Hanser: Munich, Germany, 1992.
143. Hodge, I. M. *J. Non-Cryst. Solids* **1994,** *169,* 211.
144. Geny, F.; Monnerie, L. *J. Polym. Sci., Polym. Phys. Ed.* **1977,** *15,* 1.
145. Shlesinger, M. F.; Montroll, E. W. *Proc. Natl. Acad. Sci. U.S.A.* **1984,** *81,* 1280.
146. Ngai, K. L. In *Non Debye Relaxations in Condensed Matter;* Ramarkrishnam, T. V.; Lakshoni, K. Raj., Eds.; World Scientific: Singapore, 1987, p 3.
147. Rendel, R. W.; Ngai, K. L. In *Relaxations in Complex Systems;* Ngai, K. L.; Wright, G. B., Eds.; Office of Naval Research: Arlington, VA, 1984; V9, p 309.
148. Ngai, K. L.; Rendel, R. W.; Rajagopal, R. W.; Teitler, A. K. *Ann. N. Y. Acad. Sci.* **1987,** *484,* 150.
149. *Relaxations in Complex Systems, Parts I and II;* Ngai, K. L.; Wright, G. B., Eds.; published as special editions of *J. Non-Cryst. Solids* **1991,** *131–133.*
150. *Relaxations in Complex Systems 2, Parts I and II;* Ngai, K. L.; Riande, E.; Wright, G. B., Eds.; published as special editions of *J. Non-Cryst. Solids* **1994,** *172–174.*
151. Williams, G. In *Keynote Lectures in Selected Topics of Polymer Science;* Riande, F., Ed.; CSIC: Madrid, Spain, 1996; pp 1–39 (ISBN 84–00–074726).
152. Douglas, J. F.; Hubbard, J. B. *Macromolecules* **1991,** *24,* 3163.
153. Götze, W.; Sjögren, L. *J. Phys. C., Solid State Phys.* **1987,** *20,* 879.
154. Götze, W.; Sjögren, L. *Rep. Prog. Phys.* **1992,** *55,* 241.
155. Götze, W. In *Liquids, Freezing and the Glass Transition;* Hansen, J. P.; Levesque, D.; Zinn-Justin, J., Eds.; North Holland: Amsterdam, Netherlands, 1991; p 287.
156. Götze, W. In *Amorphous and Liquid Materials;* Fritsch, G.; Jacucci, G. Martinus Nijhof: Dordrecht, Netherlands, 1987; p 34.
157. Sjögren, L.; Götze, W. *J. Non-Cryst. Solids* **1991,** *131–133,* 153.
158. Sjögren, L.; Götze, W. *J. Non-Cryst. Solids* **1994,** *172–174,* 7.
159. Götze, W.; Sjögren, L. *J. Non-Cryst. Solids* **1994,** *172–174,* 16.
160. Schmidt-Rohr, K.; Spiess, H. W. *Phys. Rev. Lett.* **1991,** *66,* 3020.
161. Schmidt-Rohr, K.; Spiess, H. W. *Multidimensional Solid State NMR and Polymers;* Academic: London, 1994.
162. Leisen, J.; Schmidt-Rohr, K.; Spiess, H. W. *J. Non-Cryst. Solids* **1994,** *172–174,* 737.
163. Rouse, P. E. *J. Chem. Phys.* **1953,** *21,* 1272.
164. Zimm, B. H. *J. Chem. Phys.* **1956,** *24,* 269.
165. Kranbuehl, D. E.; Verdier, P. H. *J. Chem. Phys.* **1972,** *56,* 3145.

166. (a) Verdier, P. H. *J. Chem. Phys.* **1966**, *45*, 2118; (b) 2122.
167. Kranbuehl, D. E.; Verdier, P. H. *J. Non-Cryst. Solids* **1994**, *172–174*, 997.
168. (a) Dubois-Violette, E.; Geny, F.; Monnerie, L.; Parodi, O. *J. Chim. Phys.* **1969**, *66*, 1865. (b) Monnerie, L. *J. Non-Cryst. Solids* **1991**, *131–133*, 755.
169. Beevers, M. S.; Williams, G. *Adv. Mol. Relax Processes* **1975**, *7*, 237.
170. Jernigan, R. L. In *Dielectric Properties of Polymers;* Karasz, F. E., Ed.; Plenum: New York, 1972; p 99.
171. Williams, G.; Cook, M. *Trans. Faraday Soc.* **1971**, *67*, 990.
172. Rosato,V.; Williams, G. *Adv. Mol. Relax. Proc.* **1981**, *20*, 233.
173. Zentel, R.; Strobl, G. R.; Ringsdorf, H. *Macromolecules* **1985**, *18*, 960.
174. Maier, W.; Meier, G. *Z. Naturforsch.* **1961**, *162*, 262.
175. Martin, A. J.; Meier, G.; Saupe, A. *Symp. Faraday Soc.* **1971**, *5*, 119.
176. Chamberlain, R. V.; Haines, D. N.; Kingsbury, D. W. *J. Non-Cryst. Solids* **1991**, *131–133*, 192.
177. Chamberlain, R. V.; Kingsbury, D. W. *J. Non-Cryst. Solids* **1994**, *172–174*, 318.
178. Wu, L.; Dixon, P. K.; Nagel, S. R.; Williams, B. D.; Carini, J. P. *J. Non-Cryst. Solids* **1991**, *131–133*, 32.
179. Götze, W.; Sjögren, L. *J. Non-Cryst. Solids* **1991**, *131–133*, 161.
180. Schonhals, A.; Kremer, F.; Schlosser, E. *Phys. Rev. Lett.* **1991**, *67*, 999.
181. Schonhals, A.; Kremer, F.; Hoffman, A.; Fischer, E. W.; Schlosser, E. *Phys. Rev. Lett.* **1993**, *70*, 3459.
182. Schonhals, A.; Kremer, F.; Stickel, F. *Phys. Rev. Lett.* **1993**, *71*, 4096.
183. Schonhals, A.; Kremer, F.; Hoffman, A.; Fischer, E. W. *Physica A* **1993**, *201*, 263.
184. Stickel, F.; Kremer, F.; Fischer, E. W. *Physica A* **1993**, *201*, 318.
185. Hoffman, A.; Kremer, F.; Fischer, E. W. *Physica A* **1993**, *201*, 106.
186. Stickel, F.; Fischer, E. W.; Schonhals, A.; Kremer, F. *Phys. Rev. Lett.* **1994**, *73*, 2936.
187. Schonhals, A.; Kremer, F. *J. Non-Cryst. Solids* **1994**, *172*, 336.
188. Kranbuehl, D. E. *J. Non-Cryst. Solids* **1991**, *131–133*, 930.
189. Johari, G. P.; Mangion, M. B. M. *J. Non-Cryst. Solids* **1991**, *131–133*, 921.
190. Cassettari, M.; Salvetti, G.; Tombari, E.; Veronesi, S.; Johari, G. P. *J. Non-Cryst. Solids* **1994**, *172–174*, 554.
191. Stephan, F.; Seytre, G.; Boiteaux, G.; Ulanski, J. *J. Non-Cryst. Solids* **1994**, *172–174*, 1001.
192. Carlini, C.; Livi, A.; Rolla, P. A.; Fioretto, D. *J. Non-Cryst.Solids* **1994**, *172–174*, 567.
193. Carlini, C.; Ciardelli, F.; Rolla, P. A.; Tombari, E. *J. Polym. Sci.* **1987**, *25*, 253.
194. Hamon, B. V. *Proc. IEE* **1952**, *99*, Pt. IV, monograph 27.
195. Roe, R. J. *J. Non-Cryst. Solids* **1994**, *172–174*, 69.
196. Wong, J.; Angell, C. A. *Glass, Structure by Spectroscopy;* Marcel Dekker: New York, 1976.
197. Cicerone, M. T.; Blackburn, F. R.; Ediger, M. D. *J. Chem. Phys.* **1995**, *102*, 471.
198. Cicerone, M. T.; Ediger, M. D. *J. Phys. Chem.* **1992**, *97*, 2156.
199. Waldow, D. A.; Ediger, M. D.; Yamaguchi, Y.; Matsushita, Y.; Noda, I. *Macromolecules* **1991**, *24*, 3147.
200. Hyde, P. D.; Evert, T. E.; Ediger, M. D. *J. Chem. Phys.* **1990**, *93*, 2274.
201. Inoue, T.; Cicerone, M. T.; Ediger, M. D. *Macromolecules* **1995**, *28*, 3425.
202. Cicerone, M. T.; Ediger, M. D. submitted to *J. Chem. Phys.* **1995**, *103*, 5684.
203. Williams, G.; Fournier, J. *J. Chem. Phys.* **1996**, *104*, 5690.
204. Jonscher, A. K. *Dielectric Relaxation in Solids;* Chelsea Dielectrics Publishers: London, 1983.

205. Dissado, L. A.; Hill, R. M. *Phys. Scr. T* **1982**, *1*, 110.
206. Dissado, L. A.; Hill, R. M. *Proc. Roy. Soc. A* **1983**, *390*, 131.
207. Dissado, L. A.; Hill, R. M. *J. Chem. Soc. Faraday Trans. II*, **1984**, *80*, 291.
208. Dissado, L. A.; Hill, R. M. *J. Mater. Sci.* **1984**, *19*, 1576.
209. Dissado, L. A.; Hill, R. M. *J. Phys. C* 1985, *18*, 3829.
210. Pathmanathan, K.; Dissado, L. A.; Hill, R. M. *J. Mater. Chem.* **1985**, *20*, 3716.
211. Dissado, L. A.; Hill, R. M. *Chem. Phys.* **1987**, *111*, 193.
212. Dissado, L. A.; Hill, R. M. *Solid State Ionics* **1987**, *22*, 331.
213. Dissado, L. A.; Hill, R. M. *J. Mater. Sci.* **1989**, *24*, 375.
214. Dissado, L. A.; Hill, R. M. *J. Appl. Phys.* **1989**, *66*, 2511.
215. Dissado, L. A.; Allison, J. M. *J. Mol. Liq.* **1993**, *56*, 295.
216. Rajagopal, A. K.; Ngai, K. L.; Rendell, R. W.; Teitler, S. *J. Stat. Phys.* **1983**, *30*, 285.
217. Ngai, K. L.; Rajagopal, A. K. J.; Rendell, R. W.; Teitler, S. *Phys. Rev. B* **1983**, *28*, 6073.
218. Ngai, K. L.; Rajagopal, A. K.; Teitler, S. *Physica A* **1985**, *133*, 213.
219. Ngai, K. L.; Rajagopal, A. K.; Rendell, R. W.; Teitler, S. *IEEE Trans. Electr. Insul.* **1986**, *21*, 313.
220. Ngai, K. L. *J. Non-Cryst. Solids* **1987**, *95*, 969.
221. Ngai, K. L.; Rajagopal, A. K.; Teitler, S. *J. Chem. Phys.* **1988**, *88*, 5086.
222. Rajagopal, A. K.; Ngai, K. L.; Rendell, R. W.; Teitler, S. *Physica A*, **1988**, *21*, 3030.
223. Rajagopal, A. K.; Ngai, K. L.; Teitler, S. *J. Chem. Phys.* **1990**, *92*, 243.
224. Rajagopal, A. K.; Ngai, K. L.; Teitler, S. *J. Non-Cryst. Solids* **1991**, *131*, 282.
225. Ngai, K. L.; Peng, S. L.; Tsang, K. Y. *Physica* **1992**, *191*, 523.
226. Ngai, K. L. *J. Chem. Phys.* **1993**, *98*, 7588.
227. Roland, C. M.; Ngai, K. L. *J. Non-Cryst. Solids* **1994**, *172*, 868.
228. Roland, C. M.; Ngai, K. L. *J. Chem. Phys.* **1995**, *103*, 1152.
229. Ngai, K. L. *J. Non-Cryst. Solids* **1991**, *131*, 80.
230. Ngai, K. L.; Rendell, R. W. *J. Non-Cryst. Solids* **1991**, *131*, 942.
231. Ngai, K. L.; Rendell, R. W.; Plazek, D. J. *J. Chem. Phys.* **1991**, *94*, 3018.
232. Ngai, K. L.; Wang, C. H.; Fytas, G.; Plazek, D. L.; Plazek, D. J. *J. Chem. Phys.* **1987**, *86*, 4768.
233. Ngai, K. L.; Mashimo, S.; Fytas, G. *Macromolecules* **1988**, *21*, 3030.
234. Böhmer, R.; Ngai, K. L.; Angell, C. A.; Plazek, D. J. *J. Chem. Phys.* **1993**, *99*, 4201.
235. Ngai, K. L.; Rendell, R. W. *J. Mol. Liq.* **1993**, *56*, 199.
236. Ngai, K. L.; Roland, C. M. *Macromolecules* **1993**, *26*, 6824.
237. Rendell, R. W.; Ngai, K. L.; Mashimo, S. *J. Chem. Phys.* **1987**, *87*, 2359.
238. Ngai, K. L. *Macromolecules* **1991**, *24*, 4865.
239. Roland, C. M.; Ngai, K. L. *Macromolecules* **1992**, *25*, 363.
240. Floudas, G.; Placke, P.; Stepanek, P.; Brown, W.; Fytas, G.; Ngai, K. L. *Macromolecules* **1995**, *28*, 6799.
241. Ngai, K. L.; Etienne, S.; Zhong, Z. Z.; Schuele, D. E. *Macromolecules* **1995**, *28*, 6423.
242. Schwarz, G. *J. Phys. Chem.* **1968**, *71*, 4021.
243. Williams, G. *Adv. Mol. Relax. Processes* **1970**, *1*, 409.
244. Müller, M.; Stadler, R.; Kremer, F.; Williams, G. *Macromolecules* **1995**, *28*, 6942.

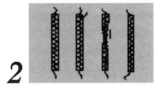

2

Broadband Dielectric Measurement Techniques

F. Kremer and M. Arndt

In this chapter we review high-accuracy dielectric measurement techniques covering the frequency range from 10^{-6} Hz up to microwave frequencies. In the first two sections we discuss frequency and time domain methods and we focus on equipment that is commercially available. In the last section a novel approach for the analysis of broadband dielectric spectra is presented.

Dielectric spectroscopy is an ideal tool to study molecular dynamics and charge transport. In its modern form it is broadband in frequency and covers the range from 10^{-6} Hz to 10^{12} Hz. To span this huge frequency window a variety of different measurement techniques have to be combined (Figure 1). The basic principles of dielectric measurement techniques are well-known (*1–15*). Our purpose for this chapter is to describe the state of the art in this field of research.

Measurement Systems in the Frequency Domain

In the frequency domain the complex dielectric function $\varepsilon^*(\omega)$ of a material is given by

$$\varepsilon^*(\omega) = \frac{C}{C_0} \tag{1}$$

where C_0 is the vacuum capacitance of a condenser, and C is the capacitance

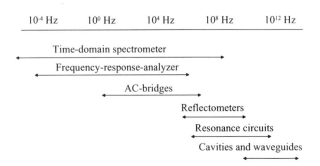

Figure 1. Summary of the various types of dielectric measurement techniques.

of the same condenser filled with the material under study. By using a sinusoidal electric field $E^*(\Omega) = E_0 \exp i\omega t$ having the angular frequency, ω, the dielectric function can be derived by measuring the complex impedance $Z(\omega)$ of the sample:

$$\varepsilon^*(\omega) = \frac{1}{i\omega Z(\omega) C_0} \tag{2}$$

To cover the frequency domain from 10^{-4} Hz to 10^{11} Hz, four different measurement systems based on different measurement techniques are employed (*16, 17*):

- frequency-response analysis (10^{-4} to 10^{11} Hz)
- impedance analysis (10^2–10^7 Hz)
- radio frequency (RF)-reflectometry (10^6–10^9Hz)
- network analysis (10^7 Hz–10^{12} Hz) (*18*)

A schematic of frequency response analysis technique is shown in Figure 2. The ac voltage $U_1(\omega)$ is applied to the sample by the generator

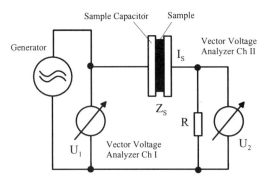

Figure 2. Schematic of frequency response analysis technique.

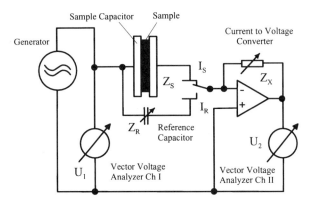

Figure 3. Schematic of the active interface in the low frequency range with active current-to-voltage converter and variable reference capacitors.

covering the frequency range from 10^{-4} to 10^7 Hz. The resistor R converts the sample current $I_S(\omega)$ into a voltage $U_2(\omega)$. The amplitude and the phases of the voltages $U_1(\omega)$ and $U_2(\omega)$ are measured by the use of two phase sensitive voltmeters (vector voltmeters). The complex sample impedance $Z_S(\omega)$ can be calculated from the measured data by

$$Z_S(\omega) = \frac{U_S(\omega)}{I_S(\omega)} = R\left(\frac{U_1(\omega)}{U_2(\omega)} - 1\right) \tag{3}$$

Commercial frequency-response analyzers are either based on digital lock-in amplifiers (e.g., Stanford SR 850, SR 830, and SR 810; frequency range: 10^{-3} to 10^5 Hz) or gain–phase analyzers (e.g., Solartron SI 1260 or SI 1255; frequency range: 10^{-5} to 10^7 Hz and TA Instruments DEA2970; frequency range: 3×10^{-3} to 10^5 Hz). For dielectric measurements with a high accuracy an active interface has to be used (*19*) (e.g., Chelsea Dielectric Interface, Novocontrol BDC).

In the Novocontrol interface the resistor R in Figure 2 is replaced by a current-to-voltage converter of variable gain (Figure 3) for low frequencies. Further improvement results from a comparison of the sample capacitor with a variable reference capacitor Z_R. If Z_x is a variable impedance that can be changed in resistance and capacitance, the sample impedance Z_S of a direct measurement is given by

$$Z_S = \frac{U_{1S}}{I_S} = -\frac{U_{1S}}{U_{2S}} Z_x \tag{4}$$

The accuracy of the measurement according to eq 4 is limited due to phase errors in the current-to-voltage converter and in the analyzer, but it

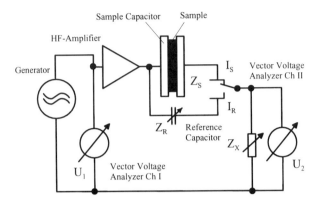

Figure 4. Schematic of the active interface in the high frequency range with passive variable impedance and variable reference capacitors.

can be improved if a reference measurement under the same conditions (especially of the current-to-voltage converter) is made for Z_R:

$$Z_R = \frac{U_{1R}}{I_R} = -\frac{U_{1R}}{U_{2R}} Z_x \tag{5}$$

Combining eqs 4 and 5 yields:

$$Z_S = \frac{U_{1S}}{U_{2S}} \frac{U_{2R}}{U_{1R}} Z_R \tag{6}$$

To reduce systematical errors the difference between the sample and the reference impedance must be minimized. In this case, the analyzer measures nearly identical values for the sample and the reference, and the accuracy is limited only by the analyzer resolution. In practice nonlinear deviations occur if the sample impedance differs more than 10% from the reference value. Therefore, the reference capacitors can be adjusted in 64 steps between 25 pF and 2 nF.

For frequencies less than 100 kHz (Figure 3) an operational amplifier is used that enables us to handle extremely low-input currents down to 40 fA. These amplifiers are limited to frequencies below 100 kHz. For frequencies between 10^5 and 10^7 Hz, a different technique of current-to-voltage conversion is used (Figure 4). The ac voltage of the analyzer generator drives a buffer amplifier that decouples the sample current from the analyzer. In contrast to Figure 3 the active current-to-voltage converter is replaced by a passive variable impedance Z_x. For ideal components and neglecting inductive effects one finds

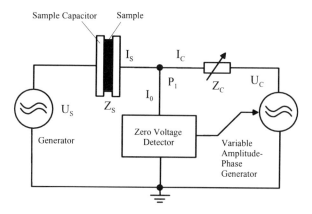

Figure 5. Schematic of an ac impedance bridge.

$$Z_S = Z_x \left(\frac{U_{1S}}{U_{2S}} - 1 \right) \tag{7}$$

With this equation the reference technique can be applied in a similar manner.

By using a Solartron SI 1260 in combination with a Novocontrol BDC active-interface resolution in $\tan \delta = \varepsilon''/\varepsilon'$, better than 10^{-4} can be achieved over the entire range of the analyzer. The impedance range is 10 to 10^{14} Ω. If a proper sample geometry is chosen, specific conductivity from 10^5 to 10^{-18} S/cm can be measured.

Figure 5 shows an ac impedance bridge that consists of the sample capacitance Z_S and the adjustable compensation impedance Z_C. On the left-hand side of the bridge, the generator drives the sample with the fixed and known ac voltage U_S, which causes the current I_S to flow into P_1. On the right-hand side of the bridge, the variable-amplitude phase generator (VAPG) feeds the current I_C through the compensation impedance Z_C into P_1. The bridge will be balanced if $I_S = -I_C$, which corresponds to $I_0 = 0$. Any deviation is detected by the zero voltage detector, which changes the amplitude and phase of the VAPG until $I_0 = 0$. In the balanced state, the sample impedance is calculated as

$$Z_S = \frac{U_S}{I_S} = -\frac{U_S}{U_C} Z_C \tag{8}$$

Several manufacturers offer automatic bridges, such as Hewlett-Packard HP 4284 (20 Hz to 1 MHz, resolution $\tan \delta < 5 \times 10^{-4}$, 1 m$\Omega$ to 100 MΩ) and HP 4192 (10 Hz to 10 MHz, resolution $\tan \delta < 10^{-3}$) and Quadtech (12 Hz to 100 kHz, resolution $\tan \delta < 10^{-5}$). For dielectric measurements a standard sample cell is required that is connected by coaxial cables to the bridge. Because of the parasitic inductances of the lines and connectors the high frequency limit is reached at about 1 MHz. Also the low-frequency limit may not be reached except for dielectric samples with high losses (tan $\delta > 1$), because the sample impedance exceeds the input impedance of the bridge. Nevertheless ac impedance bridges offer a suitable and inexpensive solution for applications, which do not need a large frequency range. As the measurement time is lower compared with frequency response analysis, ac impedance bridges are particularly suited for measurements on materials with time-dependent dielectric properties (monitoring of chemical reactions, characterization of phase transitions, etc.).

As sample geometry a disc-like capacitor arrangement is suitable for frequencies $\leq 10^6$ Hz. Typical dimensions are a thickness of approximately 50 μm and a sample diameter of 10 mm resulting in a capacitance of 69.5 pF for a value of $\varepsilon' = 5$. It is essential to measure the sample temperature in the immediate neighborhood of the sample and to keep the sample temperature as constant as possible. Both requirements are well fulfilled with the sample cells and the temperature controllers of Novocontrol.

Coaxial line reflectometry has to be employed at frequencies from 1 MHz to 10 GHz. In contrast to the low frequency techniques already described, above 10 MHz the measurement cables always contribute to the sample impedance. Above approximately 30 MHz standing waves arise at the line and a direct measurement of the sample impedance fails. This drawback can be avoided by application of microwave techniques taking the measurement line as the main part of the measured impedance into account. Therefore, precision lines with defined propagation constants are required. A schematic of the measurement technique is shown in Figure 6. The sample capacitor is used as the termination of a precision coaxial line. The complex reflection factor $r(l)$ at the analyzer end of the line depending on the sample impedance is measured with a microwave reflectometer. For this purpose, the incoming (inc.) and reflected (refl.) waves are separated with two directional couplers and are measured in amplitude, and phase r is defined as the ratio of the voltages (or electrical fields) of the reflected wave to the incoming wave on the line. It depends on the location, x, of the measurement along the line.

$$r(x) = \frac{U_{\text{refl.}}(x)}{U_{\text{inc.}}(x)} \tag{9}$$

For an ideal line $r(l)$, which is measured by the reflectometer, can be transformed to the reflection factor $r(0)$ at the sample end of the line by

$$r(0) = r(l) \exp[2l(\alpha + i\beta)] \tag{10}$$

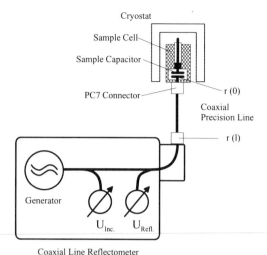

Figure 6. Schematic of a coaxial reflectometer.

where α is the damping constant and $\beta = 2\pi l/\lambda$ (where λ denotes the wavelength and l denotes the length of the line) the propagation constant of the line. From eq 10, the sample impedance is calculated as

$$Z_S = Z_0 \frac{1 + r(0)}{1 - r(0)} \tag{11}$$

where Z_0 is the wave resistance of the line. As can be seen from eq 11, the measurement range is limited to the sample impedance, Z_S, being in the range of Z_0. If this result is not the case, the reflection $r(0)$ becomes nearly 1 or -1 and the measured $r(l)$ changes marginally only.

In practice, the lines are not ideal and sophisticated calibration procedures have to be applied. Nevertheless, low loss precision lines matching the output resistance of the reflectometer are required. The line parameters α and β must be homogeneous over the whole line and also independent of temperature, because the calibration generally can only be carried out at room temperature. The same criteria mentioned for the line apply to the sample cell. Therefore, an additional calibration that eliminates the influence of internal impedances in the sample cell (Figure 7) is required. To keep the wave guide as short as possible, the cryostat with the sample is directly mounted at the front end of the analyzer. Coaxial line reflectometers cover the frequency range between 1 MHz and approximately 1 GHz. On the basis of Hewlett-Packard RF-impedance analyzers a resolution of $\tan \delta < 10^{-3}$ can be realized (HP 4291 A, frequency range 1 MHz to 1.8 GHz). Sample cells, precision extension lines, and dedicated cryostats are supplied by Novocontrol.

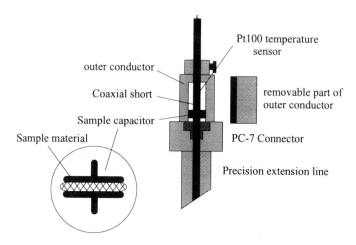

Figure 7. Schematic of an RF sample cell.

At frequencies above 1 GHz network analysis is used, in which not only the reflected wave but also the wave transmitted through a sample is analyzed in terms of phase and amplitude. This process allows extending the frequency range up to 100 GHz. But with increasing frequency and hence decreasing dimensions of coaxial lines or waveguides the calibration procedure becomes extremely cumbersome. With network analyzers based on the HP 8510 a resolution of tan $\delta < 10^{-2}$ is possible. Suitable sample cells and cryostats are offered by Novocontrol. In the frequency range between 10 GHz and 1000 GHz, oversized cavities or quasioptical spectrometers are used (*14, 15*).

Measurement Systems in the Time Domain

On the low frequency side of the dielectric spectrum (<10 Hz), time-domain techniques are often employed (*20–23*). At time $t = 0$ a voltage step U_{pol} is applied and the polarization current $I(t)$ is recorded. Analogous to eq 1, the time-dependent dielectric function is given by

$$\varepsilon(t) \ = \ \frac{C(t)}{C_0} \quad \text{and} \quad \frac{\mathrm{d}\varepsilon}{\mathrm{d}t} \ = \ \frac{I(t)}{C_0 U_{\text{pol}}} \tag{12}$$

Figure 8 shows a scheme of a time-domain spectrometer (*23*). The main item of the measurement setup is a programmable electrometer, such as Keithley 617, which is able to measure currents between 10^{-16} and 10^{-3} A. The connection between the sample and the electrometer consists of low-noise triaxial cables with an insulation resistance of 10^{15} Ω (*24*). The

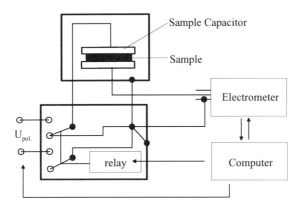

Figure 8. Schematic of a low frequency time domain spectrometer.

polarization voltage is taken from an internal voltage source of the electrometer. For depolarization a high-insulated double-relay separates the voltage source from the measurement circuit and connects the sample to the electrometer ground. To obtain the dielectric relaxation spectrum Fourier transform techniques have to be applied (*20, 25–28*). For the numerical analysis of time-domain data approximate transformations (e.g., Hamon transformation (*29*)) are used (*30, 31*). A resolution in tan $\delta < 10^{-3}$ can be realized (Keithley 617, frequency range 10^{-5} to 10 Hz).

An alternative way to measure dielectric properties in the time domain is applying a certain displacement D at $t = 0$ and acquiring the time-dependent voltage or field $E(t)$ (*32*). In this case the results refer to the dielectric modulus $M(t)$. If transformed into the frequency domain the dielectric modulus is given by $M^*(\omega) = 1/\varepsilon^*(\omega)$. Although the two quantities reflect the same microscopic effects the time scales can differ significantly. For the case of a single Debye relaxation the relation between relaxation time at constant field τ_ε and constant displacement τ_M is given by $\tau_M = \tau_\varepsilon \times \varepsilon_\infty / \varepsilon_s$ (*33*), where ε_∞ and ε_s denote the dielectric constants in the limit of high and low frequencies, respectively. For poly(vinyl acetate), one gains a factor of 6 in measurement time by recording $M(t)$ instead of $\varepsilon(t)$ (*32*).

Time-domain measurements can be extended to frequencies up to 20 GHz (*34*), but with a less accurate resolution in comparison to the network analyzer approach. In this technique a rectangular step voltage pulse is applied to the sample, and the changes in the characteristics of the pulse after the reflection from a section of a coaxial line filled with the sample being tested are monitored. The response of the sample to a fast voltage transient is recorded in the time domain and converted to a complex frequency response. Measurements are made along a coaxial transmission line with the sample mounted in a capacitive sample cell that terminates the line. The only differences between different setups are in the construction

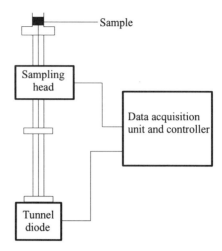

Figure 9. Schematic of a high frequency time domain spectrometer.

of the measuring cell and its location in the coaxial line, and these differences lead to different relationships between the values registered during the measurement and the dielectric characteristics of the objects under study. The limiting high and low frequencies are determined by the rise time and duration of the step pulse. The scheme of a high frequency time domain spectrometer is simple (Figure 9). The pulse of a step generator (tunnel diode) is sent to the temperature-controlled sample, and the reflected signal is measured with a sampling head and stored in a transient recorder. The sampling is made on a logarithmic time base. Complete measurement systems are offered by Dipole TDS Ltd. (*34*).

Analysis of Frequency-Domain Dielectric Data

Dielectric data can be analyzed by fitting empirical relaxation functions to the complex dielectric function $\varepsilon^*(\omega)$. The simplest case is the well-known Debye relaxation function $\varepsilon^*(\omega) - \varepsilon_\infty = \Delta\varepsilon/(1 + i\omega\tau)$. A more complex relaxation behavior like symmetric and asymmetric broadening of the relaxation peak is usually described by one parameter functions (Cole–Cole, Davidson–Cole, Fuoss–Kirkwood, and William–Watts) or two parameter functions (Havriliak–Negami, Jonscher, or Hill). For a collection, *see* the book of Jonscher (*35*).

Typically, a superposition of several such functions provides a satisfactory multiparameter fit to the experimental data. However, relating the parameters so obtained to the intrinsic physical properties of the material is not always straightforward. A further drawback of such an approach is the inherent difficulty of separating processes with comparable relaxation times. A proper choice of the number of relaxation processes is not always obvious, and additional *a priori* assumptions have to be made. An alternative way is to describe a dielectric relaxation spectrum in terms of an ensemble of Debye processes with a continuous relaxation time distribution, $g(t)$. The real and imaginary parts of the dielectric spectrum $\varepsilon^*(\omega)$ are then presented by

$$\varepsilon' = \varepsilon_\infty + \Delta\varepsilon \int \frac{g(\tau)}{1 + \omega^2\tau^2} \, d\,(\ln \tau) \tag{13}$$

$$\varepsilon'' = \Delta\varepsilon \int \frac{g(\tau)\omega\tau}{1 + \omega^2\tau^2} \, d\,(\ln \tau) \tag{14}$$

These equations are essentially Fredholm integral equations of the first kind and thus belong to the class of so-called *ill-posed* problems. A proper approach to solve such equations is the Tikhonov regularization algorithm (*36, 37*). The first attempt to apply regularization techniques to the analysis of dielectric data uses the CONTIN procedure of Provencher (*38*). A method with a significantly higher accuracy of fitting is the self-consistency method of Hohnerkamp and Weese (*39*). This technique has been applied to study dynamic exchange of glass-forming liquids confined to nanoporous materials (*40*).

Conclusion

The advantage of modern dielectric spectroscopy relies on the extraordinary width of its frequency range (10^{-6} to 10^{12} Hz). Additionally, the reduction of the required amount of sample material to the milligram level made it an attractive spectroscopic tool for samples that are obtained in small quantities only. In the recent decade studies based on dielectric spectroscopy made important contributions in a variety of fields in current research, such as scaling of the dynamic glass transition (*41–43*), molecular dynamics in polymers (*44–46*), collective and molecular dynamics in (ferroelectric) liquid crystals (*47–49*), relaxation in confining geometries (*50–53*), and relaxation and charge transport in colloidal systems (*54, 55*).

References

1. Field, R. F. In *Dielectric Materials and Applications;* von Hippel, A. R., Ed.; Wiley: New York, 1954.
2. Hill, N. E.; Vaughan, W.; Price, A. H.; Davies, M. *Dielectric Properties and Molecular Behaviour;* Van Nostrand: London, 1969.

3. Vaughan, W.; Smyth, C. P.; Powles, J. G. In *Physical Methods of Chemistry;* Weissberger, A.; Rossiter, B. W. Eds.; Wiley-Interscience, New York, 1972.
4. Blythe, A. R. *Electrical Properties of Polymers;* Cambridge University: Cambridge, England, 1979.
5. Boyd, R. H. *Methods of Experimental Physics;* Academic: Orlando, FL, 1980; Vol. 16C, Chapter 18.
6. Pochan, J. M.; Fitzgerald, J. J.; Williams, G. In *Determination of Electronic and Optical Properties,* 2nd ed.; Rossiter, B. W.; Baetzold, R. C., Eds.; Wiley: New York, 1993; Vol. 8.
7. Craig, D. Q. M. *Dielectric Analysis of Pharmaceutical Systems;* Taylor and Francis: London, 1995.
8. Gordy, W.; Smith, W. V.; Trambarulo, R. F. *Microwave Spectroscopy;* Wiley: New York, 1953.
9. Westphal, W. B. In *Dielectric Materials and Applications;* von Hippel, A. R., Ed.; Wiley: New York, 1954.
10. *Dielectric Materials and Waves;* von Hippel A. R., Ed.; Wiley: New York, 1954.
11. Cullen, A. L.; Yu, P. K. *Proc. R. Soc. Lond. Ser.* 1971, *A 325,* 483.
12. Amrhein, M.; Schulze, H. W. *Colloid Polym. Sci.* 1972, *250,* 921.
13. Gebbie, H. A.; Bohlander, R. A. *Appl. Opt.* 1972, *11,* 723.
14. Kremer, F.; Poglitsch, A.; Böhme, D.; Genzel, L. *Int. J. Infrared Millimeter Waves* 1984, II, 141.
15. Genzel, L.; Poglitsch, A.; Hässler, S. *Int. J. Infrared Millimeter Waves* 1986, *6,* 741.
16. Kremer, F.; Boese, D.; Maier, G.; Fischer, E. W. *Prog. Colloid Polym. Sci.* 1989, *80,* 129.
17. Schaumburg, G. "Dielectric Newsletter of Novocontrol," Novocontrol Company: Hundsangen, Germany, March 1994.
18. Pugh, J.; Ryan, T. *IEEE Conf. Publ.* 1979, *177,* 404.
19. Richert, R. *Rev. Sci. Instrum.* 1996, *67,* 1.
20. Mopsik, F. I. *Rev. Sci. Instrum.* 1984, *55,* 79.
21. McCrum, N. G.; Read, B. E.; Williams, G. *Anelastic and Dielectric Effects in Polymeric Solids;* Dover: New York, 1991.
22. Schiener, B.; Böhmer, R. *J. Non-Cryst. Solids,* 1995, *182,* 180.
23. Garwe, F.; Schönhals, A.; Lockwenz, H.; Beiner, M.; Schröter, K.; Donth, E. *Macromolecules* 1996, *290,* 247.
24. *Low Level Measurements,* 3rd. ed.; Keithley Instruments: Cleveland, OH, 1984.
25. Davidson W.; Wheeler, J. *J. Chem. Phys.* 1959, *30,* 1357.
26. Reddish, W. *Pure Appl. Chem.* 1962, *5,* 723.
27. Williams, G. *Polymer* 1963, *4,* 27.
28. Johnson, E.; Anderson, E. W.; Furakawa,T. *Annu. Rep. Conf. Electr. Insul. Dielectr. Phenom.* 1981, 258.
29. Hamon, V. *Proc. IEE* 1952, *99,* Part IV, Monograph 27.
30. Williams, G. *Trans. Faraday Soc.* 1962, *58,* 1041.
31. Williams, G.; Watts, D. C. *Trans. Faraday Soc.* 1970, *66,* 80.
32. Wagner, H.; Richert, R. *Polymer,* in press.
33. Fröhlich, H. *Theory of Dielectrics;* Clarendon: Oxford, England, 1958.
34. Feldman, Yu. "Dielectric Newsletter of Novocontrol," Novocontrol Company: Hundsangen, Germany, March 1995.
35. Jonscher, A. K. *Dielectric Relaxation in Solids;* Chelsea Dielectric: London, 1983.
36. Tikhonov, A. N.; Arsenin, V. Y. *Solutions of Ill-Posed Problems;* Wiley: New York, 1977.
37. Groetsch, W. *The Theory of Tikhonov Regularization for Fredholm Equations of the First Kind;* Pitman: London, 1984.

38. Provencher, W. *Comput. Phys. Commun.* **1982**, *27*, 229.
39. Hohnerkamp, J.; Weese, J. *Continuum Mech. Thermodyn.* **1990**, *2*, 17.
40. Schäfer, H.; Sternin, E.; Stannarius, R.; Kremer, F.; Arndt, M. *Phys. Rev. Lett.* **1996**, *76*, 2177.
41. Dixon, P. K.; Wu, L.; Nagel, S. R.; Williams, B. D.; Carini, J. P. *Phys. Rev. Lett.* **1990**, *65*, 1108.
42. Schönhals, A.; Kremer, F.; Schlosser, E. *Phys. Rev. Lett.* **1991**, *67*, 999.
43. Hofmann, A.; Kremer, F.; Fischer, E. W.; Schönhals, A. In *Disorder Effects on Relaxational Processes;* Richert, R.; Blumen, A., Eds.; Springer: Berlin, Germany, 1994, p 309.
44. Urakawa, O.; Adachi, K.; Kotaka, T. *Macromolecules* **1993**, *26*, 2036.
45. Schönhals, A. *Macromolecules* **1993** *26*, 1309.
46. Müller, M.; Kremer, K.; Stadler, R.; Fischer, E. W.; Seidel, U. *J. Colloid Polym. Sci.* **1995**, *273*, 38.
47. Kresse, H.; Kostromin, S. G.; Shibaev, V. P. *Liq. Cryst.* **1989**, *6*, 333.
48. Schönfeld, A.; Kremer, F.; Zentel, R. *Liq. Cryst.* **1993**, *13*, 403.
49. Groothues, H.; Kremer, K.; Schouten, P. G.; Warman, J. M. *Adv. Mater.* **1995**, *7*, 283.
50. Schüller, J.; Mel'nichenko, Yu.; Richert, R.; Fischer, E. W. *Phys. Rev. Lett.* **1994**, *73*, 2224.
51. Mel'nichenko, Yu.; Schüller, J.; Richert, R.; Ewen, B.; Loong, C. K. *J. Chem. Phys.* **1995**, *103*, 2016.
52. Stannarius, R.; Kremer, F.; Arndt, M. *Phys. Rev. Lett.* **1995**, *75*, 4698.
53. Arndt, M.; Stannarius, R.; Gorbatschow, W.; Kremer, F. *Phys. Rev. E* **1996**, *54*, 5377.
54. Myers, F.; Saville, D. A. *J. Colloid Interface Sci.* **1989**, *131*, 448.
55. Feldman, Yu.; Kozlovich, N.; Nir, I.; Garti, N. *Phys. Rev. E.* **1995**, *51*, 478.

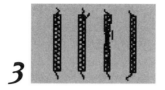

3

Dielectric Properties of Amorphous Polymers

A. Schönhals

Starting with a classification of how molecular dipoles can be attached to long flexible polymer molecules, the dielectric relaxation behavior of bulk amorphous polymers is reviewed. The peculiarities of the β-relaxation, which corresponds to localized reorientations of the dipole vector, the α-relaxation, which is related to the glass transition, and the normal-mode relaxation due to the fluctuation of the whole polymer chain are discussed in terms of the relaxation rate, the shape of the loss peak, and the relaxation strength. Relationships among the processes are described. Current research trends and open questions are discussed.

Since Debye (*1*) established the theory of dipolar relaxation, dielectric spectroscopy has proved very useful for studying the conformation and dynamics of amorphous polymers. This topic was pioneered by the classic studies of Fuoss, Kirkwood, and their collaborators (*2*), followed by the extensive work of McCrum et al. (*3*), Ishida et al. (*4*), Hedvig (*5*), Karasz (*6*), Blythe (*7*), Adachi and Kotaka (*8*), Riande and Saiz (*9*), and many, many others. In particular Williams and co-workers published many original papers as well as reviews (*see* for instance references 10 and 11). Because of recent developments in experimental technique and evaluation methods, there is a growing number of papers which

- document the dielectric behavior of well-known and new polymer systems over a large range of frequency and temperature
- interrelate dielectric, mechanical, NMR, and other relaxation processes of polymers

- compare the experimental results with the results of theories to obtain insights into the molecular factors that are responsible for the observed dielectric relaxation of amorphous polymers

This chapter is not intended to be complete either in the treated topics or in the quoted references. Rather it is my personal account of the current understanding of the dielectric relaxation processes in bulk amorphous polymers based on many earlier papers (2–11).

Dipole Moments in Polymers and Basic Relations

Dielectric spectroscopy deals with the influence of an alternating electric field $\vec{E}(\omega)$ on matter (ω, angular frequency; f, frequency; $f = 2\pi\omega$). Application of $\vec{E}$ results in a polarization $\vec{P}$ of the medium. For small electric-field strengths a linear relationship holds between $\vec{E}$ and $\vec{P}$ (12):

$$\vec{P}(\omega) = (\varepsilon^*(\omega) - 1)\varepsilon_{Vac}\vec{E}(\omega); \quad \text{with } \varepsilon^*(\omega) = \varepsilon'(\omega) - i\varepsilon''(\omega) \quad (1)$$

where ε_{Vac} denotes the permittivity in vacuum. The quantity ε^*, called the complex dielectric function (ε', real part; ε'', imaginary or loss part; $i = \sqrt{-1}$), is related by the theory of dielectric relaxation to the correlation function $\Phi(t)$ of the polarization fluctuations (12, 13):

$$\frac{\varepsilon^*(\omega) - \varepsilon_\infty}{\varepsilon_{Sta} - \varepsilon_\infty} = \int_0^\infty \left[\frac{-d\Phi(t)}{dt}\right] \exp(-i\omega t)\, dt \quad \text{with} \quad \Phi(t) = \frac{\langle \Delta P(t)\, \Delta P(0)\rangle}{\langle \Delta P(0)^2\rangle}$$

$$(2)$$

where ΔP denotes a fluctuation of the polarization around its equilibrium value, the brackets denote the averaging over an ensemble or time t, ε_∞ and ε_{Sta} are the permittivities at very high and at quasistatic frequencies, respectively.

Experiments working with sinusoidal alternating fields are called measurements in the frequency domain. Alternatively the experiments can also be carried out in the so-called time domain, where a step-like change of E is applied. In that case the dielectric behavior is discussed in terms of the time-dependent dielectric function $\varepsilon(t)$, which is directly proportional to the dipole correlation function. Therefore, the relation between frequency and time domain is given by eq 2.

From a microscopic point of view the macroscopic observable polarization P is related to the dipole density of N permanent molecular dipoles

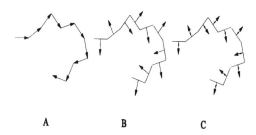

A   B   C

Figure 1. The different geometric possibilities for a location of a dipole with respect to the polymer chain.

$\vec{\mu}_i$ in a volume V.[1] In contrast to low molecular weight molecules, where the dipole moment can be well represented by a single rigid vector (*1, 12*), for long-chain molecules there are different geometric possibilities for the orientation of molecular dipole vectors with respect to the polymer backbone (Figure 1). According to Stockmayer (*15*) and Block (*16*), macromolecules with the dipoles fixed parallel to the main chain are called type-A polymers. For type-B polymers the dipole moment is rigidly attached perpendicular to the chain skeleton. Although no polymer is only of type A there are several examples of polymers like *cis*-1,4-polyisoprene (*8*) that have a dipole component both parallel and perpendicular to the chain. These polymers are also called type-A polymers, and an extensive treatment of these systems was given by Adachi and Kotaka (*8*). However, most of the synthetic macromolecules are of type B. Chain molecules having the dipoles in a more or less flexible side chain like the poly(*n*-alkylmethacrylate)s are type C.[2] A detailed discussion of each group including a list of polymers for each type can be found in references 9 and 15.

The net dipole moment per unit volume (polarization) of a polymer system is given as a vector summation over all molecular dipole types in the repeating unit, the polymer chain, and over all chains in the system:

$$\vec{P} = \frac{1}{V} \sum_{\substack{\text{all} \\ \text{chains}}} \sum_{\text{chain}} \sum_{\substack{\text{repeating} \\ \text{unit}}} \vec{\mu}_i \tag{3}$$

Because the molecular motions in dense amorphous polymer systems are controlled by very different time and length scales, different parts of

---

[1] Also, space charges that normally are responsible for conductivity contribution can contribute to P as Maxwell–Wagner–Silliars polarization (*14*) if they are partially trapped.

[2] A dipole in a side chain usually implies both a dipole component that contributes to the polarization by main-chain reorientation and a component whose reorientation is due to local side-chain motions (*16*).

the net dipole moment can be reoriented by different motional processes. Thus, the dielectric spectrum of an amorphous polymer generally shows a multiple relaxation behavior where each process is indicated by a peak in ε'' and a step-like decrease in ε' versus frequency at a fixed temperature. The dielectric relaxation behavior can also be measured isochronally at a fixed frequency versus temperature. These measurements also exhibit a loss peak, at a temperature depending on the selected frequency, and a step-like increase in ε' with increasing temperature. However, isochronal data in the temperature domain are difficult to analyze quantitatively, because the relaxation strength and the distribution of relaxation time are all generally temperature dependent. For relaxation processes ε' and ε'' are connected by the Kramers–Kronig relation (12) and so contain the same information in principle. However, the determination of ε'' from ε' requires integration over the entire frequency range, which is not possible in general. So in this account mainly the ε'' behavior will be discussed.

An ε'' peak can be fully characterized by its frequency position determined by the frequency of maximal loss f_p from which a characteristic relaxation time $\tau = 1/(2\pi f_p)$ can be obtained, its shape properties such as width and symmetry, and its relaxation strength $\Delta\varepsilon$ given by $\Delta\varepsilon = \int\varepsilon''(\omega)\,d\ln\omega$ (12). The Debye theory of dielectric relaxation, as improved by Onsager, Fröhlich, and Kirkwood (*see* reference 12 and references quoted therein), gives for the temperature dependence of $\Delta\varepsilon$ of N independently relaxing dipoles

$$\Delta\varepsilon = \varepsilon_{\text{stat},i} - \varepsilon_{\infty,i} = F_{\text{onsager}}\,g\frac{N\mu^2}{3kT} \quad \text{with} \quad F_{\text{onsager}} = \frac{1}{3}\frac{\varepsilon_{\text{stat},i}(\varepsilon_{\infty,i} + 2)^2}{2\varepsilon_{\text{stat},i} + \varepsilon_{\infty,i}}$$

$$(4)$$

where μ is the dipole moment of the moving unit, T is the temperature, k is the Boltzmann constant, $F_{\text{onsager}} \approx 1$ is the correction factor for internal field effects, and $\varepsilon_{\infty,i}$ and $\varepsilon_{\text{stat},i}$ denote the permittivities at high and at low frequencies with respect to f_p of the relaxation region under investigation. The Kirkwood correlation factor g was introduced to describe local, static correlation of dipoles. Inserting eq 3 into eq 2, the g factor can be calculated in principle for a selected system. Practically, this calculation is not possible for real, dense amorphous polymers. Actual and successful calculations were only done for several types of isolated polymer chains by using the statistical mechanics of polymers and the rotational isomeric state model (9, 17–19). Inserting eq 3 into eq 2 shows also that in addition to self-correlation terms different cross-correlation terms of dipoles such as the correlation of dipoles located in neighbored repeating units of the same or of other chains can contribute to the dielectric behavior. This topic was discussed in great detail by Williams, Cook, and Hains (20).

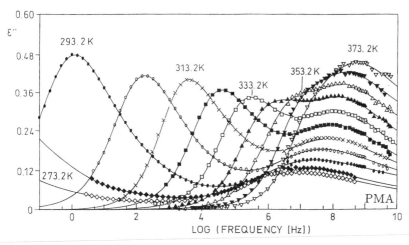

Figure 2. Dielectric loss spectra of polymethylacrylate. The symbols correspond to different temperatures. (Adapted from reference 21.)

Most of the amorphous polymers exhibit a secondary, or β-process, and a principal, or α-relaxation, located at lower frequencies or higher temperatures than the β-relaxation. As a typical example Figure 2 shows the dielectric loss spectrum for polymethylacrylate in the frequency range from 10^{-1} Hz to 10^{10} Hz displaying the high frequency, β, and the low frequency, α-processes (*21*). For type-A polymers at frequencies below the α-relaxation, a further process called α'- or normal-mode relaxation can be observed (*8*) that corresponds to the molecular motion of a whole polymer chain.

In this chapter the characteristic and fundamental peculiarities of the β-, α-, and the normal-mode relaxation of bulk amorphous polymers will be discussed in terms of the relaxation rates, the shape of the loss peak, and the relaxation strength[3] in greater detail. Special features like the dependence of the dielectric properties on the concrete chemical structure or on the tacticity cannot be described completely. For these details the reader is referred to other accounts like reference 5 or to the references cited therein.

Polymers can be also synthesized in many geometric forms other than linear chains like combs, stars, rings, or networks. The geometric form of a macromolecule will clearly influence its relaxational and dielectrical behavior. Several examples of this characteristic will also be discussed. Apart from the β-, α-, and the normal-mode processes many amorphous polymers

---

[3] Several methods like Cole–Cole plots estimate the relaxation time, the shape, and the relaxation strength for each process (12). The most convenient one is the use of model functions (*22*) (*see also* Chapter 6).

also exhibit further relaxation processes (*3*) that can be detected dielectrically (*23–25*), provided that the relaxation process involves a reorientation of the dipole moment vector.

β-Relaxation in Amorphous Polymers

Although the β-relaxation originates from localized fluctuation of the dipole vector which is well accepted, the actual molecular mechanism of this relaxation process is not yet well understood. On the one hand, according to the nomenclature developed by Heijboer (*26*), fluctuations of localized parts of the main chain, or the rotational fluctuations of side groups or parts of them, are assigned as a molecular mechanism for that relaxation process. In particular, this picture is supported by dielectric studies of model systems such as the investigation of the properties of poly(*n*-alkylmethacrylate)s in dependence on the length of the alkyl side chain (*27–30*) or more recent investigations on methyl acrylate–ethylene (*31*) and vinyl acetate–ethylene copolymers (*25*) and on blends of bisphenol-A and tetramethylbisphenol-A polycarbonate (*32*). On the other hand Goldstein and Johari (*33*) proposed that the β-relaxation is a general feature of the amorphous state because apart from amorphous polymers, β-relaxations are also observed in many other systems with very different chemical structures, including low molecular weight glass-forming liquids and rigid molecular glasses (*33, 34*). It was argued that this relaxation process is moderated to a great extent by the amorphous state. Very recently the discussion of β-relaxations was renewed by the development of the mode–mode coupling theory of the glass transition (*35, 36*). This theory will be discussed in greater detail later. In general the β-relaxation has the following properties: relaxation rate, shape of the loss peak, and relaxation strength.

Relaxation Rate

The temperature dependence of the relaxation rate is found to be Arrhenius-like: that is,

$$f_{p,\beta} = f_{\infty,\beta} \exp\left[-\frac{E_a}{kT} \right] \tag{5}$$

where $f_{\infty,\beta}$ is the preexponential factor, and E_a is activation energy. For truly activated processes $f_{\infty,\beta}$ should be of the order of magnitude 10^{12} to 10^{13} Hz. If $f_{\infty,\beta}$ is greater than this value other factors like activation entropies must contribute to the process. The activation energy E_a, related to the slope of log $f_{\infty,\beta}$ versus $1/T$, depends on both the internal rotation barriers and the environment of a moving unit. Recently the influence of the molecular environment was also discussed in terms of matrix rigidity for acrylic polymers (*36a*). Typical values for E_a are 20–50 kJ/mol. For most polymers the β-relaxation has been studied below the glass-transition temperature (T_g), but broadband studies have also shown that this relaxation exists above T_g and that the relaxation rate shows no discontinuity at T_g.

Shape of Loss Peak

In the frequency domain the dielectric β-relaxation displays a broad and in the most cases symmetric loss peak with half widths of 4–6 decades. Extracting information on the basic mechanism of motion is difficult from such broad peaks. The width of the β-peak is often explained in terms of a distribution of both the activation energy and the preexponential factor caused by a distribution of molecular environments in which the molecular motion of the β-relaxation can take place (*37*). The width of the β-peak generally decreases with increasing temperature.

Relaxation Strength

For most of the amorphous polymers having a dipole moment rigidly attached to the main chain, such as polycarbonate (*32*), polyvinyl chloride (*38, 39*), polychloroprene (*38*), or poly(ethylene terephthalate) (*40, 41*), for the relaxation strength of the β-relaxation $\Delta\varepsilon_\beta \ll \Delta\varepsilon_\alpha$ holds. Some polymers containing flexible dipolar side groups such as poly(methyl acrylate) (*21*) or higher poly(*n*-alkylmethacrylate)s (*42*) show $\Delta\varepsilon_\beta < \Delta\varepsilon_\alpha$. Exceptions are conventional poly(*n*-alkyl-methacrylate)s for which $\Delta\varepsilon_\beta > \Delta\varepsilon_\alpha$ (*3, 43, 44*). Because the main dipole moment is located in the side group, just small fluctuations of the side group can contribute significantly to the dielectric loss. On the other hand there are arguments from multidimensional NMR experiments that the main- and side-chain motions are coupled (*45*).

Generally, $\Delta\varepsilon_\beta$ increases with temperature. Because the net dipole moment is nearly independent of temperature this temperature dependence can be understood according to eq 4 by an increase of the fluctuation angle of the dipole vector or by the number of moving dipoles increasing with temperature. The suggestion that the correlation factor, which can be formally calculated from eq 4, increases with temperature seems unlikely for the β-relaxation because a reorientation of greater parts of the molecular dipole vector, which is necessary for a change of g, is not possible below T_g.

α-Relaxation in Amorphous Polymers

The α-relaxation is related to the glass transition of the system and for that reason this relaxation process is also called dynamic glass transition. The 'static' T_g, determined by differential scanning calorimetry, corresponds to a frequency of approximately 10^{-3} Hz. Of course this frequency is dependent on the cooling and heating rate. A dynamic glass transition temperature from dielectric experiments can be defined as the temperature of maximum loss at the selected frequency. In general the α-relaxation and the related glass-transition phenomenon are not well understood, and the real microscopic description of the relaxation remains an unsolved and current problem of polymer science. This statement is also true for low molecular weight liquids. For an overview, see reference 46 or 47.

It seems clear that the glass transition is a cooperative phenomenon and most workers would agree that for polymers the α-process corresponds to the micro-Brownian segmental motions of chains. Most workers would further agree that these motions are related to conformational changes like gauche trans transitions that lead to a rotation of a dipole around the chain that is rigidly attached perpendicular to the chain skeleton. Compared with an isolated low molecular weight molecule, this elementary motional process for an isolated polymer chain—called local chain motion—is very complex and involves many degrees of freedom. One proposal for the mechanism is the crankshaft motion in which the tails of the chains are in the same position in the initial and final states such as the Shatzki crankshaft (48) or three-bond motion (49–51). The current understanding of the local chain motion is based on ideas of Helfand and Hall (52) and Skolnick and Yaris (53), who described this process as damped diffusion of conformational changes along the chain. These model considerations are in agreement with most experimental results as well as computer simulations (54–56). A recent rediscussion of this type of motion was given by Ediger and Adolf (57).

For bulk amorphous polymers this motional process will take place in dense environments built by other chains. So intermolecular interactions contribute to that relaxation process in addition to intramolecular ones. For most amorphous polymers the α-relaxation has the following peculiarities: relaxation rate, shape of the loss curve, and relaxation strength.

Relaxation Rate

The temperature dependence of the relaxation rate of the α-relaxation, $f_{p,\alpha}$, shows a curved trace in a plot log $f_{p,\alpha}$ versus T^{-1}. This dependence can be well described by the Vogel–Fulcher–Tammann–Hesse (VFTH) (58–60) equation

$$\log f_{p\alpha} = \log f_{\alpha\infty} - \frac{A}{T - T_0} \tag{6}$$

where $\log f_{\alpha,\infty}$ ($f_{\alpha,\infty} \approx 10^{10}-10^{13}$ Hz) and A are constants, and T_0 is the so-called ideal glass transition or Vogel temperature, which is generally 30–70 K below T_g. Equation 6 indicates that the α-relaxation is not an activated process. An apparent, temperature-dependent activation energy can be calculated formally from eq 6. However, near the glass transition this apparent activation energy is much greater than the binding energy for a C–C bond and has therefore no physical or chemical meaning.

Several models were proposed to understand the physical nature of the VFTH equation and of T_0. In the free-volume approach of Cohen and Turnball (for an overview, *see* reference 61), the fractional free volume becomes zero at T_0. Adam and Gibbs (*62*) treated for the first time the α-relaxation as a cooperative process. This cooperativity sets in well above T_g and increases with decreasing temperature. In this kinetic approach (*62*) and in the fluctuation model of Donth (*63, 64*), the volume of the cooperatively rearranging region, defined as the smallest volume element that can relax to a new configuration independently, diverges at $T = T_0$. The thermodynamic lattice theory of the glass transition of DiMarzio and Ann (*65*) predicts that at T_0 a second-order phase transition takes place and overcomes the Kauzman paradox (*66*). Although the physical meaning of T_0 is up to now not yet clear the universality of the VFTH equation near T_g suggests that T_0 is a significant temperature for the dynamics of the glass transition. At higher temperatures ($T \approx T_g + 100$ K) and higher frequencies, a change may occur in the temperature dependence of $\log f_{p,\alpha}$ from VFTH behavior to another behavior (*67, 68*).

Equation 6 can be transformed into the Williams–Landel–Ferry (WLF) relation (*69*)

$$\log \frac{f_{p,\alpha}(T)}{f_{p,\alpha}(T_r)} = - \frac{C_1(T - T_r)}{C_2 + T - T_r} \tag{7}$$

where T_r is a reference temperature, $f_{p,\alpha}(T_r)$ is the relaxation rate at this temperature, and C_1 and $C_2 = T_r - T_0$ are the so-called WLF parameters. It has been claimed that these parameters should have universal values if $T_r = T_g$ is chosen (*69*). However, these universal values were found experimentally to be only rough approximations.

Because the α-relaxation is related to the glass transition, $f_{p,\alpha}$ shows the same dependence on molecular weight and chain architecture as T_g (for an overview, *see* references 64 and 69). For linear chains and stars, $f_{p,\alpha}$ increases slightly with molecular weight (M_w) below the critical value M_c ($M_c \approx 10^4$) and is independent of M_w above M_c. For networks, $f_{p,\alpha}$ generally increases with increasing cross-linking density. The unusual dependence of T_g on molecular weight known for ring polymers was recently investigated by a dielectric study on the molecular dynamics on cyclic and linear poly(dimethyl siloxanes) (70).

Shape of Loss Peak

In general the α-process is well defined in the frequency domain and shows a relatively broad (the width ranges in dependence on structure from 2 to 6 decades) and asymmetric peak. Several functions like the Cole–Cole (71), Cole–Davidson (72, 73), and the Fuoss–Kirkwood functions (74) in the frequency domain are able to describe broad symmetric or asymmetric loss peaks. The most general one is the model function of Havriliak and Negami (75, 76) (HN function), which reads

$$\varepsilon^*(\omega) - \varepsilon_\infty = \frac{\Delta\varepsilon}{(1 + (i\omega\tau_{HN})^{\beta_{HN}})^{\gamma_{HN}}} \tag{8}$$

τ_{HN} is a characteristic relaxation time related to the peak frequency f_p (77), and β_{HN} and γ_{HN} are fractional shape parameters with $0 < \beta_{HN}$; $\beta_{HN}\gamma_{HN} \leq 1$ related, with respect to τ_{HN}^{-1}, to the limiting low and high frequency slopes of log ε'' versus log ω:

$$\frac{\partial \log \epsilon''}{\partial \log \omega} = \beta_{HN} = m \qquad \text{for } \omega << \frac{1}{\tau_{HN}};$$

$$\frac{\partial \log \epsilon''}{\partial \log \omega} = -\beta_{HN}\gamma_{HN} = -n \qquad \text{for } \omega >> \frac{1}{\tau_{HN}}$$

This nomenclature was first introduced by Jonscher (78). The connection between the HN parameters m and n with the Debye, the Cole–Cole, and the Cole–Davidson function is given in Figure 3. The HN function is equivalent for $\beta_{HN} = \gamma_{HN} = 1$ to the Debye function, for $\gamma_{HN} = 1$ to the Cole–Cole-function, and for $\beta_{HN} = 1$ to the Cole–Davidson function. Therefore, the HN function can be regarded as a combination of the Cole–Cole and Cole–Davidson functions.

In the time domain, the dielectric behavior is very often described by the Kohlrausch–Williams–Watts (stretched-exponential, KWW) function (79)

$$\Phi(t) = \exp[-(t/\tau_{KWW})^{\beta_{KWW}}] \quad \text{with} \quad 0 < \beta_{KWW} \leq 1 \tag{9}$$

because no simple analytical model function with four shape parameters (similar to eq 8) is known for time-dependent permittivity $\varepsilon(t)$. The rela-

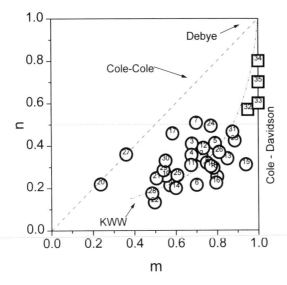

Figure 3. Shape parameters n versus m of the β-relaxation peak for different polymers (○) and low molecular weight glass-forming liquids (□). Dashed line represents the Cole–Cole function, the y-axis at m = 1 represents the Cole–Davidson function, and the dashed double dot line is calculated according to the KWW function. The dotted line at n = 0.5 is the upper limit for the local chain motion according to the Hall–Helfand model (52).

tionships of the KWW function to the Cole–Davidson and to the HN functions were discussed previously (*80, 81*). A fast numerical procedure to transform the HN function into the time domain is given in reference 82. Figure 3 also displays the connections between the stretched exponential function and the HN shape parameters *m* and *n*. The KWW line was obtained by calculating ε(*t*) for different values of β_{KWW} according to eq 9 and subsequently fitting these data by a numerical Fourier transformation of the HN function into the time domain (*83*). It follows from Figure 3 that every value of β_{KWW} corresponds to special combination of *m* and *n*.

Experimentally the width of the α-peak depends on such factors as temperature or cross-linking density. It becomes narrower with increasing temperature and broadens dramatically with cross-linking. The HN shape parameters *m* and *n* are plotted in Figure 3 for many polymers and several low molecular weight glass-forming liquids from which one can conclude (*82*):

- No correlation exists between the shape parameters *m* and *n* (as by the KWW function) so that at least two shape parameters are necessary to describe the data in general. A more extended discussion of this issue for more than 1000 materials recently was given by Havriliak and Havriliak (*84*).

- For polymeric materials *n* is found to be $0 < n \leq 0.5$ (*82*). Moreover both parameters have a different dependence on temperature (*85, 86*), crystallization (*83, 87*), and cross-linking density (*88, 89*).

The width of the α-peak is generally not assumed to be caused by a distribution of relaxation times due to local heterogeneities. Rather this broad, asymmetric loss peak is an intrinsic feature of the dynamics of glass-forming systems. Because the α-relaxation is only poorly understood, there is no molecular model that starts from first principles to calculate the correlation function for dense amorphous systems up to now. One way to find out what is important for the construction of appropriate models is to carry out many complementary experiments, where the word *complementary* refers to variations in the chemical structure and the use of different experimental methods. Modern techniques like multidimensional NMR spectroscopy (*45, 90*) or scattering studies may become very important for complementary experiments. Such considerations and investigations have led to several models having a more or less phenomenological basis that try to explain the nonexponential character of the α-peak. An overview and a discussion of several of these models like small-step rotational diffusion, barrier models, or defect diffusion are discussed in an article by Williams (*10*) or in reference 12. Montroll and Bendler (*91*) extended the defect diffusion picture by incorporation of a waiting time distribution. This approach results in stretched exponentials with stretching parameters different from 0.5 depending on the waiting time distribution. Recently Matsuoka (*92*) presented a new model that explicitly incorporates intermolecular cooperativity. Also the coupling scheme introduced by Ngai (*93*) starts from the picture of intermolecular cooperativity. A recent discussion of the coupling scheme can be found in reference 94. This model, which has been applied to many polymeric systems (*95–99*), gives a KWW-type correlation function where β_{KWW} should decrease phenomenologically with increasing cooperativity. Moreover the coupling model predicts a relationship between the shape of the relaxation function and the VFTH equation, which has been tested experimentally (*99*).

None of these models is able to predict two shape parameters in accordance to eq 8 and Figure 3. For the first time Dissado and Hill (*100*) presented a cluster approach for the structure and the dynamics of imperfect materials that predicts, for the relaxation function, two power laws in accordance with eq 8. The high frequency parameter *n* is related to intra-cluster motions, and *m* is related to intercluster motions (*100*). A model that was designed for the dynamics in dense polymer systems was published (*101*). This model starts from the point of view that the local chain dynamics must also be the elementary motional mode in dense systems. Of course in dense melts that motional process will be influenced by the neighbored chains. So the process of local chain motion changes to a hindered diffusion of conformational changes. Furthermore this motional process takes part in a fluctuating environment. The model predicts for the relaxation function two shape parameters, where n $(0 < n \leq 0.5)$ is related to a hindered local chain motion, and m $(0 < m \leq 1)$ is related to the cooperative motion of a larger environment. Several experimental results support these considerations (83–89).

Relaxation Strength

For the α-process, $\Delta\varepsilon_\alpha$ increases with decreasing temperature. However, Figure 4a, where the product $T\Delta\varepsilon$ is plotted versus T for several typical polymers, shows that this increase is much stronger than the temperature dependence predicted by eq 4. From a practical point of view, therefore, eq 4 can be used only for a rough estimation of the dipole moment. This very strong temperature dependence of $\Delta\varepsilon$ near T_g is found both for polymers (102–105) and for low molecular weight molecules (67, 68, 105, 106), and cannot be explained by the small increase of the density with decreasing temperature. Also, the modelling of this behavior by a temperature-dependent g factor remains formal because g was introduced to describe direct correlations between dipoles like association (12). It seems more likely that this temperature dependence results from an influence of cross-correlation terms to $\langle \mu^2 \rangle$, which becomes more important for decreasing temperatures. Therefore, the reorientation of a selected test dipole is more and more influenced by its molecular environment with decreasing temperature. Surprisingly, a plot of $\Delta\varepsilon$ versus log $f_{p\alpha}$ (Figure 4b) gives a straight correlation. The same behavior is also found for a lot of low molecular glass-forming systems (107). To decide if this dependence is or is not a general behavior of the dynamic glass transition requires additional studies. However, because the temperature dependence of both $f_{p\alpha}$ and $\Delta\varepsilon$ can be understood in the framework of cooperativity, the correlation given in Figure 4b gives additional support for the point of view that cooperativity increases with decreasing temperature.

Relationship of the β- and α-Relaxation: The (αβ)-Process

Because log $f_{p\alpha}$ increases more rapidly with increasing temperature than log $f_{p\beta}$, the α- and β-processes merge at higher temperatures and do form the so-called $(\alpha\beta)$-process. Or in other words, with decreasing temperatures the α- and β-processes will separate from each other in a certain tempera-

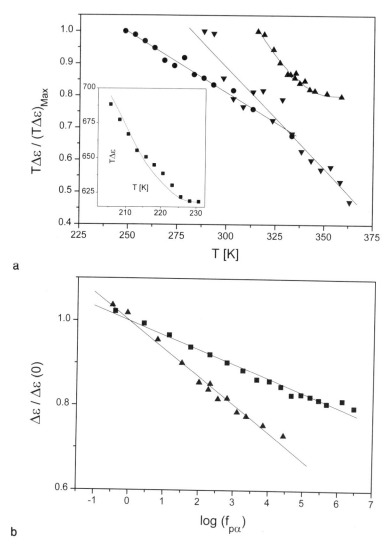

a

b

Figure 4. (a) Values of $T\Delta\varepsilon$ *reduced by the value* $(T\Delta\varepsilon)_{max}$ *at the lowest measured temperature for polyvinyl acetate data taken from Mashimo and Nozaki (103)* (▲), *polymethylacrylate* (▼) *and polyvinylisobutyl ether* (●) *data taken from Hofmann (152). The inset shows unreduced data for polypropylene glycol* (■), $M_w = 4000$, *data taken from reference 105. (b) Values of* $\Delta\varepsilon$ *versus* $\log f_{p\alpha}$ *reduced to the value at* $\log f_{p\alpha} = 0$ *for polyvinyl acetate (s) data taken from Mashimo and Nozaki (103) and for polypropylene glycol (n) data taken from reference 105. Lines are linear regressions to the data.*

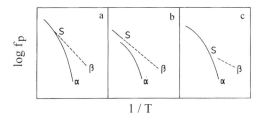

Figure 5. Three scenarios for the splitting of the α- and β-processes.

ture region called splitting region S. Figure 5 gives three different patterns for this separation:

1. Goldstein and Johari suggested the classical true merging of the α- and β-relaxations at high temperatures, whereas $\log f_{p(\alpha\beta)}$ should have an activated temperature dependence with the same activation energy as the low temperature β-process (Figure 5a). On the one hand, experiments support that picture (*33, 34*). On the other hand, recent dielectric investigations have shown that $\log f_{p(\alpha\beta)}$ cannot be described by an Arrhenius law in general (*108*) even at very high temperatures compared to T_g.

2. From theoretical considerations (*109*) a separate onset of the α-relaxation (Figure 5c) is predicted and found for poly(n-butylmethacrylate) (*110*) and other poly(n-alkylmethacrylate)s (*30*). At present it is still an open question if this behavior is characteristic for all polymers with bulky side chains.

3. Figure 5c is expected if the α-relaxation in a certain temperature range is a precondition for the β-process. This scenario is found for many materials including amorphous polymers (*37, 105, 111–114*).

As an example Figure 6 shows the temperature dependence of $\log f_{p\alpha}$ and $\log f_{p\beta}$ over 14 decades of frequency for poly(propylene glycol) (PPG) with a molecular weight of 4000 (*105*). In the second part of Figure 6 for PPG, $\Delta\varepsilon_\alpha$ is plotted versus $\log f_{p\alpha}$ in such a way that the frequency scales can be compared. Obviously in the same frequency (or temperature) range where the β-relaxation separates from the α-process the slope of $\Delta\varepsilon_\alpha$ versus $\log f_{p\alpha}$ shows a drastic change. The line through the relaxation rates of the β-relaxation is a fit of the Arrhenius equation to the data, which yield reasonable parameters ($\log f_{\infty\beta} = 12.4$ Hz and $E_a = 27.5$ kJ/mol). Whether this behavior is or is not general requires additional investigations especially in the high frequency range. Provided that this behavior is characteristic for glass-forming materials, a new possibility is opened to define another

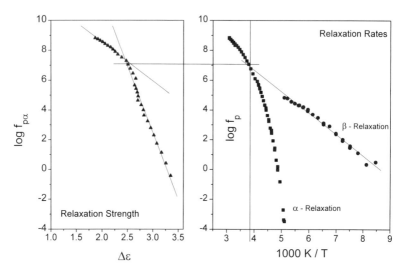

*Figure 6. Relaxation behavior of polypropylene glycol: (a) relaxation rates for the α-
and β-relaxations versus 1/T; (b) $\Delta \varepsilon_\alpha$ versus log $f_{p\alpha}$. The line through the data points
of the β-relaxation results from a fit of the Arrhenius equation to the data. The other
lines are guides for the eyes.*

specific temperature for the α-relaxation (*107*) besides the Vogel temperature T_0. This temperature can be identified with the temperature where
the cooperativity of the α-relaxation sets in. In the same temperature range,
a change in the diffusion behavior (*115–117*) has been observed.

The discussion about the microscopic origin of the glass transition has
been stimulated by the application of the mode–mode coupling theory
(MCT) to the supercooled state (*35, 36*). The MCT describes in the hydrodynamic limit the slowing down of the dynamics of the density fluctuation
by a nonlinear feedback mechanism that causes a progressive increase of
the viscosity as temperature decreases. In its idealized version (neglecting
so-called hopping processes) the MCT predicts two types of relaxation dynamics: a fast process and a slow process that freezes according to

$$f_p \sim \left(\frac{T - T_c}{T_c} \right)^{-\gamma_{MCT}} \tag{10}$$

for $T > T_c$ and therefore this process is classified as an α-relaxation (*5*).
T_c is a critical temperature where a dynamic phase transition from an ergodic ($T > T_c$) to a nonergodic behavior ($T < T_c$) takes place. γ_{MCT} is a
fit parameter. Mainly scattering experiments (*118–120*) have given support
to this picture, and the evaluation of the data shows that T_c should be about
1.2 T_g (*115*). In addition to other difficulties of the MCT, especially in

explaining the universal WLF behavior, a new technique for the analysis of broadband data (*108*) has shown that in general the dielectric data cannot be described by eq 10.

The fast process was assigned to the β-relaxation, and furthermore scaling laws were predicted for the shape of the α- and β-processes by the MCT (*35*). The analysis of the dielectric behavior of different amorphous polymers delivers only partial agreement with the predictions of the MCT. Especially, T_c, which is essential for the MCT, cannot be extracted without ambiguity (*21, 39, 41, 105, 121*). The current widely accepted opinion is that the dielectric β-relaxation, sometimes called Goldstein–Johari β-relaxation, cannot be described by the MCT (*46*).

Dielectric Normal-Mode Relaxation

For polymers having a dipole component parallel to the chain backbone at frequencies below the α-relaxation, a further process can be observed called normal-mode relaxation (*8*). The molecular dipole vectors that are parallel to the repeating unit are summed up over the chain, and therefore the normal-mode relaxation is related to both the geometry and the dynamics of the macromolecule (*8, 9*). Of course the observed behavior depends on the connection of the molecular dipoles (head to tail, etc.) (*8*), which is related to chemical constitution.

The theory of this relaxation process (*8, 122, 123*) predicts that the contribution of this relaxation process is proportional to the fluctuation of the end-to-end vector of the polymer chain $<r(0)r(t)>$. For oligomeric, nonentangled ($M < M_c$) melts, the mean fluctuation of the end-to-end vector $<r(0)r(t)>$ can be calculated by using the bead-spring model in the free-draining limit, introduced by Rouse (*124*) for dilute solutions, because the hydrodynamic interactions (*125*) are screened out in undiluted melts (*126*):

$$\langle r(0)r(t)\rangle/\langle r^2\rangle = (8/\pi^2) \sum (1/p^2) \exp(-t/\tau_p) \tag{11}$$

$$\tau_p = \frac{\zeta N^2 b^2}{3\pi^2 kTp^2} \quad \text{or} \quad \tau_p = \frac{12M\eta}{\pi^2\rho RTp^2} \qquad p = 1,3,5\ldots \tag{12}$$

where τ_p is the relaxation time for the p-th normal mode, N is the number of chain segments with bond length b, ζ is the monomeric friction coefficient, and η is the zero-shear viscosity. Equation 12 predicts $\tau_1 \sim N^2 \sim M^2$.

For entangled chains ($M > M_c$) de Gennes (*126*) and Doi and Edwards (*127*) assume in the Reptation picture that a Rouse-like chain is confined in a tube with diameter d. The functional form of eq 11 is maintained with a different relaxation time:

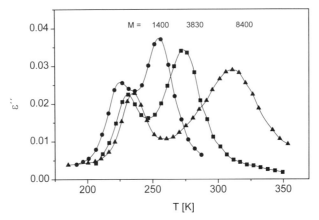

Figure 7. Values of ε″ versus T *at a frequency of 10^4 Hz for* cis-1,4-polyisoprene *samples of* M_w *1400* (●), *3830* (■), *and 8400* (▲). *(Adapted from reference 128.)*

$$\tau_p = \frac{\zeta N^3 b^4}{3\pi^2 d^2 kT p^2} \qquad p = 1,3,5 \ldots \qquad (13)$$

From that equation it follows that τ_1, also called tube disengagement time, is proportional to M_w^3. Figure 7 shows ε″ versus T for *cis*-1,4-polyisoprene samples of different molecular weights at a frequency of 10^4 Hz (*128*). The low temperature (high frequency) peak that shows only a weak dependence on M_w is assigned to the α-relaxation, whereas the high temperature peak caused by the normal-mode process depends strongly on the molecular weight.

Relaxation Rate

The relaxation rate f_{pn}, which can be extracted from the peak frequency of the normal-mode relaxation, corresponds to τ_1 which is the first Rouse or Reptation mode. Figure 8 shows the molecular weight dependence of τ_1 at a fixed temperature for *cis*-1,4-polyisoprene taken from reference 123. At low molecular weights, clearly a Rouse behavior of $\tau_1 \sim M_w^2$ is fulfilled. Around M_c a crossover from a slope of 2 to a slope of 3.7 can be observed. Although this slope of 3.7 is greater than the value predicted by the original Reptation picture, it agrees well with mechanical data (*69, 92*) and recent improvements of the Reptation theory. Furthermore, by using τ_1 and the end-to-end distance $\langle r^2 \rangle$ of the chain a self-diffusion coefficient, D_{self}, can be calculated from the dielectric experiment via $D_{self} \approx \langle r^2 \rangle / \tau_1$. On the other hand, D_{self} can be measured by direct methods like field-gradient NMR spectroscopy (*129*). Independently determined quantities display the same

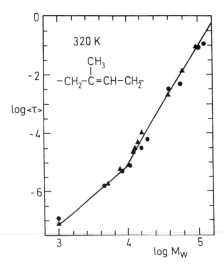

Figure 8. Mean relaxation time $\tau_1 = 1/(2\pi f_{pn})$ for the normal-mode relaxation of cis-1,4-polyisoprene versus molecular weight at 320 K. (Adapted from Boese and Kremer (123).)

temperature dependence (*129*). This result lends further support to the argument that the dielectric normal-mode relaxation is really due to the molecular motion of the whole polymer chain. A more refined discussion of the temperature dependence of the normal-mode relaxation time will be given in the next section.

The work on dielectric normal-mode relaxation on star-branched polymers was pioneered by Stockmayer and Burke (*130*) followed by investigations on multi-arm *cis*-polyisoprene stars by Boese, Kremer, and Fetters (*131, 132*). The main result is that for stars the relaxation time of the normal-mode relaxation is delayed by a factor of four. This result can be understood by theoretical considerations treating the dynamics of a polymer chain that is fixed at one end (*133*). Moreover the experimental results show that the molecular motion of one arm is nearly independent of the other arms (*131, 132*). Whether this is also the case if the star is built up with arms of a different chemical structure remains an open question. If, in networks of type-A polymers, the normal-mode relaxation can (*134*) or cannot (*135*) be observed dielectrically depends on the given molecular structure.

Shape of Loss Peak

Equation 11 predicts that the shape parameters for the normal-mode peak should be $m = 1$ and $n = 0.7$ (*128*) independent of molecular weight.

Experimentally, for samples with very narrow distribution of molecular weights the loss peaks are much broader than predicted by the theory (*8, 123, 128*). This result is true for molecular weights below and above M_c. (Dielectric normal-mode spectroscopy was used to extract the molecular weight distribution (*136*).) The same result is obtained also by means of mechanical spectroscopy (*137*). Moreover the shape of the loss peak seems to be dependent on molecular weight (*128, 138*). There are several treatments help understand this experimental fact. One model is based on the coupling scheme (*97*). Ngai and co-workers also presented a Focker–Planck approach to the chain dynamics of polymers and to this problem (*139*). Geyler and Pakula (*140*) carried out computer simulations for cooperative relaxation in dense macromolecular systems. The calculated end-to-end vector correlation function that shows a strong nonexponential behavior depends on molecular weight. Very recently Schweitzer (*141*) published a scaling approach to improve the Rouse and Reptation picture and to handle this problem.

Relaxation Strength

For type-A polymers the whole dipole moment parallel to the chain skeleton is obtained by integration over the chain length. Therefore, the relaxation strength of the normal-mode relaxation $\Delta\varepsilon_n$ is proportional to the mean end-to-end vector of the chain $\langle r^2 \rangle$. If the dipole moment parallel to the repeating unit and the molecular structure is known from $\Delta\varepsilon_n$, the mean end-to-end vector $\langle r^2 \rangle$ can be estimated quantitatively by using eq 4. For *cis*-1,4-polyisoprene the molecular weight dependence of $\langle r^2 \rangle$ obtained by dielectric spectroscopy is in agreement with that obtained by scattering results (*142*). At least changes of the geometric chain dimensions caused by temperature or by mircophase separation in diblock-copolymers (*143*) can be detected.

Relationship of α-Relaxation and Normal-Mode Relaxation

Equation 12 gives a relationship between the relaxation time for the normal-mode relaxation and the monomeric friction coefficient, which is related to the α-relaxation (*69*). For that reason the relaxation rate of the normal-mode relaxation f_{pn} should have a WLF-like temperature dependence, and moreover both processes should have the same Vogel temperature. Indeed f_{pn} shows a temperature dependence according to the WLF equation (*8,*

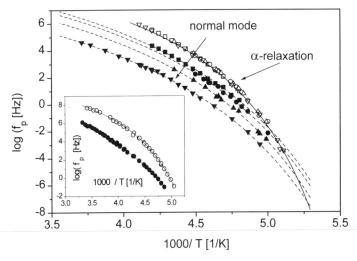

Figure 9. Relaxation rates of the α-relaxation (open symbols) and of the normal-mode relaxation (solid symbols) for polypropylene glycol with different molecular weights: □, 810; ○, 1000; △, 2000; ▽, 4000. Lines are fits of the WLF equation to the data. The inset shows the same behavior for a cis-1,4-polyisoprene sample of $M_w = 1000$.

122, 123, 128) as is demonstrated in Figure 9 for PPGs with different molecular weights. This plot shows that the relaxation rate of the α-relaxation has a much stronger temperature dependence compared to that of the normal-mode relaxation, and both processes will merge at low frequencies. (The analysis of the so-called δ-relaxation in side-chain liquid crystalline polymers displayed a similar behavior (*144, 145*).) The inset of Figure 9 displays the same behavior for a *cis*-1,4-polyisoprene melt with a molecular mass of 1000. Because this behavior is found for different polymers (*128, 146*) and because the analysis of creep data by a master-curve construction (*147, 148*) gave similar results, this behavior should be regarded as a general feature of polymer dynamics. As a consequence of this result the use of free volume corrections that must be made to extract monomeric friction coefficient should be rediscussed (*69, 149*).

The first attempt to explain this behavior was made on the basis of the coupling scheme by Ngai et al. (*98, 150*). Especially, this model can describe data at low frequencies. A recent extension of these investigations to higher frequencies has shown that at high temperatures the relaxation rates of both processes have a similar temperature dependence but decouple at low temperatures (*105*). This behavior is more clearly demonstrated by the inset of Figure 10, where the ratio of the relaxation rates of the α-relaxation and of the normal-mode process is plotted versus inverse temperature. Because the α-relaxation corresponds to a rotational movement of dipoles around the chain skeleton and the normal-mode process is related to trans-

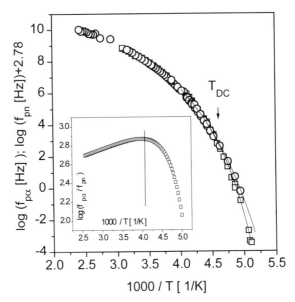

Figure 10. Decoupling of the relaxation rates of the α-relaxation (□) and the normal-mode relaxation (○) for polypropylene glycol with $M_w = 4000$; the relaxation rate of the normal-mode relaxation was shifted by 2.78 to higher frequencies. The lines are guides for the eyes. The inset displays the ratio of the relaxation rates of the α-relaxation and of the normal-mode process.

latorial diffusion, the observed behavior is similar to the decoupling of rotational and translatorial diffusion found first by Sillescu and co-workers (*116*) and later by other groups (*117*). Up to now there is no good model to describe these experimental observations and further experimental investigations are necessary. There is experimental evidence that this decoupling depends on the molecular weight of the diffusing molecule (*151*) and that this phenomenon must be related to the cooperative nature of the α-relaxation (*144*).

Acknowledgment

This chapter is dedicated to E. Schlosser, who introduced me to dielectric spectroscopy.

I am grateful to F. Kremer (Leipzig), E. W. Fischer (Mainz), E. Donth (Halle), K. L. Ngai (Washington, DC), E. Schlosser, and H. E. Carius (Berlin) for helpful discussions.

References

1. Debye, P. "Polar Molecules," Chemical Catalog Co.: New York, 1929 (reprinted by Dover: New York).

2. Fuoss, R. M.; Kirkwood, J.G. *J. Am. Chem. Soc.* **1941,** *63,* 385.
3. McCrum, N. G.; Read, B. E.; Williams, G. *Anelastic and Dielectric Effects in Polymeric Solids;* Wiley: New York, 1967 (reprinted by Dover: New York, 1991).
4. Ishida, Y.; Yamafuji, K. *Kolloid Z.* **1961,** *177,* 97; Ishida, Y.; Matsuo, M.; Yamafuji, K. *Kolloid Z.* **1969,** *180,* 108; Ishida, Y. *J. Polym. Sci., Part A-2* **1969,** *7,* 1835.
5. Hedvig, P. *Dielectric Spectroscopy of Polymers;* Adam Hilger: Bristol, England, 1977.
6. *Dielectric Properties of Polymers;* Karasz, F. E., Ed.; Plenum: New York, 1972.
7. Blythe, A. R. *Electrical Properties of Polymers;* Cambridge University: Cambridge, England, 1979.
8. Adachi, K; Kotaka, T. *Prog. Polym. Sci.* **1993,** *18,* 585.
9. Riande, E.; Saiz, E. *Dipole Moments and Birefringence of Polymers;* Prentice Hall: Englewood Cliffs, NJ, 1992.
10. Williams, G. *Adv. Polym. Sci.* **1979,** *33,* 60
11. Williams, G. In *Comprehensive Polymer Science;* Allen, G.; Bevington, J. C., Eds.; Pergamon: Oxford, England, 1989; Vol II.
12. Böttcher, C. J. F. *Theory of Dielectric Polarization;* Elsevier: Amsterdam, Netherlands, 1973, Vol. I; Böttcher, C. J. F.; Bordewijk, P. Theory of Dielectric Polarization; Elsevier: Amsterdam, Netherlands, 1978, Vol. II.
13. Cook, M.; Watts, D. C.; Williams, G. *Trans. Faraday Soc.* **1970,** *66,* 2503.
14. Sillars, R. W. *J. Inst. Elect. Eng.* **1937,** *80,* 378.
15. Stockmayer, W. *Pure Appl. Chem.* **1967,** *15,* 539.
16. Block, H. *Adv. Polym. Sci.* **1979,** *33,* 94.
17. Debye, P.; Bueche, F. *J. Chem. Phys.* **1951,** *7,* 589.
18. Wolkenstein, M. V. *Configurational Statistics of Polymeric Chains;* Wiley Interscience: New York, 1963.
19. Flory, P. J. *Statistical Mechanics of Chain Molecules;* Hanser Publishers: Munich, Germany, 1988.
20. Williams, G.; Cook, M.; Hains, P. J. *J. Chem. Soc., Faraday Trans. II* **1972,** *68,* 1045.
21. Kremer, F.; Hofmann, A.; Fischer, E. W. *Am. Chem. Soc. Polym. Prepr.* **1992,** *33,* 96.
22. Schlosser, E.; Schönhals, A. *Colloid Polym. Sci.* **1989,** *267,* 963.
23. Pathmanathan, K.; Johari, G. P. *J. Polym. Sci.* **1987,** *25,* 379
24. Johari, G. P. *Polymer* **1986,** *27,* 866.
25. Buerger, D. E.; Boyd, R. H. *Macromolecules* **1989,** *22,* 2659.
26. Heijboer, J. In *Molecular Basis of Transitions and Relaxations;* Meier, D. J. Gordon and Branch: New York, 1978.
27. Tetsutani, T.; Kakizaki, M.; Hideshima, T. *Polym. J.* **1982,** *14,* 305.
28. Tetsutani, T.; Kakizaki, M.; Hideshima, T. *Polym. J.* **1982,** *14,* 471.
29. Gomes Ribelles, J. L.; Diaz Calleja, R. J. *J. Polym. Sci. Polym. Phys. Ed.* **1985,** *23,* 1297.
30. Garwe, F.; Schönhals, A.; Beiner, M.; Schröter, K.; Donth, E. *Macromolecules* **1996,** *29,* 247.
31. Buerger, D. E.; Boyd, R. H. *Macromolecules* **1989,** *22,* 2649.
32. Katana; Kremer, F.; Fischer, E. W.; Plaetscke, R. *Macromolecules* **1993,** *26,* 3075.
33. Johari, G. P.; Goldstein, M. *J. Chem. Phys.* **1970,** *53,* 2372.
34. Johari, G. P. *J. Chem. Phys.* **1973,** *28,* 1766.
35. Götze, W.; Sjögren, L. *Rep. Prog. Phys.* **1992,** *55,* 241.
36. Götze, W. In *Liquids, Freezing and the Glass Transition;* Hansen, J. P.; Levesque, D.; Zinn-Justin, J., Eds.; North-Holland: Amsterdam, Netherlands, 1991. (a) Havrilak, S.; Havriliak, S. *Polymer* **1992,** *33,* 938.
37. Wu, L. *Phys. Rev. B* **1991,** *43,* 9906.

38. Matsuo, M.; Ishida, Y.; Yamafuji, K.; Takayanagi, M; Irie, F. *Kolloid-Z. Z. für Polym.* **1965**, *201*, 7.
39. Colmenero, J.; Arbe, A.; Alegria, A. *Physica A* **1993**, *201*, 447
40. Coburn, J.; Boyd, R. H. *Macromolecules* **1986**, *19*, 2238.
41. Hofmann, A.; Kremer, F.; Fischer, E. W. *Physica A* **1993**, *201*, 106.
42. Williams, G.; Watts, D.nC. *Trans. Faraday Soc.* **1971**, *67*, 2793.
43. Williams, G.; Edwards, D. A. *Trans. Faraday Soc.* **1966**, *62*, 1329.
44. Sasabe, H; Saito, S. *J. Polym. Sci., Part A-2* **1968**, *6*, 1401.
45. Kulik, A. S.; Beckham, H. W.; Schmidt-Rohr, K.; Radloff, D.; Pawelzik, U.; Boeffel, C.; Spiess, H. W. *Macromolecules* **1994**, *27*, 4746.
46. "Proceedings of the 2nd Conference on Relaxation in Complex Systems"; Ngai, K. L.; Riande, E.; Wright, G. *J. Non.-Cryst. Solids* **1994**, 172–174.
47. *Disorder Effects on Relaxational Processes;* Richert, R.; Blumen, A., Eds.; Springer-Verlag: Berlin, Germany, 1994.
48. Shatzki, T. F. *J. Polym. Sci.* **1962**, *57*, 496; *Polym. Prepr.* **1965**, *6*, 646.
49. Verdier, P. H.; Stockmayer, W. *J. Chem. Phys.* **1962**, *36*, 227.
50. Jones, A.; Stockmayer, H. *J. Polym. Sci. Polym. Phys. Ed.* **1977**, *15*, 847
51. Valeur, B.; Jarry, J.-P.; Geny, F.; Monnerie, L. *J. Polym. Sci. Polym. Phys. Ed.* **1975**, *13*, 667.
52. Hall, C. K; Helfand, E. *J. Chem. Phys.* **1982**, *77*, 3275.
53. Skolnick, J.; Yaris, R. *Macromolecules* **1982**, *15*, 1041.
54. Dejean de la Batie, R.; Lauprêtre, F.; Monnerie, L. *Macromolecules* **1988**, *21*, 2045.
55. Viovy, J. L.; Monnerie, L.; Brochon, J. C. *Macromolecules* **1983**, *16*, 1845.
56. Helfand, E.; Wassermann, Z. R.; Weber, T. A. *Macromolecules* **1980**, *13*, 526.
57. Adolf, D. A.; Ediger, M. D. *Macromolecules* **1991**, *24*, 5834; *Macromolecules* **1992**, *25*, 1074.
58. Vogel, H. *Phys. Z.* **1921**, *22*, 645.
59. Fulcher, G. S. *J. Am. Chem. Soc.* **1925**, *8*, 339.
60. Tammann, G.; Hesse W. *Z. Anorg. Allg. Chem.* **1926**, *156*, 245.
61. Grest, G. S.; Cohen, M. H. *Adv. Chem. Phys.* **1981**, *48*, 455.
62. Adam, G.; Gibbs, J .H. *J. Chem. Phys.* **1965**, *43*, 139.
63. Donth, E. *J. Non-Cryst. Solids* **1982**, *53*, 325.
64. Donth, E. *Relaxation and Thermodynamics in Polymers: Glass Transition;* Akademie-Verlag: Berlin, Germany, 1992.
65. DiMarzio, E. A. *Ann. N.Y. Acad. Sci.* **1981**, *371*, 1.
66. Angell, C. A.; Rao, K. J. *J. Chem.* **1972**, *57*, 470.
67. Dixon, P. *Phys. Rev. B* **1990**, *42*, 8179.
68. Schönhals, A.; Kremer, F.; Hofmann, A.; Fischer, E. W.; Schlosser, E. *Phys. Rev. Lett.* **1993**, *70*, 3459.
69. Ferry, J. D. *Viscoelastic Properties of Polymers*, 3rd ed.; J. Wiley and Sons: New York, 1980.
70. Kirst, K. U.; Kremer, F.; Pakula, T.; Hollinghurst, J. *Colloid Polym. Sci.* **1994**, *272*, 1420.
71. Cole, K. S.; Cole, R. H. *J. Chem. Phys.* **1941**, *9*, 341.
72. Davidson, D. W.; Cole, R. H. *J. Chem. Phys.* **1950**, *18*, 1417.
73. Davidson, D. W.; Cole, R. H. *J. Chem. Phys.* **1951**, *19*, 1484.
74. Fuoss, R. M.; Kirkwood, J. G. *J. Am. Chem. Soc.* **1941**, *63*, 385.
75. Havriliak, S.; Negami, S. *J. Polym. Sci.* **1966**, *C14*, 99.
76. Havriliak, S.; Negami, S. *Polymer* **1967**, *8*, 161.
77. Schlosser, E.; Kästner, S.; Friedland, K.-J. *Plaste und Kautschuk* **1981**, *28*, 77.
78. Jonscher, A. *Dielectric Relaxation in Solids;* Chelsea Dielectric: London, 1983.

79. Williams, G.; Watts, D. *Trans. Farad. Soc.* **1970,** *66,* 80.
80. Lindsey, C. P.; Patterson, G. D. *J. Chem. Phys.* **1980,** *73,* 3348.
81. Alvarez, F.; Alegria, A.; Colmenero, J. *Phys. Rev. B* **1991,** *44,* 7306.
82. Schönhals, A. *Acta Polym.* **1991,** *42,* 149.
83. Schönhals, A.; Schlosser E. *J. Non.-Cryst. Solids* **1991,** *131–133,* 1161.
84. Havriliak, S.; Havriliak, S. J. *J. Non.-Cryst. Solids* **1994,** *172–174,* 297.
85. Schönhals, A.; Kremer, F.; Schlosser, E. *Phys. Rev. Lett.* **1991,** *67,* 999.
86. Fioretto, D.; Livi, A.; Rolla, P. A.; Socino, G.; Verdini, L. *J. Phys.: Condens. Matter* **1994,** *6,* 1.
87. Matsuoka, S.; Ishida, Y. *J. Polym. Sci.* **1966,** *C14,* 247.
88. Glatz-Reichenbach, J. K. W.; Sorriero, L. J.; Fitzgerald, J. J. *Macromolecules* **1994,** *27,* 1338.
89. Schlosser, E.; Schönhals, A. *Colloid Polym. Sci.* **1989,** *267,* 133.
90. Schmidt-Rohr, K.; Spiess, H. W. *Phys. Rev. Lett.* **1991,** *66,* 3020.
91. Montroll, E. W.; Bendler, J. T. *J. Statictical Phys.* **1984,** *34,* 129.
92. Matsuoka, S. *Relaxation Phenomena in Polymers;* Hanser Publishers: Munich, Germny, 1992; Matsuoka, S.; Quan, X. *Macromolecules* **1991,** *24,* 2770.
93. Ngai, K. L. *Comments Solid State Phys.* **1979,** *9,* 127; Ngai, K. *Comments Solid State Phys.* **1979,** *9,* 141.
94. Ngai, K. L. In *Disorder Effects on Relaxational Processes;* Richert, R.; Blumen, A., Eds.; Springer-Verlag: Berlin, Germany, 1994.
95. Rendell, R. W.; Ngai, K .L.; Mashimo, S. *J. Chem. Phys.* **1987,** *87,* 3259.
96. Ngai, K. L.; Mashimo, S.; Fytas, G. *Macromolecules* **1988,** *21,* 3030.
97. Ngai, K. L; Rendell, R. W. *Macromolecules* **1987,** *20,* 1066.
98. Ngai, K. L.; Schönhals, A.; Schlosser, E. *Macromolecules* **1992,** *25,* 4915.
99. Plazek, D. J.; Ngai, K. L. *Macromolecules* **1991,** *24,* 1222.
100. Dissado, L. A.; Hill, R. M. *Proc. R. Soc. London Ser. A* **1983,** *390,* 131.
101. Schönhals, A.; Schlosser, E. *Colloid Polym. Sci.* **1989,** *267,* 125.
102. Alegria, A.; Colmenero, J.; del Val, J. J; Barandiaran J. M. *Polymer* **1985,** *26,* 913.
103. Nozaki, R.; Mashimo, S. *J. Chem. Phys.* **1987,** *87,* 2271.
104. Schlosser, E.; Schönhals A. *Polymer* **1991,** *32,* 2135.
105. Schönhals, A.; Kremer, F. *J. Non.-Cryst. Solids* **1994,** *172–174,* 336.
106. Stickel, F.; Kremer, F.; Fischer, E. W. *Physica A* **1993,** *201,* 318.
107. Schönhals, A. In *Non-Equilibrium Phenomena in Supercooled Fluids, Glasses, and Amorphous Materials;* Giordono, M.; Leporini, D.; Tos, M. P., Eds.; World Scientific: Singapore, 1996.
108. Stickel, F.; Fischer, E W.; Richert, R. *J. Chem. Phys.* **1995,** *102,* 6257; Stickel, F.; Fischer, E. W.; Schönhals, A.; Kremer, F. *Phys. Rev. Lett.* **1994,** *73,* 2936
109. Schulz, M.; Donth E. *J. Non.-Cryst. Solids* **1994,** *168,* 186.
110. Garwe, F.; Schönhals, A.; Beiner, M.; Schröter, K.; Donth, E. *J. Phys.: Condens. Matter* **1994,** *6,* 6941.
111. Matsuoka, S.; Johnson, G. E. *ACS-PMSE Prepr.* **1994,** *70,* 369.
112. Pschorn, U; Rössler, E.; Sillescu, H.; Kaufmann, S.; Schaefer, D.; Spiess, H. W. *Macromolecules* **1991,** *24,* 389.
113. Rössler, E. *Phys. Rev. Lett.* **1992,** *69,* 1620.
114. Murty, S. S. N. *J. Mol. Liq.* **1989,** *44,* 51.
115. Rössler, E. *Phys. Rev. Lett.* **1990,** *65,* 1595.
116. Lohfink, M.; Sillescu, H. *Proceedings of the 1st Tohwa University International Symposium;* American Physical Society: College Park, MD, 1991.
117. Blackburn, F. R.; Cicerone, M. T.; Hietpas, G.; Wagner, P. A.; Ediger, M. D. *J. Non-Cryst. Solids* **1994,** *172–174,* 256.

118. Frick, B.; Farago, B.; Richter, D. *Phys. Rev. Lett.* **1990**, *64*, 2921.
119. Li, G.; Du, W.M.; Sakai, A.; Cummins, H. Z. *Phys. Rev. A* **1192**, *46*, 3343.
120. Sidebottom, D. L.; Bergman, R.; Börjesson, L.; Torell, L. M. *Phys. Rev. Lett.* **1992**, *68*, 3587.
121. Hofmann, A.; Kremer, F.; Fischer, E. W.; Schönhals, A. In *Disorder Effects on Relaxational Processes;* Richert, R.; Blumen, A., Eds.; Springer-Verlag: Berlin, Germany, 1994.
122. Adachi, K; Kotaka, T. *Macromolecules* **1988**, *21*, 157.
123. Boese, D.; Kremer, F. *Macromolecules* **1990**, *23*, 829.
124. Rouse, P. E. *J. Chem. Phys.* **1953**, *21*, 1272.
125. Zimm, B. H. *J. Chem. Phys.* **1956**, *24*, 269.
126. de Gennes, P.-G. *Scaling Concepts in Polymer Physics;* Cornell University : Ithaca, NY, 1980.
127. Doi, M.; Edwards, S. F. *The Theory of Polymer Dynamics;* Clarendon: Oxford, England, 1986.
128. Schönhals, A. *Macromolecules* **1993**, *26*, 1309.
129. Fleicher, G.; Appel, M. *Macromolecules* **1995**, *28*, 7281; Appel, M.; Fleischer, J.; Kärger, G.; Chang, I.; Fujara, F.; Schönhals, A. *Colloid Polym. Sci.,* in press.
130. Stockmayer, W. H.; Burke, J .J. *Macromolecules* **1969**, *2*, 647.
131. Boese, D.; Kremer, F.; Fetters, L. J. *Makromol. Chem., Rapid Commun.* **1988**, *9*, 367.
132. Boese, D.; Kremer, F.; Fetters, L. J. *Macromolecules* **1990**, *23*, 1826.
133. Grest, G. S.; Kremer, K.; Milner, S. T.; Witten, T. A. *Macromolecules* **1989**, *22*, 1904.
134. Steemann, P. A. M.; Nusselder, J. J. H. *ACS-PMSE Prepr.* **1994**, *70*, 281.
135. Kirst, U. Ph.D. Thesis, University of Mainz, 1993.
136. Fodor, J. S.; Hill, D. A. *J. Phys. Chem.* **1994**, *98*, 7674.
137. Winter, H. H. *J. Non.-Cryst. Solids* **1994**, *172–174*, 1158.
138. Adachi, K.; Kotaka, T. *J. Mol. Liq.* **1987**, *36*, 75.
139. Ngai, K. I..; Peng, S. L.; Skolnick, J. *Macromolecules* **1992**, *25*, 2184.
140. Pakula, T.; Geyler, S. *Macromolecules* **1987**, *21*, 2909.
141. Schweitzer, K. S. *Phys. Scripta* **1993**, *T49*, 99.
142. Adachi, K.; Nishi, I.; Itoh, S.; Kotaka, T. *Macromolecules* **1990**, *23*, 2550.
143. Stühn, B.; Stickel, F. *Macromolecules* **1992**, *25*, 5306.
144. Attard, G. S.; Moura-Ramos, J. J.; Williams, G. *J. Polym. Sci. Polym. Phys. Ed.* **1987**, *25*, 1099.
145. Schönhals, A.; Geßner, U.; Rübner, J. *Marcomol. Chem. Phys.* **1995**, *196*, 1671.
146. Schönhals, A.; Schlosser, E. *Phys. Scripta* **1993**, *T49*, 233.
147. Plazek, D. J.; O'Rourke, M. *J. Polym. Sci. A2* **1971**, *9*, 209.
148. Plazek, D. J. *Polym. J. (Jpn.)* **1980**, *12*, 43.
149. von Meerwall, E.; Grigsby, J.; Tomich, D.; van Antwerp, R. *J. Polym. Sci. Polym. Phys. Ed.* **1982**, *20*, 1037.
150. Ngai, K. L.; Plazek, P. J.; Deo, S. S. *Macromolecules* **1987**, *20*, 3047.
151. Schönhals, A. Dielectric Newsletter, 1995.
152. Hofmann, A. Ph.D. Thesis, University of Mainz, 1993.

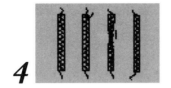

4

Dielectric Spectroscopy of Semicrystalline Polymers

Richard H. Boyd and Fuguo Liu

Polymers rarely crystallize very completely and are thus two-phase systems consisting of crystals and a residual uncrystallized amorphous fraction. The amorphous fraction exhibits the counterparts of the processes occurring in wholly amorphous polymers, namely a glass–rubber relaxation region and often at least one subglass process. The glass transition process is very significantly perturbed by the presence of the crystal phase but subglass processes are not. These observations are to be rationalized in terms of the effective length scales involved in comparison with the interlamellar spacing. Crystal phase relaxations are often not observed dielectrically because of selection rules but the well-known α-crystalline process in polyethylene can be studied and interpreted in detail.

This chapter is meant to be an introduction into some of the issues that arise in the study of semicrystalline polymers by means of dielectric spectroscopy. No attempt is made to provide a comprehensive coverage of the considerable literature that exists. For that matter, many of the details of the important issues are skipped over to present the essentials. Two related and much more comprehensive reviews of both dielectric and mechanical relaxation processes in semicrystalline polymers have previously been made (1, 2) and are still useful. Many of the details can be pursued through these reviews. The most important point made is that semicrystalline polymers are two-phase systems consisting of both amorphous domains and crystals and that both phases can have relaxation processes. The plan is to first describe what relaxation processes are found in the amorphous fraction and to contrast these with their counterparts in wholly amorphous polymers. Detailed interpretation requires knowledge of the properties of the individual phases. The problem of determining phase dielectric constants from the semicrystalline composite values is thus considered. Then, the interpretation of the amorphous-phase relaxations in terms of dipolar correlation functions and relaxation width measures is taken up. Finally, crystal-phase relaxations are considered. The relaxation in dipole decorated polyethylene (PE) is examined in this context.

Experimental

Because this chapter is largely tutorial in nature it is important that the experiments chosen to illustrate some of the principles stressed be particularly clear and consistent. To this end, some data that were especially taken for the purpose and were not previously published are used.

Dielectric constant and loss were determined for poly(ethylene terephthalate) (PET), poly(vinyl acetate) (PVAc), and (oxidized) linear polyethylene (LPE). Two PET samples were studied. They were obtained from DuPont as Mylar film. One was an amorphous film and the other was a semicrystalline sample. The degree of crystallinity of the semicrystalline sample from density measurements was 50%. The PVAc was obtained from Polysciences. It was compression-molded into disks. The oxidized LPE was a sample similar to those previously described (3, 4). It was also compression-molded into disks. All of the specimen electrodes were created by evaporative gold coating and included a guard ring. The details of the PVAc measurements were reported more fully elsewhere (5).

The dielectric measurements were carried out by using an Imass time-domain dielectric spectrometer. The instrument performs the transform to the frequency domain. The frequency range covered was 10^{-3} Hz to 10^4 Hz in most cases.

Morphology of Semicrystalline Polymers

Polymer chains that contain long sequences of identical repeat units are capable of aggregation into the three-dimensional repeating arrays that characterize the crystalline state (6). Nearly always the packing is such that the array consists of bundles of parallel chains. Each of the chains is in an extended conformation that can be regarded as a helix. Thus, the unit cell is determined by the packing details transverse to the helix axis (the a, b directions) and the nature of the repeat units along the helix axis (the c

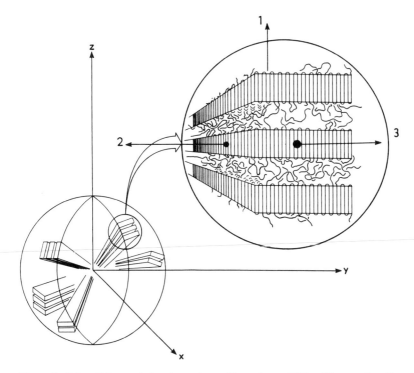

Figure 1. Spherulitic morphology in semicrystalline polymer. Ribbon-like crystals radiate from the primary nucleation site. Locally (enlarged region) the ribbons form a lamellar structure with alternating crystalline and amorphous layers.

direction). Because of their great chain length polymers find it difficult to form large crystals kinetically. In fact, in the case of polymers cooled quiescently from the melt they find it kinetically advantageous to fold back and forth across a growing crystal face of relatively small dimensions in the chain direction (*see* face 3, Figure 1). In the lateral direction on the growth face (face 2, Figure 1) the crystal dimensions are much larger. The crystal can continue to grow for considerable distances (face 3, Figure 1). A typical crystal is therefore ribbon- or lamella-like. These ribbons radiate outward from a primary nucleation site but locally they are packed, one over another, into stacks (Figure 1). Because the crystal thickness in the chain direction (face 1, Figure 1) on the growing face is less than the chain length, a given chain can crystallize onto the faces of several nearby growing crystals. This process means that it is difficult for the material to crystallize completely. Significant numbers of noncrystallized chain segments may be trapped between the stacked crystal sheaths. Thus, quiescently cooled crystalline polymers are obviously *polycrystalline.* They contain radial arrays (*spherulites*) of many crystal ribbons emanating from many primary nucleation sites. However, and perhaps of more importance for the discussion at hand, they are also *semicrystalline*. That is, they contain considerable amounts of uncrystallized material.

Many other morphologies beyond the spherulitic one described previously can be achieved (7–9). Those resulting from orientation under solid-state drawing, melt-spinning, and crystallization from solution are examples. However, the extent of crystallization typically is incomplete. A mixture of relatively well-formed crystals coexisting with a residual uncrystallized amorphous fraction results.

Signatures of Relaxations in Semicrystalline Polymers

The two-phase picture described previously gives rise to several important questions. The first is whether the relaxation processes that occur in typical *completely* amorphous polymers also occur in the amorphous fraction in semicrystalline polymers. To a large extent these processes do occur in the amorphous fraction environment. Derivative of the first question, then, is whether or to what extent these amorphous-phase processes are altered or modified in the semicrystalline situation. Because the thickness of the amorphous interlayer is typically quite small (~50–100 Å) it would not be surprising for such modification to occur. It turns out that the answer to this question depends strongly on the type of process. Finally, there is the obvious prospect of relaxation processes occurring within the crystals themselves. This result is also observed, but not universally, and in fact, for several reasons, observation is rather uncommon in dielectric spectroscopy. The description of crystal phase processes is deferred to a separate section.

Glass Transition Region in Semicrystalline Polymers

The glass transition region in *wholly* amorphous polymers has a universal and distinct signature, or set of characteristics.

Wholly Amorphous Polymers

The shapes of the dielectric constant and loss curves when plotted versus log frequency or in terms of the loss plotted against dielectric constant in an Argand diagram are quite distinct. The loss is skewed toward high frequency and the curves are fit very well by the Havriliak–Negami (HN) function (10). In the time domain, the dielectric constant is fit satisfactorily by the Kohlrausch–Williams–Watts (KWW) stretched exponential function (11–13). The fit is not as close as for the HN frequency domain cases but is accomplished with one less parameter.

The signature also includes the effects of temperature. Central relaxation times from HN function or KWW function fits to data at a number of temperatures display Vogel–Fulcher (VF) (*14, 15*) or Williams–Landel–Ferry (WLF) (*16*) behavior when plotted as $-\log \tau$ (relaxation time) versus $1/T$ (temperature). Alternatively, $\log f_{max}$ (frequency at which loss factor is maximum) values from isochronal or isothermal scans may be used to show this behavior. The characteristic curvature in these plots is common to all vitrifying materials whether polymeric or not.

Fortunately, the effect of the semicrystalline environment on the glass transition process can be directly assessed experimentally in a few cases. That is, sometimes the same polymer can be prepared in a completely amorphous condition as well as in a semicrystalline state. PET is a good example. A totally amorphous glass can be made by quenching the melt to room temperature. Degrees of crystallinity of up to ~50% are achieved by annealing the glass near or above the glass transition temperature (T_g).

In Figure 2 the loss curves versus log frequency for PVAc and amorphous PET are compared at temperatures corresponding to the glass transition region for each. PVAc is, of course, a very typical and much studied

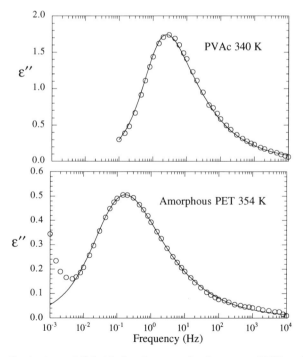

Figure 2. Comparison of dielectric loss factor vs. log frequency of PVAc with that of amorphous PET. The curves are fits of the HN function.

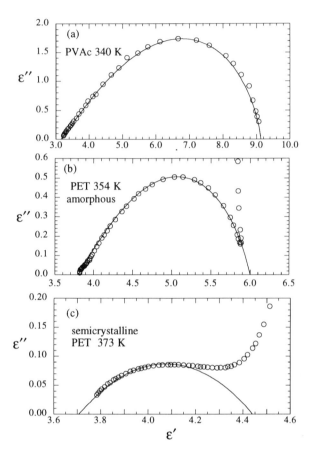

Figure 3. Argand complex plane plots of dielectric loss factor vs. dielectric constant for PVAc, amorphous PET, and semicrystalline PET. For panels (a) and (b) the curves are fits of the HN function. For panel (c) the fit is for the Cole–Cole function.

amorphous polymer. The results for PET are in every respect similar. The high frequency tail in the loss curve is evident in both. In Figure 3 the Argand diagrams display the same behavior. In both Figures 2 and 3, the curves shown are HN function fits to the data. Figure 4 illustrates the VF–WLF nature of the process in amorphous PET.

As an aside, the upturn in loss at low frequency for PET, Figure 2, is due to conductance loss. That is, migration of charged impurities, as opposed to the dipolar reorientation associated with the loss peak, gives rise to an unbounded increase in loss as frequency decreases (*see* Appendix).

Effect of Crystal-Phase Presence

Some of the effects of the presence of the crystal phase on the amorphous phase glass transition process may be seen in Figure 5. The dielectric con-

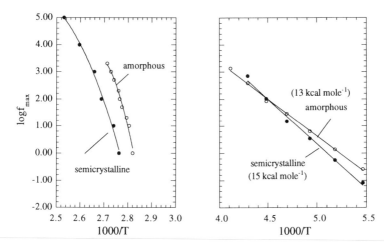

Figure 4. Relaxation maps. The maximum in dielectric loss factor vs. log f curves: that is, log f_{max}, plotted against reciprocal temperature. The left-hand panel compares amorphous and semicrystalline PET in the α glass transition region. The right hand panel is for the β subglass region.

stant and loss curves versus log frequency of amorphous PET are compared with those of a 50% crystalline specimen. A dramatic change has taken place. The loss curve is much broader and is displaced some 20 K to higher temperature when in the same frequency window. The temperature dependence of the loss peak location, as evidenced in log f_{max} versus $1/T$ remains VF/WLF in nature (Figure 4).

The conductance that affected only the loss component, ε'', in the amorphous polymer now causes a low frequency upturn in the real component, ε', as well. This effect is a consequence of the two-phase nature of the system and is an example of the classic Maxwell–Sillars–Wagner interfacial polarization effect (*17*) (*see* Appendix).

The increment in the dielectric constant, ε', that defines the relaxation strength is considerably smaller than that found in the completely amorphous polymer. In part this effect can come from the dilution effect of the dielectrically inactive crystal phase. However, reference to Figure 3 shows that the increment for the wholly amorphous case is ~2.2, whereas it is only about 0.7 for the semicrystalline polymer. Because the degree of crystallinity is 50%, a reduction in relaxation strength beyond the dilution effect is suggested. This result is confirmed by plotting the *strength* of the relaxation, $\varepsilon_R - \varepsilon_U$ (where ε_R is the limiting low frequency relaxed dielectric constant and ε_U its high frequency unrelaxed counterpart) versus degree of crystallinity (Figure 6). The data shown are from a study of PET over a wide range of crystallinities (*18*). The strength does indeed decrease with increasing crystallinity as expected for a process occurring in the amorphous

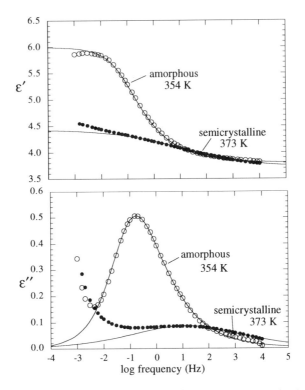

Figure 5. Dielectric constant (upper panel) and loss (lower panel) vs. log frequency for amorphous and semicrystalline PET. For amorphous PET the curves are fits of the HN function. For semicrystalline PET the fit is for the Cole–Cole function.

fraction. In addition, however, the point for the wholly amorphous sample falls above the regression line for the semicrystalline specimens. This result would appear to indicate that the amorphous phase in the semicrystalline environment has a lower dielectric relaxation strength than in the wholly amorphous environment.

To summarize, the signature of the glass transition region in semicrystalline polymers, in comparison with this region in wholly amorphous polymers, consists of great broadening of the process in the frequency or time domain, some shift in location (to higher temperature or lower frequency), reduction in relaxation strength beyond the dilution effect, and the onset of Maxwell–Wagner–Sillars interfacial conductance polarization.

Subglass Relaxations in Semicrystalline Polymers

When a vitrifying liquid is cooled well below T_g to form a glass, there is the temptation to think of molecular motions that could give rise to dipolar

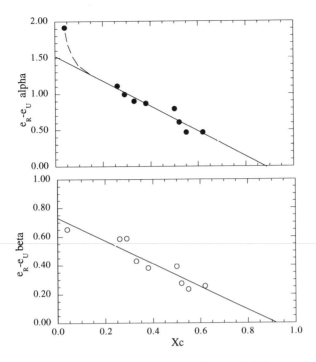

Figure 6. Dielectric relaxation strength: that is, the relaxed dielectric constant, ε_R, minus the unrelaxed value, ε_U, plotted against degree of crystallinity, X_c, for PET. The upper panel is for the α glass transition region. The lower one is for the β subglass region. The solid lines are linear regressions. For the α process the lowest crystallinity point, for an effectively wholly amorphous specimen, is not included in the regression. The dashed curve serves as a guide to connect this point to the rest of the data.

relaxation processes as being frozen out. However, very often this is not the case. It is quite common to find relaxations in the glassy state that can be well characterized by dielectric spectroscopy. Most amorphous polymers display one or more such relaxation processes below T_g. These relaxations are characterized by extreme breadth in frequency or time, Arrhenius behavior of log f_{max}, and modest activation energy. The strengths of the processes can vary widely among various polymers, all the way from barely detectable to an important fraction of the total relaxation strength in the equilibrium melt above T_g.

The question at hand is whether and to what extent the semicrystalline environment affects subglass processes in the amorphous fraction. Again PET serves as an excellent example for it shows a prominent subglass process (β-relaxation) that can be studied in both the wholly amorphous and semicrystalline polymer.

Figure 7 shows the dielectric constant and loss versus log frequency in

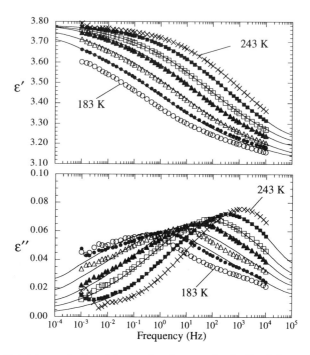

Figure 7. Dielectric constant and loss vs. log frequency in the β subglass region for amorphous PET. The temperatures are at 10 degree intervals between the limits shown. The curves are fits of the Cole–Cole function.

amorphous PET at a number of temperatures. The loss process is exceedingly broad (*see* Figure 2). It is so broad in fact that the actual shape with respect to whether it is skewed to high or low frequency is not certain. It does sharpen with increasing temperature, and the symmetric Cole–Cole function appears to fit the data reasonably well at higher temperatures (Figure 7). The parameters for the fits are listed in Table I.

Figure 8 compares the β-processes in the wholly amorphous and semicrystalline conditions. A comparison of complex plane plots is made in Figure 9. The temperature dependencies of log f_{max} are compared in Figure 4. Other than a decrease in strength the β-process in the semicrystalline environment is very similar to that in the wholly amorphous case. The peak occurs in the same frequency region with similar breadth. That the reduction in strength is attributable to dilution by the crystal phase is shown in Figure 6. All of the specimens appear to fall on the same regression line.

In summary, the subglass process is extremely broad in frequency or time and has Arrhenius temperature dependence with modest activation energy. Unlike the case of the glass transition process, the subglass process is largely insensitive to the presence of the crystal phase. These effects appear to be general for a wide variety of polymers.

Table I. Parameters from Curve Fitting of Dielectric Data for Polymers Measured in This Work[a]

Polymer	T (K)	ε_U	ε_R	α	β	τ_0 (s)
Amorphous PET						
	193	3.09	3.80	0.21		0.2
	203	3.09	3.80	0.22		0.03
	213	3.12	3.80	0.243		0.007
	223	3.13	3.79	0.259		0.002
	233	3.14	3.78	0.286		0.0005
	354	3.80	6.00	0.675	0.56	2.0
Crystalline PET						
	193	3.18	3.66	0.192		0.2
	203	3.18	3.67	0.196		0.04
	213	3.19	3.66	0.212		0.01
	373	3.71	4.44	0.298		0.01
PVAc						
	340	3.15	9.13	0.872	0.49	0.12
	351	3.20	8.89	0.875	0.51	0.012

[a]α is the width parameter of the Cole–Cole equation or the symmetric broadening parameter of the HN equation, β is the skewing parameter of the HN equation, and τ_0 is the central relaxation time. If no β entry is present the Cole–Cole equation was invoked.

Phase Properties from Specimen Measurements

Detailed interpretation of the mechanism underlying a relaxation process occurring in one of the phases, amorphous or crystalline, requires knowledge of the properties of that phase. For example, if the Kirkwood–Onsager correlation factor is to be computed for an amorphous phase process it is implied that the dielectric constant of the amorphous phase is known. Yet dielectric measurements are unavoidably carried out on a composite semicrystalline specimen. Thus, some sort of mixing rule or equation must be available to "back out" the dielectric constant of the phase in question. In the previous discussion centering on plotting relaxation strengths against degree of crystallinity (Figure 6) it was implied that the specimen dielectric constant is a *linear* function of degree of crystallinity. This need not be the case and in fact in general it is not.

The general subject of the relation of the properties of physical mixtures to those of the components is a very complicated one. These relations are obviously highly dependent on the properties in question. Also, the closer together the properties of the components are the more accurate an approximate mixing rule or equation becomes. Much of the theoretical work on mixtures has taken the form of deriving bounds on the composite properties. That is, sometimes it can be shown that the actual behavior must lie between certain limiting cases.

In the dielectric case a common situation is where one of the phases,

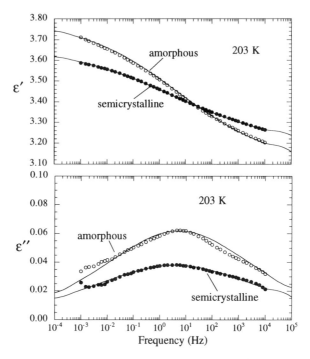

Figure 8. Dielectric constant and loss vs. log frequency in the β subglass region compared for semicrystalline and amorphous PET. The curves are fits of the Cole–Cole function.

possessing a relaxation process, takes on a relaxed value whereas the second phase possesses no relaxation and takes on its unrelaxed value. In semicrystalline polymers these values may be quite different. Fortunately, however, they are often not so different but they can be treated adequately by an appropriate mixing equation (*19*).

Mixing Equations for Lamellar Structures

As indicated previously, a common morphology encountered is that of stacks of crystalline lamellae separated by amorphous interlayers. In this case the dielectric constant of the mixture, ε, is given in the horizontal (H) in plane directions (faces 2 and 3) of Figure 1 by

$$\varepsilon_H = v_1\varepsilon_1 + v_2\varepsilon_2 \tag{1}$$

and in the normal or vertical (V) direction (face 1) by

$$1/\varepsilon_V = v_1/\varepsilon_1 + v_2/\varepsilon_2 \tag{2}$$

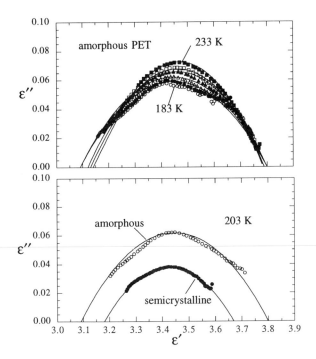

Figure 9. Argand complex plane plots of dielectric loss factor vs. dielectric constant in the β subglass region. Upper panel is for amorphous PET at 10 degree intervals. The lower panel compares amorphous PET with semicrystalline PET at the same temperature. The curves are fits of the Cole–Cole function.

where ε_1 and v_1 are the dielectric constant and volume fraction, respectively, of the amorphous phase, subscript 1, and subscript 2 refers to the crystal-phase counterparts. Presumably in a quiescently crystallized specimen the amorphous phase itself is nearly isotropic, and ε_1 has the same value in both the H and V directions. If the relaxation is in the amorphous phase, ε_2 for the crystal phase is an unrelaxed value and therefore nearly isotropic. In the actual measurements on the macroscopically isotropic specimen neither the H or V direction prevails, but rather some average of them will result. However ε_H and ε_V in eq 1 and eq 2 can be regarded as upper and lower bounds, respectively, to the mixture dielectric constant. These bounds are not very close, however, and are not very useful when ε_1 and ε_2 differ considerably (Figure 10). Much better bounds can be found (*19*) by constructing a local, direction-dependent, dielectric tensor, **e**, for the lamellar material from ε_H and ε_V in eqs 1 and 2 as

$$\mathbf{e} = \begin{pmatrix} \varepsilon_V & 0 & 0 \\ 0 & \varepsilon_H & 0 \\ 0 & 0 & \varepsilon_H \end{pmatrix} \tag{3}$$

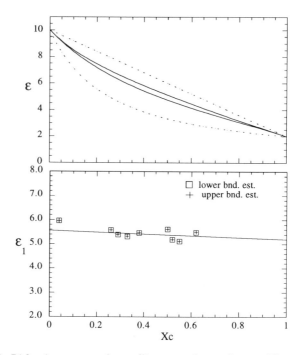

Figure 10. Dielectric constant of crystalline–amorphous mixtures. The upper panel shows the calculated dielectric constant of an example mixture ($\varepsilon_1 = 10$, $\varepsilon_2 = 2$) as a function of crystallinity, X_c. The dashed curves are for the simple "parallel" and "series" equations, eqs 1 and 2, respectively. The solid curves are for the lamellar morphology bounds, eqs 4 and 5. The lower panel shows the dielectric constant of the amorphous phase, ε_1 back-calculated from experimental values for semicrystalline mixtures. The lamellar bound equations are used, eqs 4 and 5. The experimental data for PET and for the relaxed dielectric constant of the combined α and β processes shown in Figure 6. The solid line is a linear regression that excludes the lowest crystallinity, essentially wholly amorphous specimen.

Averaging of ε over all orientations of the lamellar normals results in the following relation for the mixture dielectric constant

$$\langle \varepsilon \rangle = (\varepsilon_V + 2\varepsilon_H)/3 \tag{4}$$

Conversely averaging ε^{-1} over lamellar orientations leads to

$$\langle \varepsilon^{-1} \rangle = (1/\varepsilon_V + 2/\varepsilon_H)/3 \tag{5}$$

The average $\langle \varepsilon \rangle$ in eq 4 forms an upper bound to the dielectric constant for the mixture, and $1/\langle \varepsilon^{-1} \rangle$ in eq 5 forms a lower bound. The bounds computed by using eqs 4 and 5, with eqs 1 and 2 used to define ε_V and ε_H, are much tighter than those from eqs 1 and 2 directly (Figure 10).

In the usual application, the measured specimen dielectric constant is identified with $\langle \varepsilon \rangle$ in eq 4 or $1/\langle \varepsilon^{-1} \rangle$ in eq 5. Then for each case, the phase dielectric constant, ε_1, is back calculated by using known values of the degree of crystallinity and ε_2. The dielectric constant, ε_2, for the dielectrically inactive crystalline phase usually can be identified with the measured unrelaxed constant, ε_U, for the specimen. If ε_1 is less than approximately $5\,\varepsilon$, this procedure yields two values for ε_1 that are sufficiently close together. As an example, Figure 10 also shows the result of calculating the amorphous-phase dielectric constants, ε_1, for PET from the specimen constants, ε, and $\varepsilon_2 = \varepsilon_U$. The data (*18*) are for the same series of semicrystalline specimens as in Figure 6. The unrelaxed constant, ε_U, refers to the specimen constants below both the α- and β-relaxations. The specimen constants, ε, are at 363 K and are the relaxed values of the specimen constants above both the α- and β-processes. First, the upper and lower bound estimates of ε_1 are indistinguishable. Second, the values of the amorphous-phase constant, ε_1, are essentially independent of crystallinity in the semicrystalline specimens. However, the wholly amorphous specimen seems to have a slightly higher ε_1 value than inferred from the regression line for the semicrystalline samples. This result is in accordance with the results in the upper panel of Figure 6 and as discussed previously under the effect of crystal-phase presence on the glass transition process.

Interpretation of Amorphous-Phase Relaxation Processes

A dielectric relaxation process has two aspects. The first is the strength as represented by $\varepsilon_R - \varepsilon_U$. This is an equilibrium property and is subject to treatment by statistical mechanics. The other aspect of relaxation is the dynamic character: that is, the location and shape of the process in time and temperature. Some comments on both of these aspects are made.

Amorphous-Phase Orientation Correlation Functions

As demonstrated, reliable values of the amorphous-phase relaxed dielectric constants can be determined from the specimen measurements. It is then very informative to convert these to dipolar orientational correlation functions. These functions give some insight into the nature of equilibrium or statistical mechanical average behavior of the amorphous chain dipoles and the effect of the crystalline environment on them.

Correlation Functions from Kirkwood–Onsager Theory

The determination of correlation functions is accomplished by the Kirkwood–Onsager–Fröhlich equation (20)

$$\varepsilon_R - \varepsilon_U = \left(\frac{3\varepsilon_R}{2\varepsilon_R + \varepsilon_U}\right)\left(\frac{\varepsilon_U + 2}{3}\right)^2 \frac{4\pi\overline{N}}{3kT} g\mu^2 \qquad (6)$$

where ε_R and ε_U are the relaxed and unrelaxed phase dielectric constants, respectively, $\overline{N}$ is the dipole number density, μ is the dipole moment, k is the Boltzmann constant, T is temperature, and g is the desired correlation function. Equation 6 is formulated by using electrostatic units for μ, that is with the 4π term and no κ_0 permittivity of free-space constant, so that the Debye unit as 10^{-18} esu can be used; kT and dipole number density should thus be in centimeter–gram–second units.

The correlation function responds in general to two types of structural influences. A commonly presented formulation is

$$g = 1 + \Sigma_j\langle\cos \gamma_j\rangle \qquad (7)$$

where γ_j is the angle a jth neighbor dipole makes with a typical chosen dipole, and the sum is over all of the neighbors. The $\langle\ \rangle$ brackets denote a statistical mechanical average. In an amorphous polymer melt this averaging and summation are dominated by correlations arising from conformational effects between nearby dipoles in the same chain. This application is one of the central applications of rotational isomer state theory applied to polymers (21). Implicit in this is the assumption that the central or chosen dipole eventually relaxes so that its own spatial orientations with respect to a fixed reference frame are random. However, in a system where the dipoles may be confined, such as by attachment of a chain to a crystal, such three dimensional randomness may not be possible. This spatial restriction causes *self-correlation* for the dipole (22). The intramolecular conformationally driven correlation of nearby dipoles leads to a correlation function value that can be either greater or less than the uncorrelated value of $g = 1$. Spatial-confinement-induced self-correlation always contributes to g being less than unity. It is not possible to resolve the experimental g values into these two types of contributions. However, there are circumstances where it is clear which type the dominant contributions represent.

In Figure 11 the dipolar correlation functions, g, for amorphous and semicrystalline PET are compared. The g factors for the amorphous phase in the semicrystalline specimen were determined from amorphous-phase relaxed and unrelaxed dielectric constants (18). The g factors from the unrelaxed constants were in turn derived from experimental specimen values by using the lamellar morphology equations above (eqs 4 and 5). The

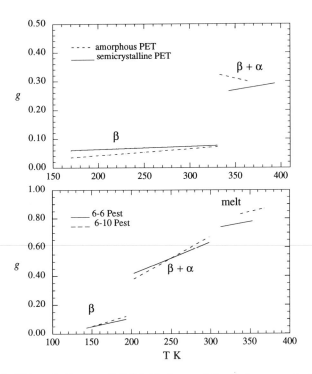

Figure 11. Kirkwood–Onsager–Fröhlich dipolar correlation factors, g, vs. temperature. The upper panel is for the α- and β-relaxation regions in PET. The lower panel is for 6-6 and 6-10 linear aliphatic polyesters (PEST).

α- and β-process correlation functions for both the free amorphous phase and for the semicrystalline environment are considerably less than one. This primarily reflects the intramolecular conformational correlation effect. The conformation of the ester groups across a given aromatic ring tends to be trans with respect to each other. This conformation results in partial cancellation of the moments of the adjacent group and a negative contribution to the correlation factor. However, the *g* values for the semicrystalline environment are noticeably smaller than those for the wholly amorphous situation and increase with temperature. This result can be attributed to the self-correlation confinement effect of the crystal phase on the available chain configurations.

Semicrystalline aliphatic polyesters are an example where the intramolecular conformational correlation is considerably reduced and thus the effects of the crystalline environment are more transparent (*23, 24*). These polymers are the analogs of linear aliphatic polyamides (nylons) and the same numbering convention may be invoked. The reduction in conformational correlation compared to PET is simply the result of the conformational flexibility of the several aliphatic C–C bonds separating the polar groups. The comparison of the semicrystalline environment with the wholly amorphous state is in this case affected by making measurements on melts of the crystalline samples. Results for 6-6 and 6-10 linear aliphatic polyesters (PESTS) are also shown in Figure 11. The correlation factors, g, for the melt are much higher than for PET and appear to be approaching unity. The g values for 6-10 PEST are higher than for the 6-6, no doubt due to the greater number of aliphatic bonds separating the polar groups. In the solid, the α- and β-process g values are considerably smaller and more temperature dependent. This is attributable to the immobilizing effect of the crystalline phase on amorphous chain configurations.

Significance of Correlation Function Behavior

The principal finding with respect to the effect of crystal-phase presence is that the correlation function is reduced and is significantly temperature dependent and g increases with temperature. The reduction in g can be accommodated by the notion that some amorphous-phase dipoles are totally unable to participate in relaxation by virtue of chain attachments to crystal surfaces. Effectively, this result would mean that g for a polymer without intramolecular conformational correlation is the *fraction of amorphous-phase dipoles able to relax*. However, the temperature dependence of g suggests that this fraction increases with temperature, such as the α and β curves in Figure 11. Thus, the concept arises that energy differences exist between different spatial orientations of a given dipole. If a dipole has two positions, 1 and 2, available that are separated by energy ΔU, then the correlation factor is given by (20, 22, 25)

$$g = 2(1 - \cos \gamma)P_1 P_2 \tag{8}$$

where P_1 and P_2 are the probabilities of being in states 1 and 2, respectively, and are thus given by

$$P_1 = 1/(1 + e^{-\Delta U/kT}) \tag{9}$$

$$P_2 = e^{-\Delta U/kT}/(1 + e^{-\Delta U/kT}) \tag{10}$$

where state 1 is of lower energy and γ is the angle through which the dipole reorients in going from state 1 to 2. Equation 8 leads to g increasing with temperature toward the value dictated by the reorientation angle. Experimentally, the typical increase in g with temperature is more gradual than given by eq 8. However if a *distribution* of site energy differences is invoked the experimental behavior can be accommodated.

In the previous picture the immobilizing effect of the crystal phase is described by the phenomenon of dipoles in the vicinity of the crystal surfaces finding themselves in "traps." That is, they are now in situations where the conformation reorientations in the free amorphous state that have no intermolecular energy penalty now have such penalties. Because little is known a priori of the actual energy penalties this picture is mainly conceptual or perhaps a means to deduce something about the trap distribution from experiments.

For the β-subglass processes in all cases, wholly amorphous or semicrystalline, the measured g functions are very small and of the order of 0.1 at low temperature. They are also temperature dependent and increase with temperature. These low values could possibly be attributed to the result of only a fraction of the dipoles finding themselves in environments in the glass where they are able to relax, to limited angular reorientational possibilities (i.e., the cos γ term in eq 8), or to a distribution of energy differences existing between orientational states. All of these factors could arise because of the intermolecular packing in the glass and its inherent variations or heterogeneity. The experiments themselves give little clue as to the detailed interpretation. It would appear that molecular simulations will be a promising entry into resolution of this quandary (*26*).

Dynamics of Amorphous-Phase Relaxations

From the previous section on signatures, the amorphous-phase glass transition region (α-process) broadens greatly when the crystal phase is present. This result is displayed more quantitatively in Figure 12 where Cole–Cole width parameters are plotted versus temperature for the α-process in PET and aliphatic polyesters. Little is available from theory with respect to interpretation of this effect. However, it is interesting to view the phenomenon in terms of an effective chain segment length involved in the relaxation in an unconstrained amorphous phase well above T_g. This segment length appears to be of the order of several persistence lengths (*21, 27*) for equilibrium chain properties. That is, it is probably somewhat greater than the segment length required for bond vectors in the same chain to lose correlation. If a free chain is constrained at widely separated points such as in a lightly cross-linked rubber, the melt viscosity, which depends on the ability of entire chains to undergo transport, is drastically perturbed and becomes infinite. However, there is little effect on dynamic relaxation processes such as dipolar relaxation. The effective segment length involved in dipolar relaxation is shorter. There is little effect on T_g. However as the space between

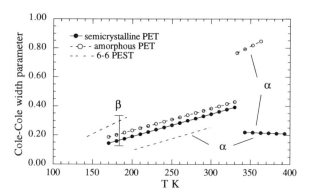

Figure 12. Cole–Cole width parameters vs. temperature for the α- and β-relaxation regions in PET and aliphatic polyesters. For amorphous PET the α-process parameter shown is the symmetric broadening parameter. The skewing parameter values may be found in Coburn and Boyd (18).

confinements is decreased the effective segment length for relaxation is approached. This result causes some of the modes for relaxation to be unavailable and others to become slower. The glass temperature increases.

In the semicrystalline environment the thickness of the amorphous interlayer also approaches the segment length for relaxation. Chains entering the crystal surface from the amorphous phase behave somewhat like a cross-linked rubber and segments comprising loops and folds at the surface are even more seriously confined.

This qualitative picture has relevance for the nature of the dynamics of subglass processes as well. As was seen, such processes are relatively unaffected by the presence or absence of a crystal phase. A corollary of the concept of an effective segment length for dynamics in an unconstrained melt above T_g is that well below T_g in the glass there are no dynamic processes on this length scale. This corollary implies that, whatever the molecular details of the dynamic processes underlying subglass motions may be, the segment length involved is short. This result is entirely consistent with the lack of effect of crystal-phase presence. The length scale for subglass motions is too short to be seriously perturbed by the confinements introduced by the crystal surfaces. The localized nature of the process is also consistent with the modest activation energies observed.

The extreme width of subglass relaxations is very possibly connected with the heterogeneity of packing in glasses. That is, relatively poorly packed regions could be expected to have relaxation times associated with localized processes that are shorter than those associated with well-packed regions. Some recent molecular simulation results indicate that the spatial scale and degree of heterogeneity are appropriate (26).

Much speculation has taken place with respect to the molecular details of main-chain motions that could be highly localized (*2, 28*). Certain conformational sequences such as *crankshafts* might be relatively favorable with respect to allowed motion. However, generally speaking, when the likely populations of specialized conformational sequences are assessed there are not enough of them to explain even the weak strengths or small *g* factors associated with many main-chain subglass processes. In many chains of simple geometry, a number of localized conformational transitions are possible. Again, molecular simulations are likely to be helpful in the interpretation process for this area.

Further Interpretation, Three Phases

A further interpretation that has sometimes been used concerns accommodating the effects of the crystal phase on the amorphous-phase glass–rubber relaxation region by invoking an additional phase. In one version this third phase is assumed to be an immobilized interfacial layer between the crystal and amorphous phases. For example, if the correlation factor, *g*, in a semicrystalline polymer is found to be reduced compared to a wholly amorphous counterpart due to immobilization effects by the crystals, the fractional reduction in *g* can be used to compute the fraction of the third phase, an immobilized or interfacial amorphous phase (*29*). This concept does have conceptual appeal in that it succinctly summarizes the effects of crystal immobilization on the *strength* descriptor of the processes in amorphous component. It also recognizes the intuitive feeling that the most serious effects of crystal presence should be near the crystal surfaces. However it has serious shortcomings that compromise its usefulness. Implied in this description is that the perturbing effect of the crystals is confined to the immobilized interfacial phase, and therefore the remaining amorphous phase is not perturbed. However only one α- or T_g-process is experimentally observed. In this version of the three-phase model this observation occurs in the unperturbed amorphous fraction. The immobilized or interlayer phase has no experimental signature and is assigned by the reduction in the correlation factor value for the unperturbed amorphous phase. However, as emphasized in the earlier sections, the most dramatic effect of crystal-phase presence is on the experimentally observed *dynamic* descriptors of the glass–rubber process. The *entire* relaxation spectrum is shifted and greatly broadened. Thus, dynamically, the perturbing effects of crystal presence propagate throughout the amorphous fraction. Because both the strength and dynamic descriptors belong to the same process, it is misleading to conceptually localize the perturbing effects to a separate phase.

Another version of the three-phase model is the opposite of the previous version. It ascribes the experimentally observed relaxation process in semicrystalline polymers to the interfacial layer and not to the bulk of the amorphous phase (*30, 31*). In the case of polymers like PET where both the wholly amorphous and semicrystalline versions are available this interpretation is obviously not true. The strengths of the processes are too similar to admit this possibility. The effects of crystallization on strength, discussed previously (e.g., Figures 6, 10, and 11), while observable are too small to attribute the relaxation process entirely to an interfacial region. In the case of the linear aliphatic polyesters (Figure 11), which cannot be quenched to the wholly amorphous state, the *g* factors can be compared to the melt. Again they are too large to be ascribed to a limited interfacial region. Thus, the use of the interfacial phase concept has been limited to polymers that cannot be quenched to the completely amorphous state for comparison or where quantification of the strengths is difficult.

Undoubtedly, the properties of the amorphous phase are very different near the crystal surfaces. However, the effects of the crystals on the amorphous-phase glass transition region are more serious and interesting than invocation of a third-phase interfacial region can accommodate.

Crystal-Phase α-Process in Polyethylene

Crystalline-phase relaxation processes are observed in *mechanical* relaxation spectroscopy in several polymers of simple chain structure (*1, 28*). These include polyethylene (PE), polypropylene (PP), poly(tetrafluoroethylene) (PTFE), and poly(oxymethylene) (POM). However, of these only in PE has this type of process been found dielectrically. The reason for this difference is connected with the requirement for dielectric activity. Polymer chains consisting of aliphatic carbon atoms with tetrahedral bonding geometry have cancellation of bond moments in the total molecular moment if the two substituents are the same. Thus, PTFE and PE are effectively nonpolar and dielectrically inactive. In PP the geometry is distorted enough from ideal tetrahedra that dielectric activity is strong enough for measurements to be made. POM is quite dielectrically active. However, the requirements for a crystal-phase process are more stringent. For activity to occur, dipolar reorientation must take place. In the crystal a screw axis motion will allow a chain to move along the *c* axis and to come back into crystallographic register. No change in energy will occur within the crystal between the initial and final states. No intensity penalty will occur of the sort associated with ΔU in eq 8. However, no change in moment will occur, either. Such screw axis motions will be dielectrically silent. This result is analogous to vibration spectroscopy where selection rules based on changes of dipole moment with vibration govern the observance of IR bands. Mechanically such motions do lead to release of the crystal constraints on the amorphous phase and softening of the amorphous phase. The softening effect gives rise to mechanical activity (*2*).

In PE, dielectric activity can be introduced by *decorating* it with a few dipoles. This process can happen adventitously simply by processing a PE

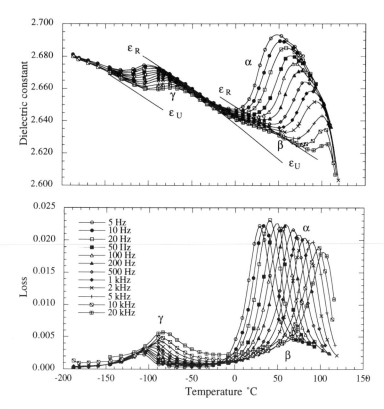

Figure 13. Dielectric constant and loss in dipole decorated (slightly oxidized) LPE. Approximate relaxed and unrelaxed dielectric constants for the β- and γ-processes are indicated on the figure. (Reproduced with permission from reference 4. Copyright 1994 Butterworths–Heinemann (Elsevier).)

melt. Oxidation results in a few ketone groups, perhaps a few per thousand CH_2 units, being introduced along the chain. Relaxation processes can then be studied dielectrically (*3, 4, 28*). The highest temperature process, labeled as α, as seen in the data of Graff and Boyd (*4*) in Figure 13, is prominent and is crystal phase in origin. The amorphous-phase glass–rubber relaxation region, now labeled β, is very broad but can be observed and the low temperature subglass process, γ, is also prominent.

The reason for the dielectric activity of the crystalline α-process is connected with the small number of carbonyl groups. A single group is likely to be the only one in a chain segment spanning a crystal. A screw axis rotation in PE causes the chain to rotate by 180° and advance by $c/2$. When a screw axis rotation takes place it causes the dipole to rotate by 180°, thus causing observable dielectric activity.

An isothermal frequency scan taken for the present work in the α-region is shown in Figure 14 along with an Argand plot. The process is

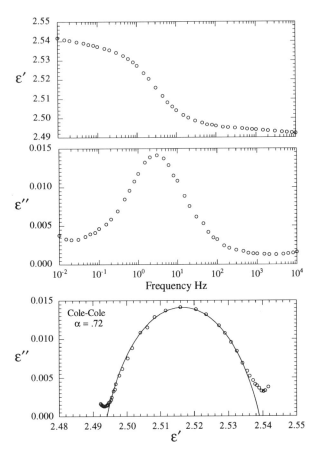

Figure 14. Dielectric constant and loss vs. log frequency and Argand diagram for slightly oxidized LPE in the α-crystal-phase relaxation region.

exceedingly narrow in frequency and is approaching single relaxation time behavior. It is perhaps the narrowest process to be observed in a bulk polymer relaxation.

It is of interest to pursue the details of how a chain turns around in the crystal (the screw axis rotation), which would seem impossible as a rigid motion along a segment spanning an entire crystal. The activation energy would be prohibitively high for a long segment in a thick crystal because an energy contribution would be associated with the crystal mismatch of each CH_2 group. This question has occupied the attention of a number of researchers for many years (*20, 32–36*). The most salient experimental observation is that the measured relaxation times become longer as crystal thickness increases (*3, 20*). However, the effect saturates because in very thick crystals, a limiting, relatively slow increase of relaxation time with thickness appears to be approached. Figure 15 shows a plot of experimental

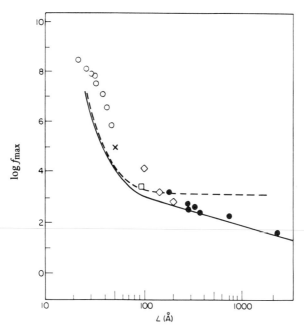

Figure 15. Dependence of the location of the α-crystal phase in LPE, as measured by log f_{max}, on crystal thickness, L. The points are experimental and include a variety of LPE and branched polyethylene (BPE) specimens as well as ketone-containing paraffins. See Ashcraft and Boyd (3) and Mansfield and Boyd (36) for data sources. The curves are calculated from a detailed atomistic model based on the motion of the smooth twist of Figure 16 (36). The solid curve contains a correction for possible trapping of the twist that the dashed curve does not. (Reproduced with permission from reference 2. Copyright 1985 Butterworths–Heinemann (Elsevier)).

log f_{max} values versus lamellar thickness in dipole decorated PE (*36*). Data for paraffin hosts that contain ketones are included as well.

This behavior has been interpreted in terms of a nonrigid chain rotation (*20, 32–36*). A twist in the chain is postulated to occur and this twist propagates across the crystal (Figure 16). This process has been modelled in atomistic detail (*36*). The most efficient twist contains about 12 CH_2 groups. This twist once formed can move through the surrounding lattice with little or no hindrance. The activation energy then becomes the energy necessary to create the twist. This energy can be calculated as a function of lamellar thickness. In a two-site model the relaxation time, τ, and f_{max} are given by

$$2\pi f_{max} = 1/\tau = 2k_{rate} \tag{11}$$

where k_{rate} is the rate constant for barrier crossing. In terms of absolute rate theory, f_{max} and k_{rate} at T_{max} are related to the barrier as

$$\ln f_{max} = \ln k_{rate}/\pi = \ln(kT_{max}/\pi h) + \Delta S^*/k - \Delta H^*/kT_{max} \tag{12}$$

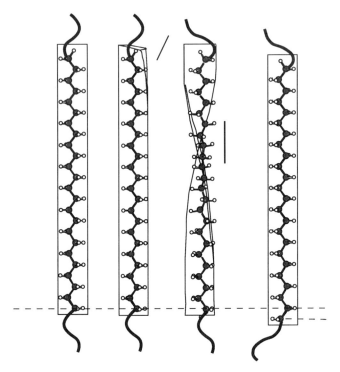

Figure 16. Source of the α-relaxation in PE. A chain turns around in the crystal by a twist starting at the crystal surface. Then, the twist propagates across the crystal. In doing so it advances by half a unit cell. If a carbonyl is attached to the chain it turns around and dielectric activity results. Mechanical activity results from the advance of the chain and its effect on the amorphous interlayer.

where ΔH^* and ΔS^* are the enthalpy and entropy of activation, respectively. Both the activation enthalpy and entropy are accessible to modeling. Figure 15 shows calculated log f_{max} results in comparison with the experimental values (*36*). The model captures the physical situation rather well.

Appendix: Effects of Ionic Conductance on Measured Dielectric Constant and Loss

Homogeneous Systems

In frequency-domain measurements the dielectric constant can be considered as a complex number, $\varepsilon^*(\omega)$, that is a function of the frequency ω. A parallel plate capacitor filled with a material of dielectric constant $\varepsilon^*(\omega)$ will have an impedance Z^* given by

$$1/Z^* = i\omega \frac{\kappa_0 A}{d} \varepsilon^*(\omega) \tag{A-1}$$

where κ_0 is the permittivity of free space, A is the area of the plates, and d is the distance of their separation (37). The real and imaginary parts of ε^* are thus determined as

$$\varepsilon' = \mathrm{Re}\left(Z^{*-1} \frac{d}{i\omega\kappa_0 A}\right) = \frac{d}{\omega\kappa_0 A} \mathrm{Im}(Z^{*-1}) \tag{A-2}$$

$$\varepsilon'' = -\mathrm{Im}\left(Z^{*-1} \frac{d}{i\omega\kappa_0 A}\right) = \frac{d}{\omega\kappa_0 A} \mathrm{Re}(Z^{*-1}) \tag{A-3}$$

where "Re" means "real part of" and "Im" means the imaginary counterpart. A given measuring apparatus will typically provide, in one form or another, the real and imaginary parts of Z^* so that ε' and ε'' can be calculated from A-2, A-3.

The migration of free charges will be associated with a specific conductance, σ. In the absence of electrode polarization effects the conduction acts as a purely resistive shunt in parallel to the dielectric filled capacitor. The resistance of the shunt, R, is given by $1/R = A\sigma d$. The total measured impedance, Z^*, of the dielectric filled capacitor together with the resistive shunt is given by

$$1/Z^* = i\omega \frac{\kappa_0 A}{d} \varepsilon^*(\omega) + A\sigma/d \tag{A-4}$$

Because the measuring apparatus delivers only the real and imaginary parts of Z^*, eqs A-2 and A-3 must still be used to compute ε' and ε''. In this case, *apparent* values of ε' and ε'' are now obtained. The conductance term in eq A-4 is real so that (cf. eq A-3) it will only contribute to the apparent loss, ε''(app). The value of ε''_{app} is given through application of eq A-3 to eq A-4 as

$$\varepsilon''_{app} = \varepsilon''(\omega) + \frac{\sigma}{\omega\kappa_0} \tag{A-5}$$

Thus, the conductance causes the apparent loss to increase without limit as frequency decreases.

Interfacial Polarization in Heterogeneous Systems

The case of Maxwell–Wagner–Sillars polarization (17) in a two-phase system is now considered. One of the phases (1) is conductive and has conductance, σ_1, and the other (2) is not. An explicit relevant case that can be easily treated is that of lamellar geometry of the phases. In the direction normal to the lamellae the impedances of the phases are added in series to obtain the sample impedance, or $Z^* = Z_1^* + Z_2^*$ where subscript 1 and 2 refer to the phases. Use of eq A-1 leads to

$$1/\varepsilon^* = \frac{v_1}{\varepsilon_1^* + \dfrac{\sigma_1}{i\omega\kappa_0}} + \frac{v_2}{\varepsilon_2^*} \tag{A-6}$$

where advantage has been taken of the fact that the measuring area A is the same for both phases and thus the phase volume fractions, v_1 and v_2 are given by $d_1(d_1 + d_2)$ and $d_2(d_1 + d_2)$, respectively. For simplicity in further discussion suppose that neither of the phases shows dipolar orientation loss, and therefore $\varepsilon_1^* = \varepsilon_1'$, $\varepsilon_2^* = \varepsilon_2'$. Rationalization of eq A-6 leads to

$$\varepsilon' = \frac{\varepsilon_1'\varepsilon_2'}{v_1\varepsilon_2' + v_2\varepsilon_1'} + \frac{\left(\varepsilon_2'/v_2 - \dfrac{\varepsilon_1'\varepsilon_2'}{v_1\varepsilon_2' + v_2\varepsilon_1'}\right)}{1 + x^2} \tag{A-7}$$

and

$$\varepsilon'' = \frac{x\left(\varepsilon_2'/v_2 - \dfrac{\varepsilon_1'\varepsilon_2'}{v_1\varepsilon_2' + v_2\varepsilon_1'}\right)}{1 + x^2} \tag{A-8}$$

where

$$x = \frac{\omega\kappa_0}{v_2\sigma_1} (v_1\varepsilon_2' + v_2\varepsilon_1') \tag{A-9}$$

The variables ε' and ε'' have the frequency dependence (through x) of a single relaxation time Debye process. At infinite frequency ε' approaches the value expected for both phases having no conductance. However, as frequency decreases the dielectric constant increases toward the limiting value ε_2'/v_2. The loss has a maximum at $x = 1$. Thus, the presence of conductance in one of the phases gives rise to an apparent loss process. These equations were derived for the special case of the normal direction in lamellar morphology. However, the only general requirement for the effect is that the conducting phase be isolated from a direct dc path between the electrodes. Thus any noncontinuous suspension of a conducting phase in a nonconducting matrix will show such a dispersion process. Very often the maximum in the loss occurs at frequencies far below the experimental range investigated. This result causes only the increase in both dielectric constant and loss because frequency decreases are observed in the usual case.

Acknowledgment

We are grateful for the financial support of our work in the dielectrics area received from the Polymers Program, Division of Materials Research, National Science Foundation. We thank J. C. Coburn of DuPont for furnishing the PET samples and R. B. Thayer for making the PVAc measurements. Figure 1 originated in the Ph.D. Dissertation of J. C. Coburn, University of Utah, 1984.

References

1. Boyd, R. H. *Polymer* **1985,** *26,* 323.
2. Boyd, R. H. *Polymer* **1985,** *26,* 1123.
3. Ashcraft, C. R.; Boyd, R. H. *J. Polym. Sci., Polym. Phys. Ed.* **1976,** *14,* 2153.
4. Graff, M. S.; Boyd, R. H. *Polymer* **1994,** *35,* 1797.
5. Thayer, R. B. B.Sc. Dissertation, Department of Materials Science and Engineering, University of Utah, 1989.
6. Tadokoro, H. *Structure of Crystalline Polymers;* Wiley-Interscience: New York, 1979.
7. Bassett, D. C. *Principles of Polymer Morphology;* Cambridge University: New York, 1981.
8. Woodward, A. E. *Understanding Polymer Morphology;* Hanser/Gardner: Cincinnati, OH, 1995.
9. Woodward, A. E. *Atlas of Polymer Morphology;* Hanser: Munich, Germany, 1988.
10. Havriliak, S.; Negami, S. *Polymer* **1967,** *8,* 161.
11. Williams, G.; Watts, D. C. *Trans. Faraday Soc.* **1970,** *66,* 80.
12. Williams, G.; Watts, D. C.; Dev, S. B.; North, A. *Trans. Faraday Soc.* **1971,** *67,* 1323.
13. Kohlrausch, R. *Pogg. Ann.* **1847,** *12,* 393.
14. Vogel, H. *Phys. Z.* **1921,** *22,* 645.
15. Fulcher, G. S. *J. Am. Chem. Soc.* **1925,** *8,* 339, 789.
16. Williams, M. L.; Landel, R. F.; Ferry, J. D. *J. Am. Chem. Soc.* **1955,** *77,* 3701.
17. Smyth, C. P. *Dielectric Behavior and Structure;* McGraw-Hill: New York, 1955; Chapter 2.
18. Coburn, J. C.; Boyd, R. H. *Macromolecules* **1986,** *19,* 2238.
19. Boyd, R. H. *J. Polym. Sci., Polym. Phys. Ed.* **1983,** *21,* 505.
20. Fröhlich, H. *Theory of Dielectrics,* 2nd ed.; Oxford University: Oxford, England, 1958; Chapter 2.
21. Flory, P. J. *Statistical Mechanics of Chain Molecules;* Wiley-Interscience: New York, 1969.

22. Smith, G. D.; Boyd, R. H. *Macromolecules* **1991**, *24*, 2731.
23. Boyd, R. H.; Aylwin, P. A. *Polymer* **1984**, *25*, 330.
24. Boyd, R. H.; Hasan, A. A. *Polymer* **1984**, *25*, 347.
25. Hoffman, J. D.; Pfeiffer, H. G. *J. Chem Phys.* **1954**, *22*, 132.
26. Smith, G. D.; Boyd, R. H. *Macromolecules* **1992**, *25*, 1326.
27. Boyd, R. H.; Phillips, P. J. *The Science of Polymer Molecules;* Cambridge University: New York, 1993.
28. McCrum, N. G.; Read, B. E.; Williams, G. *Anelastic and Dielectric Effects in Polymeric Solids;* Wiley: New York, 1967.
29. Cebe, P.; Huo, P. P. *Thermochim. Acta* **1994**, *238*, 229.
30. Popli, R.; Glotin, M.; Mandelkern, L.; Benson, R. S. *J. Polym. Sci., Polym. Phys. Ed.* **1984**, *22*, 407.
31. Han, B.; Wendorff, J.; Yoon, D. *Macromolecules* **1985**, *18*, 718.
32. Tuijman, C. A. F. *Polymer* **1963**, *4*, 315.
33. Booij, H. C. *J. Polym. Sci., Polym. Symp.* **1967**, *16*, 1761.
34. Hoffman, J. D.; Williams, G.; Passaglia, E. *J. Polym. Sci., Polym. Symp.* **1966**, *14*, 173.
35. Williams, G.; Lauritzen, J. I.; Hoffman, J. D. *J. Appl. Phys.* **1967**, *38*, 4203.
36. Mansfield, M. L.; Boyd, R. H. *J. Polym. Sci., Polym. Phys. Ed.* **1978**, *16*, 1227.
37. Boyd, R. H. In *Methods of Experimental Physics;* Fava, R. A., Ed.; Academic: Orlando, FL, 1980; Vol. 16, Part C.

Modeling and Techniques

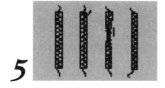

5

Calculation of Dipole Moments and Correlation Parameters

Richardo Diaz-Calleja and Evaristo Riande

Calculated dipole moments are correlated to conformation of chain molecules. Conformational energies and the location of rotational states are correlated with polymer properties, particularly in the case of polymers with polar side groups. A connection is shown between relaxation strength and dipolar correlation coefficients.

Macromolecular chains are made of thousands and even hundreds of thousands of skeletal bonds of the type that can rotate, so that the chains may adopt an almost unlimited number of spatial conformations. This conformational versatility is responsible for the unique physical properties exhibited by polymeric materials, and consequently the interpretation and detailed understanding of chain conformations have become a subject of great interest in polymer science. Thus, in the last 30 years a variety of conformation-dependent properties have been measured and interpreted within the framework of statistical mechanical methods, specifically the rotational isomeric state model. Among the conformation-dependent properties most successfully treated, several stand out (*1–3*), such as the mean-square end-to-end distance, $\langle r^2 \rangle$; mean-square dipole moment, $\langle \mu^2 \rangle$; molar Kerr constant, K_m; and stress optical coefficient, C. Because the bonds vary much more in polarity and polarizability than they do in length, dipole moments and Kerr constants are often more sensitive to structure than other more classical conformation-dependent properties such as the unperturbed dimensions.

Dipole Moments

A dipole can be defined as a physical entity in which a positive charge $(+q)$ is separated by a relatively short distance (r) from an equal negative charge $(-q)$. The dipole moment is a vectorial quantity pointing toward the positive charge whose magnitude is given by $\mu = qr$. The unit of dipole is the debye (D) $(1\ D = 10^{-18}$ esu cm in the centimeter–gram–second system and 3.336×10^{-30} C m in Système International units). Accordingly, the dipole moment of two unit electronic charges separated by 1 Å is 4.8 D. The dipole moment of a molecule can be taken as the vectorial sum of bond dipoles. However, because the bond moments measure the electrical asymmetry of certain sections of a molecule, they are affected by their immediate environment, and consequently, group dipoles rather than bond dipoles should be used in the evaluation of the dipole moment of polyatomic molecules.

The response of polymer chains in an electric field involves the orientation polarization, caused by the permanent dipoles of the chains, and the atomic and electronic polarizations arising, respectively, from the distortions of the electronic clouds and the changes in position of the nucleus of the atoms produced by the electric field. The experimental determination of dipole moments of polymers is based on the cavity model developed by Debye for gases. In this model it is assumed that each molecule inside the dielectric experiences the action of the macroscopic field increased by the electric field arisen from the charges on the surface of a cavity surrounding it. This approach leads to the following relationship for the square dipole moment μ^2 of gases (4):

$$\frac{\varepsilon - 1}{\varepsilon + 2} = \frac{4\pi\rho N_A}{M}\left(\frac{\mu^2}{3kT} + \alpha_d\right) = P_0 + P_d \tag{1}$$

where ε is the dielectric permittivity, α_d is the induced polarizability; ρ and M are, respectively, the density and molecular weight of the gas; T is the absolute temperature; k is the Boltzmann constant; N_A is Avogadro's number; and P_0 and P_d represent the orientation and induced polarizations, respectively, P_d being the sum of the atomic (P_a) and electronic (P_e) polarizations. In most cases P_a is relatively small in comparison with P_e, so that the electronic polarization can be obtained directly by setting $\mu = 0$ in eq

1. By further using the Maxwell relationship, $\varepsilon = n^2$, where n is the refractive index of the gas, one finds

$$P_e = \frac{n^2 - 1}{n^2 + 2} \tag{2}$$

Combination of eqs 1 and 2 leads to the Debye equation

$$\frac{\varepsilon - 1}{\varepsilon + 2} - \frac{n^2 - 1}{n^2 + 2} = \frac{4\pi\rho N_A \mu^2}{3kMT} \tag{3}$$

This equation was suitable to determine the dipole moment of gases at ordinary pressures, but it fails for pure liquids. The limitations of the Debye equation led Onsager to consider an additional polarization arising from the reaction field that acts on the dipole as a result of the electric displacement caused by its own presence. Onsager's (5) approach gives

$$\frac{\varepsilon - 1}{\varepsilon + 2} - \frac{\varepsilon_\infty - 1}{\varepsilon_\infty + 2} = \frac{3\varepsilon(\varepsilon_\infty + 2)}{(2\varepsilon + \varepsilon_\infty)(\varepsilon + 2)} \frac{4\pi\rho N_A \mu^2}{3kMT} \tag{4}$$

that differs from the Debye relation in the term $3\varepsilon(\varepsilon_\infty + 2)/[(2\varepsilon + \varepsilon_i)(\varepsilon + 2)]$. Here ε and ε_∞ are the dielectric permittivities at frequency 0 and ∞ respectively.

The Onsager equation converges to the Debye expression when ε approaches to ε_∞, as occurs with gases at atmospheric pressure or below. Improvements in the Onsager equation were further carried out by Kirkwood (6), who considered that rotations of the dipoles in the cavity are hindered by interactions with the dipoles of neighboring molecules. Pursuing this reasoning, Fröhlich (7) obtained the expression

$$\frac{\varepsilon - 1}{\varepsilon + 2} - \frac{\varepsilon_\infty - 1}{\varepsilon_\infty + 2} = \frac{3\varepsilon(\varepsilon_\infty + 2)}{(2\varepsilon + \varepsilon_\infty)(\varepsilon + 2)} \frac{4\pi\rho N_A g \mu_0^2}{3kMT} \tag{5}$$

where $g = 1 + \Sigma_{ij} \langle \cos \gamma_{ij} \rangle$ and $\langle \cos \gamma_{ij} \rangle$ represents the average of the cosine of the angle formed by the i and j molecules, extended to all the orientations. Theories that treat intermolecular interactions explicitly at all stages avoiding the introduction of cavities have also been reported and details of the approach can be found in reference 8.

Intermolecular interactions among polar molecules are severely reduced if they are separated by nonpolar molecules, as occurs in dilute solutions, because in this case the electric behavior of the solutions reflects that of the solute in the gaseous state. Consequently, dielectric measurements on dilute solutions of polymers in nonpolar solvents are a method commonly used to determine the dipole moments of molecular chains, by using Debye's approach for gaseous systems. Long chain molecules are continuously changing spatial conformation so that the dipole moments measured are average values $\langle \mu^2 \rangle$. Application of Debye's equation to polymer solutions, in which the weight fraction of solute is w_2, leads to

$$\frac{4}{3} \pi N_A \left[\alpha_{a2} - \alpha'_{a2} + \frac{\langle \mu^2 \rangle}{3kT} \right] = \frac{3M_2}{\rho_1} \left[\frac{1}{(\varepsilon_1 + 2)^2} \frac{d\varepsilon}{dw_2} - \frac{1}{(n_1^2 + 2)^2} \frac{dn^2}{dw_2} \right]$$

(6)

where the subindexes 1 and 2 refer to the solvent and solute, respectively; α_{a2} is the atomic polarizability of the solute, and α'_{a2} represents a fictitious polarizability that is related to the atomic polarizability of the solvent α_{a1} by the equation

$$\alpha'_{a2} = \alpha_{a1} \frac{V_2}{V_1}$$

(7)

where V_2 and V_1 are the partial molar volume of the solute and solvent, respectively. Careful experiments carried out in nonpolar substances suggest that the atomic polarizability may be one-tenth or even less of the electronic polarizability α_e. Because α_e is in turn much smaller than the orientation polarizability for polar molecules, the assumption $\alpha_{a2} \approx \alpha'_{a2}$ can further be made and leads to eq 5, the so-called Guggenheim (9) and Smith (10) equation

$$\langle \mu^2 \rangle = \lim_{w_2 \to 0} \frac{27kTM}{4\pi\rho N_A(\varepsilon_1 + 2)^2} \left[\frac{d\varepsilon}{dw_2} - 2n_1 \frac{dn_1}{dw_2} \right]$$

(8)

which gives the mean-square dipole moments of isolated chains.

Excluded Volume Effects

In contrast to other conformation-dependent properties such as the unperturbed dimensions $\langle r^2 \rangle_0$, the dipole moments of chain molecules can be measured over a wide interval of chain lengths, ranging from oligomers to high molecular weight polymers. Another additional bonus, indicated before, is that skeletal bonds change much more in polarity than they do in length. As a result, $\langle \mu^2 \rangle$ is much more sensitive to the chemical structure than $\langle r^2 \rangle_0$.

Excluded volume effects on the experimental determination of $\langle \mu^2 \rangle$ depend on the type of dipoles associated with the chains. Stockmayer (11) classified the molecular chains in three groups: A, B, and C. Type-A polymers are those in which the dipole moments of the polar bonds are parallel to the chain contour so that

$$\langle \boldsymbol{\mu}\cdot\mathbf{r}\rangle = \text{constant} \times \langle r^2\rangle \tag{9}$$

Therefore, the dipole moments of these hypothetical chains would undergo the same excluded volume effects as the molecular dimensions. Type-B polymers have the dipole moments rigidly attached to the main chain and perpendicular to the chain contour. In this case a large number of different values of the resulting dipole moment are compatible with a fixed end-to-end distance displacement of the molecular chains. Accordingly, $\langle \boldsymbol{\mu}_\perp\cdot\mathbf{r}\rangle = 0$. On the basis of a crude model without rigorous theoretical reasoning, Benoit and Marchal (*12*) suggested that for this type of molecule, excluded volume effects on the dipole moments should vanish.

By using a multivariate Gaussian distribution for $\boldsymbol{\mu}$ and r_{ij}, Nagai and Ishikawa (*13*) related the coefficients $\alpha_\mu^2 = \langle \mu^2\rangle/\langle \mu^2\rangle_0$ and $\alpha_r^2 = \langle r^2\rangle/\langle r^2\rangle_0$ (the subindex 0 indicates unperturbed conditions) by the expression

$$\alpha_\mu^2 - 1 = \frac{\langle r\cdot\mu\rangle_0^2}{\langle r^2\rangle_0\langle \mu^2\rangle_0} (\alpha_r^2 - 1) \tag{10}$$

This equation suggests that if $\langle \mathbf{r}\cdot\boldsymbol{\mu}\rangle_0 = 0$, $\alpha^2\mu = 1$, irrespective of the value that α_r^2 can take owing to excluded volume effects. However, if $\mathbf{r}$ is parallel to $\boldsymbol{\mu}$, $\langle \mathbf{r}\cdot\boldsymbol{\mu}\rangle_0^2/\langle r^2\rangle_0\langle \mu^2\rangle_0 = 1$ and $\alpha^2\mu = \alpha_r^2$. Although pure type-A polymers are not known, there are chains in which the dipoles have a component parallel, $\mu_\parallel$, and another perpendicular, $\mu_\perp$, to the chain contour. Here $\langle \mathbf{r}\cdot\boldsymbol{\mu}\rangle \sim r_n^2$, and consequently, the dipole moments are perturbed by long range interactions. Because for any polymer $0 \le \langle \mathbf{r}\cdot\boldsymbol{\mu}\rangle_0^2/\langle r^2\rangle_0\langle \mu^2\rangle_0 \le 1$, it may be expected that $\alpha^2 \le \mu_r^2$.

Dipoles that are attached to flexible side groups are called type-C polymers in Stockmayer's nomenclature. As occurs with type-B polymers, the dipole moment of these chains is not sensitive to excluded volume effects. In general $\langle \mathbf{r}\cdot\boldsymbol{\mu}\rangle_0 = 0$ for chains containing symmetry planes, symmetry axes, or symmetry centers. Most of the conventional polymers present some of these symmetries, and consequently their mean-square dipole moment should be insensitive to excluded volume effects. However, polyesters and polyamides derived from α,ω-hydroxy acids and α,ω-amino acids stand out among the polymers in which $\langle \mathbf{r}\cdot\boldsymbol{\mu}\rangle_0 \ne 0$. According to the Ishikawa and Nagai (*13*) treatment, the dipole moment of these polymers and others, such as polypropylene oxide, polypropylene sulfide, and poly(*cis*-isoprene), should be sensitive to excluded volume effects so that this physical property should be determined in Θ solvents.

The validity of eq 10 was investigated some time ago by Mattice and co-workers (*14*, *15*) by using simulation methods. The principal conclusions of these investigations were that the combination $\langle \mathbf{r} \cdot \boldsymbol{\mu} \rangle_0 = 0$ and $\alpha_\mu^2 \neq 1$ holds whenever the dipole moment associated with bond i has a component perpendicular to the planes of bonds i and $i - 1$. For symmetric chains in which the dipole moment of the all trans conformation is zero, the calculations predict that $\alpha_\mu^2 < 1 < \alpha_r^2$, even though $\langle \mathbf{r} \cdot \boldsymbol{\mu} \rangle_0 = 0$. Whatever the merits of these calculations are, the experimental results at hand seem to suggest that the excluded volume effects on the dipole moments are not important for most flexible polymers (*16–20*).

Theoretical Calculations

A given conformation of a molecular chain is defined by the skeletal bond lengths l, backbone bond angles γ, and the rotational angle φ about each bond. By assigning a reference frame at each skeletal bond, as indicated in Figure 1, the matrix $\mathbf{T}_i$ associated with the bond i that refers to a vector in the reference frame of $1 + 1$ bond to that of bond i is (*1*):

$$\mathbf{T}_i = \begin{pmatrix} \cos \theta & \sin \theta & 0 \\ \sin \theta \cos \varphi & -\cos \theta \cos \varphi & \sin \varphi \\ \sin \theta \sin \varphi & -\cos \theta \sin \varphi & -\cos \varphi \end{pmatrix}_i \quad (11)$$

where $\theta = (180 - \gamma)$ is the supplement bond angle. Then, the dipole moment associated with a given conformation of a chain with n skeletal bonds can be expressed in an explicit way by

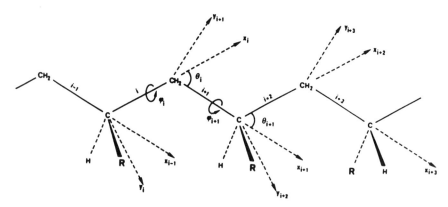

Figure 1. Segment of a polymer chain in all-trans ($\phi = 0$) conformation. Coordinate systems affixed to each skeletal bond are indicated.

$$\mu = \sum_{i=1}^{n} \mathbf{m}_i = \mathbf{m}_1 + \mathbf{T}_1\{\mathbf{m}_2 + \mathbf{T}_2[\mathbf{m}_3 + \mathbf{T}_3(\mathbf{m}_4 + \ldots)]\ldots\} \quad (12)$$

where $\mathbf{m}_i$ represents the dipole associated with bond i. An easier and more elegant way to calculate μ is provided by the matrix-multiplication scheme developed by Flory and co-workers (1, 21). According to this scheme, eq 12 can be written as

$$\mu = \mathbf{P}_1 \left(\prod_{i=2}^{n-1} \mathbf{P}_i \right) \mathbf{P}_n \quad (13)$$

where $\mathbf{P}_i$ is the generator matrix given by

$$\mathbf{P}_i = \begin{pmatrix} \mathbf{T} & \mathbf{m} \\ 0 & 1 \end{pmatrix}_i \quad (14)$$

which includes the contribution of the bond to the dipole of the conformation as well as the geometric terms of the conformation associated with bond i. In eq 13, $\mathbf{P}_1 = [\mathbf{T}_1\, \mathbf{m}_1]$ and $\mathbf{P}_n = \mathrm{column}[\mathbf{m}_n\, 1]$. In the same way, the square dipole moment of a given conformation can be written as

$$\mu^2 = \sum_{i=1}^{n} \mathbf{m}_i \cdot \sum_{j=1}^{n} \mathbf{m}_j = \sum_{i=1}^{n} m_i^2 + 2\sum\sum_{i<j} \mathbf{m}_i\mathbf{m}_j$$

$$= \sum_{i=1}^{n} \mu_i^2 + \sum\sum_{i \neq j} \mathbf{m}_i^T \mathbf{T}_i \mathbf{T}_{i+1} \ldots \mathbf{T}_{j-1}\mathbf{m}_j \quad (15)$$

where the superindex $\mathbf{T}$ in m^T means the transpose vector. By using the generator matrix (21)

$$\mathbf{M}_i = \begin{pmatrix} 1 & m^T\mathbf{T} & m^2 \\ 0 & \mathbf{T} & \mathbf{m} \\ 0 & 0 & 1 \end{pmatrix}_i \quad (16)$$

Equation 15 can be expressed in the matrix multiplication scheme by

$$\mu^2 = \mathbf{M}_1 \left(\prod_{i=2}^{n-1} \mathbf{M}_i \right) \mathbf{M}_n \quad (17)$$

where $\mathbf{M}_1 = \mathrm{row}[1\, \mathbf{m}_1^T\mathbf{T}\, m_1^2]$ and $\mathbf{M}_n = \mathrm{column}[m_n^2\, \mathbf{m}_n\, 1]$ convert the matrix product into a scalar.

Statistical Averages

The statistical average of μ^2, represented by $\langle\mu^2\rangle$, depends on the number of conformations allowed to the chain. The value of this quantity can be written as

$$\langle\mu^2\rangle = \left\langle \sum_i^n \mathbf{m}_i \sum_{j=1}^n \mathbf{m}_j \right\rangle = \sum_{i=1}^n m_i^2 + \left\langle \sum\sum_{i\neq j} \mathbf{m}_i\cdot\mathbf{m}_j \right\rangle$$

$$= \sum_{i=1}^n m_i^2 + \left\langle \sum\sum_{i\neq j} m_i m_j \cos\gamma_{ij} \right\rangle \qquad (18)$$

For real chains the rotational states of each bond are assumed to be correlated with those of its first neighbors. According to the rotational isomeric state model, if the rotational states $\alpha, \beta, \ldots \nu$ for each bond are permitted, the statistical weight of each allowed conformation associated with a pair of bonds $i - 1$ and i is written as a matrix of $\nu \times \nu$ dimensions defined as

$$\mathbf{U}_i = \begin{pmatrix} u_{\alpha\alpha} & u_{\alpha\beta} & \cdot\cdot & u_{\alpha_\nu} \\ u_{\beta\alpha} & u_{\beta\beta} & \cdot\cdot & u_{\beta_\nu} \\ \cdot & \cdot & \cdot\cdot & \cdot \\ \cdot & \cdot & \cdot\cdot & \cdot \\ u_{\nu\alpha} & u_{\nu\beta} & \cdot\cdot & u_{\nu\nu} \end{pmatrix}_i \qquad (19)$$

where $u_{\eta\gamma;i} = \exp[-E_{\eta\gamma;i}/RT]$ is the statistical weight or Boltzmann factor for a conformation of energy $E_{\eta\gamma}$ in which the bond $i - 1$ is in the rotational state η and the bond i in the rotational state γ. The rotational partition function of the chain, Z, is given by (*1, 21*)

$$Z = \mathbf{U}_1 \left[\prod_{i=2}^{i=n-1} \mathbf{U}_i \right] \mathbf{U}_n \qquad (20)$$

where $\mathbf{U}_1 = \text{row}[1\ 0\ 0.\ ^\nu.0]$ and $\mathbf{U}_n = [1\ 1.^\nu.1]$ convert the matrix product in the scalar Z. By defining the supermatrix (*21*)

$$\mathbf{G}_i = (\mathbf{U}_i \otimes \mathbf{E}_5)\mathbf{A}_i \qquad (21)$$

where $\mathbf{E}_5$ is the identity matrix of order 5, the symbol $\otimes$ denotes the direct product, and $\mathbf{A}_i$ is a pseudodiagonal matrix given by

$$\mathbf{A}_i = diag[\mathbf{M}(\varphi_\alpha), \mathbf{M}(\varphi_\beta)\ldots, \mathbf{M}(\varphi_\nu)] \qquad (22)$$

the mean-square dipole moment can finally be written as

$$\langle \mu^2 \rangle = Z^{-1} \mathbf{G}_1 \left[\prod_{i=2}^{n-1} \mathbf{G}_i \right] \mathbf{G}_n \tag{23}$$

where $\mathbf{G}_1$ and $\mathbf{G}_n$ are, respectively, the first row and the last column of the matrices $\mathbf{G}$ corresponding to the first and last bonds.

The supergenerator matrix $\mathbf{G}_i$ contains the matrix $\mathbf{U}_i$ that accounts for the statistical weights of the conformational states of bond i, a coordinate transformation matrix $\mathbf{T}_i$, which depends on the skeletal bond angle and the rotational angles associated with this bond, and the specific contribution of bond i to the dipole moment of the chains. In other words, $\mathbf{G}_i$ matrices embody the geometric and energetic parameters that describe the conformations of the chains. The knowledge of the energies associated with the conformational states of the chains is paramount to calculate the mean-square dipole moments of polymers.

Conformational Energies

The location of the rotational states and the establishment of their relative energies are obtained, whenever possible, by direct spectroscopic or thermodynamic measurements, on small molecules having structural features similar to those of the chains under investigation. When information of this kind is not available, use is also made of semiempirical conformation energy calculations that provide supplementary information useful in accessing the overall conformation energy surface as a function of skeletal rotational angles.

The force field or effective potential for a chain whose conformations are defined by the rotational angles $\varphi_1, \varphi_2, \varphi_3, \ldots \varphi_n$ is given by (22)

$$
\begin{aligned}
\mathbf{E}(\varphi_1, \varphi_2, \ldots, \varphi_n) \\
= \sum (1/2) K_b [b - b_0]^2 + \sum (1/2) K [\gamma - \gamma_0]^2 \\
+ \sum (1/2) K_\varphi (1 - \cos n\varphi) + \sum_{i<j} \sum_j \left[\frac{B_{ij}}{r_{ij}^{12}} - \frac{C_{ij}}{r_{ij}^6} + \frac{q_i q_j}{\varepsilon r_{ij}} \right]
\end{aligned}
\tag{24}
$$

The first term accounts for the bond-stretching energy along the bond b, the parameter b_0 and K_b representing, respectively, the minimum energy bond length and the force constant. The contribution of the distortions of the skeletal bonds to the potential is given by the second term, where γ_0 is the minimum energy skeletal bond angle. The third term represents the torsional contribution about the bonds where K_φ is the height barrier energy separating rotational states. The fourth term includes both the van der

Waals interactions of nonbonded atoms and the coulombic contribution of the residual charges to the potential. The attractive parameters C_{ij} in the Lennard–Jones potential function are usually obtained from the Slater–Kirkwood equation (23), whereas the repulsive terms B_{ij} are assigned to minimize the potential V_{ij} for a given pair of atoms when their distance r_{ij} is set equal to the sum of the corresponding van der Waals radii. The average energy for a potential well associated with the pair of bonds $i - 1$ and i is given by

$$\langle E(\varphi_{i-1},\varphi_i)\rangle = \frac{\sum\sum E(\varphi_{i-1},\varphi_i)\,\exp\left(-\dfrac{E(\varphi_{i-1},\varphi_i)}{RT}\right)}{\sum\sum \exp\left(-\dfrac{E(\varphi_{i-1},\varphi_i)}{RT}\right)} \qquad (25)$$

In the theoretical evaluation of conformation-dependent properties use is often made of the conformational energies corresponding to two consecutive skeletal bonds, which can be calculated by means of eq 25 assuming that the rest of the bonds remain in trans conformation. Although this semiempirical method permits the prediction of reliable values for the conformational energies of n-alkanes, including polyethylene, the reliability of this approach is dubious in systems in that sulfur, oxygen, and halogen atoms are present in the chains. In these cases, more reliable values of the conformational energies can be obtained by critical analysis of the configurational properties that are sensitive to a particular conformation.

Dipolar Correlation Coefficient

For a freely jointed chain of n skeletal bonds there is no correlation between the orientations, that is, any value of γ_{ij} from 0 to 2π has the same probability of occurrence and the average of its cosine in eq 18 vanishes. Consequently, this equation can be written as

$$\langle\mu^2\rangle = \sum_{i=1}^{n} m_i^2 = nm^2 \qquad (26)$$

where $m^2 = \sum_{i=1}^{n} m_i^2/n$. We should note that as occurs with the molecular dimensions, $\langle\mu^2\rangle \sim n$. The intramolecular dipolar correlation coefficient g_{intra}, also named dipole moment ratio, can be defined as the ratio between the actual value of the mean-square dipole moment and that of a freely joined chain ($\langle\mu^2\rangle/nm^2$). Accordingly,

$$g_{intra} = \frac{\langle\mu^2\rangle}{\sum_1^n m_i^2} = 1 + \left\langle\sum\sum_{i\neq j}\cos\gamma_{ij}\right\rangle \qquad (27)$$

if all dipole entities have the same value. Because for a freely jointed chain $g_{intra} = 1$, the larger the departure of this coefficient from 1, the higher the correlation between the dipole entities of the chains. Dipole moments obtained from dielectric measurements carried out on the bulk involve both intramolecular and intermolecular correlations. The global dipolar correlation coefficient, g, is given by the following expression (24)

$$ g = 1 + \left\langle \sum_{i \neq j} \sum \cos \gamma_{ij} \right\rangle^{\text{intra}} + \left\langle \sum_{k \neq l} \sum \cos \gamma_{kl} \right\rangle^{\text{interm}} \tag{28} $$

Because the dipolar correlation decreases rapidly as the distance between the correlating entities increases, it is expected that for polymer chains $g \cong g_{intra}$.

Dependence of Dipolar-Correlation Coefficient and Dipole Moments on Structure of Polymer Chains

Poly(oxides or ethers) and Poly(sulfides or thioethers)

Linear polyethers with structural unit $[(CH)_2]_z$—O— are an important and interesting kind of polymers to obtain information on the relationship between structure and property in molecular chains. The dipole moment associated with CH_2—O bonds lies along the bond and its value is 1.07 D (25, 26), whereas the dipole associated with the CH_2—CH_2 bond can be considered to be zero. The experimental values of the dipolar correlation coefficient for different polyoxides are shown in Table I

The statistics of these chains can be described by means of a 3×3 rotational states scheme, and rotational angles are located in the vicinity of 0° (trans, t), 120° (gauche positive, g^+), and −120° (gauche negative, g^-). Values of about 110° were used for the skeletal bond angles of these chains.

The first member of the series, polymethylene oxide (PMO), has an anomalous high melting point and it is insoluble in most organic solvents. Information about the conformational characteristics of this polymer was obtained from the critical interpretation of the dipole moment of dimethoxy methane (CH_3—O—CH_2—O—CH_3), a low molecular weight analog of the polymer. From the values of g_{intra} calculated as a function of the energy of gauche states about CH_2—O bonds, $E_{\sigma\eta}$, one finds that agreement between theoretical and experimental results is obtained assuming

Table I. Experimental Values of g_{intra} for Several Polyoxides at 25 °C

Polyoxide	g_{intra}	Refs.
PMO		
$-CH_2O-$	0.30	27
PEO		
$-(CH)_2O-$	0.51	12, 28, 29
Poly(trimethylene oxide) (P3MO)		
$-(CH_2)_3O-$	0.41	30
Poly(tetramethylene oxide) (P4MO):		
$-(CH_2)_4O-$	0.50, 0.53	16, 26
Poly(hexamethylene oxide) (P6MO):		
$-(CH_2)_6O-$	0.54	29
Isotactic poly(propylene oxide)		
$-CH(CH_3)CH_2O-$	0.50, 0.45	31, 32
Poly(3,3-dimethyl oxetane)		
$-CH_2C(CH_3)_2CH_2O-$	0.20, 0.25	33, 34
Poly(2-methyl oxetane)		
$-OCH(CH)_3(CH_2)_2-$	0.35	35
Poly(3-methyl oxetane)		
$-OCH_2CH(CH_3)CH_2-$	0.35	36
Poly(3-methyl tetrahydrofuran)		
$-OCH_2CH(CH_3)(CH_2)_2-$	0.53	37

that $E_{\sigma\eta}$ has a value of 1.4 kcal mol^{-1} below that of the alternative trans states (26). Calculations of $E_{\sigma\eta}$ using Buckingham potential functions give for this quantity a value of only 0.3 kcal mol^{-1} below that of the corresponding trans states. The difference between the value of $E_{\sigma\eta}$ calculated from semiempirical potential functions and that obtained from the analysis of conformation-dependent properties, customarily called *gauche effect*, is about 1.1 Kcal mol^{-1}. The dipolar correlation coefficient of PMO, calculated by using the energy indicated previously, amounts to 0.1. This value is significantly lower than 0.30, the experimental result obtained at 200 °C from dielectric measurements on the melt (27).

The next member of the family of linear polyoxides, polyethylene oxide (PEO), is a semicrystalline polymer that melts at 67 °C and is readily soluble in nonpolar solvents. The experimental value of g_{intra} at 30 °C is 0.51 (12, 28, 29). The intramolecular energies required in the theoretical evaluation of the dipole moments of the chains were calculated by using Buckingham potential functions. Gauche states about CH_2—O and CH_2—CH_2 bonds have energies of 0.9 ($E_{\sigma''}$) and 0.5 ($E_{\sigma'}$) kcal mol^{-1}, respectively, above those of the alternative trans states. Moreover rotations of the same sign about the two consecutive OCH_2—CH_2—OCH_2 bonds produce pentane type CH_2••••O interactions whose energy, E_ω, is 0.56 kcal mol^{-1} above that of the tt conformation. The rotational states about CH_2—CH_2 and CH_2—O bonds are located at 0, $\pm120°$, and 0, $\pm110°$, respectively. The

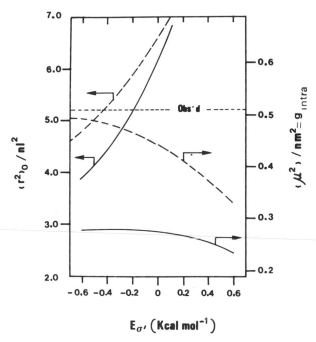

Figure 2. Characteristic ratio ($\langle r^2 \rangle / nl^2$) and the dipolar correlation coefficient g_{intra} ($\langle \mu^2 \rangle / nm^2$) for PEO at 30 °C and 20 °C, respectively, as a function of the energy $E_{\sigma'}$, associated with gauche states about C—C bonds. Solid lines: $E_{\sigma''} = 1.3$ kcal/mol, $E_\omega = 0.4$ kcal/mol, $\phi_g(C—O) = \pm 100°$, $\phi_g(C—C) = \pm 120°$. Dashed lines: $E_{\sigma''} = 0.9$ kcal/mol, $E_\omega = 0.4$ kcal/mol, $\phi_g(C—O) = \phi_g(C—C) = \pm 120°$. Horizontal lines represent experimental values. (Reproduced from reference 26. Copyright 1976 American Chemical Society.)

value of g_{intra} for PEO chains, calculated by using the set of statistical parameters previously given, is only 50% of the experimental result (*26*). Agreement between the theoretical and experimental results can only be reached by increasing the gauche population about CH_2—CH_2 bonds. In this case, not only the dipole moments of the chains but also other conformation-dependent properties can be reproduced, as can be seen in Figure 2 where both g_{intra} (= $\langle \mu^2 \rangle / nm^2$) and the characteristic ratio C_n(= $\langle r^2 \rangle_0 / nl^2$) are represented as a function of the conformational energy $E_{\sigma'}$ about CH_2—CH_2 bonds. It can be seen in this figure that the only way of reproducing both the characteristic ratio and the dipolar correlation coefficient for this polymer is to assume that $E_{\sigma'}$ has a value nearly 0.5 kcal mol^{-1} below that of the corresponding trans state (*26*). Here the gauche effect is only slightly lower than that obtained for PMO.

Conformational energies arising from first-order interactions between methylene groups and oxygen atoms were also investigated for polyoxides

Table II. Gauche Effect (Kcal/mol^{-1}) for Conformations Involving Oxygen Atoms in Various Polyoxides

Polyoxide	Bond	$E_{exp}-E_{calc}$
PMO	OC–OC	1.1
PEO	OC–CO	1.0
P3MO[a]	OC–CC	0.3
P4MO[a]	OC–CC	0.2

[a] See Table I for abbreviations.
SOURCE: Data from reference 26.

in which the number of methylene groups in the repeating unit z is larger than 2 (*16, 26, 29, 30*). In general, the difference between the values obtained from the analysis of the dipole moment and those calculated by using semiempirical potential functions decreases as z increases. As can be seen in Table II, the difference $E_{exp} - E_{calc}$ is maximum for PMO and it is nearly zero for polyoxides in which z > 4. Other polymers such as poly(propylene oxide) (*31, 32*), symmetric (*33, 34*) and asymmetric (*35, 36*) polyoxetanes, and poly(3-methyltetrahydrofuran) (*37*) have also been studied, and by comparing the theoretical and experimental dipole moments information was obtained about the conformational energies associated with the rotational states of different skeletal bonds in these polymers.

The reliability of the energies obtained from the critical analysis of the conformation-dependent properties of PMO and PEO can be checked by means of the theoretical evaluation of the dipolar correlation coefficient of the polyformals with repeating unit —O—CH$_2$—O(CH$_2$)$_z$—. These polymers, which can be considered alternating copolymers of oxymethylene and oxyalkylene segments, have dipole moments strongly dependent on the value of z. As indicated before, there is little information on chain conformations for the first member of the series, PMO, owing to its high melting point (210 °C), which renders this polymer insoluble in most organic solvents. The second member of the series, poly(1,3-dioxolane) (z = 2) (PXL), has a relatively low melting point and is readily soluble in a number of common organic solvents. Moreover PXL is an alternating copolymer of methylene oxide and ethylene oxide so that the study of this polymer can give information on the statistics of PMO as well. Bonds of type a and e in the repeating unit of PXL (Figure 3) are similar to those

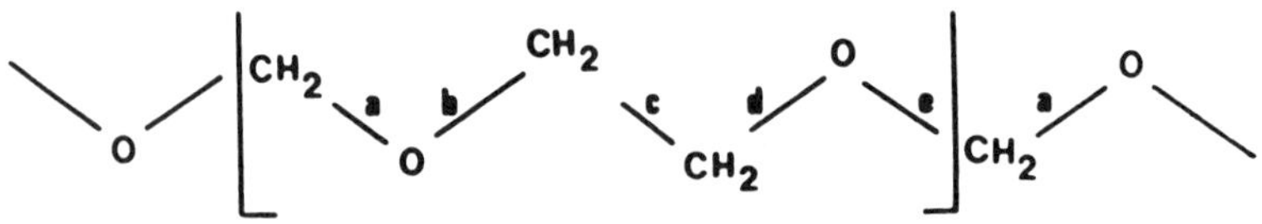

Figure 3. Repeating unit of poly(1,3-dioxolane) in all-trans conformation.

of PMO whereas bonds of type c, d, and b are similar to those of PEO. The statistical weight matrices corresponding to the skeletal bonds of the repeating unit of PXL are (*38*)

$$U_a = U_e = \begin{pmatrix} 1 & \sigma_\eta & \sigma_\eta \\ 1 & \sigma_\eta & 0 \\ 1 & 0 & \sigma_\eta \end{pmatrix}; \qquad U_b = U_d = \begin{pmatrix} 1 & \sigma'' & \sigma'' \\ 1 & \sigma'' & \omega\sigma'' \\ 1 & \omega\sigma'' & \sigma'' \end{pmatrix}$$

$$U_c = \begin{pmatrix} 1 & \sigma' & \sigma' \\ 1 & \sigma' & \sigma'\omega \\ 1 & \sigma'\omega & \sigma' \end{pmatrix} \tag{29}$$

Values of g_{intra}, calculated by using the conformational energies that give a good description of the conformation-dependent properties of PEO, are represented as a function of E in Figure 4. Good agreement between theory and experiment is found when the conformational energy associated with gauche states about the acetalic CH_2—O bonds is about 1.2 kcal mol^{-1} below that of the alternative trans states. This value is in very good agreement with the value obtained for this energy from the analysis of the dipole moments of dimethoxymethane, but is in clear disagreement with the value

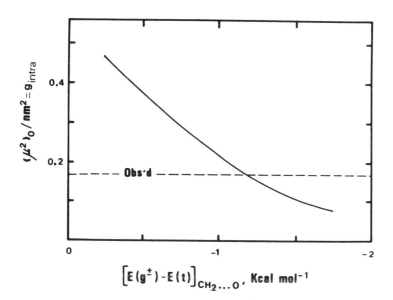

Figure 4. Dependence of intramolecular dipolar correlation coefficient g_{intra} *($\langle\mu^2\rangle$/ nm^2) for PXL on the conformational energy of gauche states about bonds of type a and e (Figure 3) in which the interacting species are CH$_3$ groups and O atoms (T = 30 °C). (Reproduced from reference 38. Copyright 1978 American Chemical Society.)*

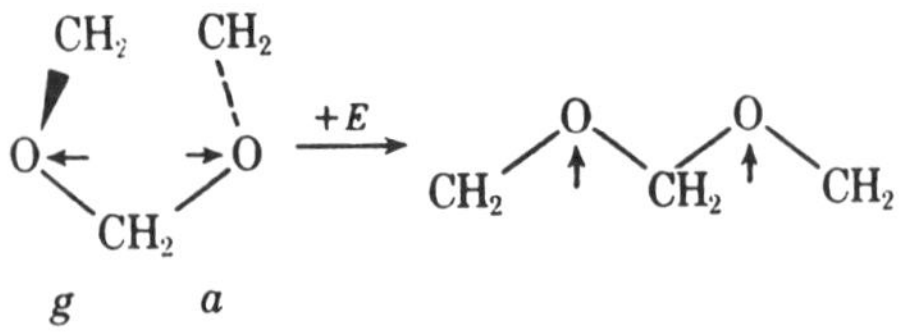

Figure 5. Dipole orientation for tt and g^+g^+ conformations in the CH_2O—CH_2—OCH_2 acetalic residue.

of -0.3 kcal mol^{-1} calculated for this energy by using semiempirical potential functions.

The theoretical value of the temperature coefficient of the dipole moments of PXL chains, expressed as d $\ln\langle\mu^2\rangle/dT$, amounts to 6.1×10^{-3} K^{-1}, in excellent concordance with the experimental result (6.0×10^{-3} K^{-1}). This is one of the highest values reported for the temperature coefficient of any configuration-dependent property. The relatively low value of the dipolar correlation coefficient and its temperature coefficient may be explained in simple molecular terms by the large population of gauche states about the acetalic skeletal bonds. Gauche states of the same sign about two consecutive acetalic bonds place the dipole moments in nearly antiparallel direction and hence the low value of g_{intra}. An increase in temperature causes an increase in the fraction of tt conformations in which the dipoles are in parallel direction, and consequently, the dipole moments of PXL chains exhibit a very large temperature dependence (Figure 5). The nature of the oxyalkylene segments in the polyformals does not seem to affect the polarity of these polymers, as can be seen in the third column of Table III where the experimental values of the dipolar correlation coefficient at 30 °C are represented for PMO ($z = 1$) (27), PXL ($z = 2$) (38), poly(1,3-dioxopane) ($z = 4$) (39), poly(1,3-dioxocane) ($z = 5$) (40), and poly(1,3-dioxonane) ($z = 6$) (40). The fourth column of this table shows

Table III. Experimental and Theoretical Dipolar Correlation Coefficients for
Several Polyformals at 30 °C

Polymer	Repeating Unit	$(g_{intra})_{exp}$	$(g_{intra})_{calc}$
POM	$-CH_2O-$	0.3^a	0.12
PXL	$-CH_2O(CH_2)_2O-$	0.17	0.17
PXP	$-CH_2O(CH_2)_4O-$	0.16	0.16
PXC	$-CH_2O(CH_2)_5O-$	0.17	0.18
PXN	$-CH_2O(CH_2)_6O-$	0.18	0.18

NOTE: PXP is poly(1,3-dioxopane), PXC is poly(1,3-dioxocane), and PXN is poly(1,3-dioxonane).
^{a} Data from reference 27.

the values calculated for g_{intra} by using the value of -1.2 kcal mol^{-1} for the energy associated gauche states about CH_2O—CH_2O bonds in polyformals. The very good agreement between theory and experiment supports the reliability of the value obtained for the energy associated with gauche states about acetalic bonds from the critical analysis of the dipole moment ratio of PXL chains.

The schematic substitution of the oxygen atoms in aliphatic polyethers for sulfur atoms leads to the series of polythioethers with repeat unit —$S(CH_2)_z$—. The first member of the series, poly(methylene sulfide) (PMS), has not yet been studied, largely because its rather high melting temperature (257 °C) and its instability above the melting point render it difficult to handle experimentally. The second member of the series, poly(ethylene sulfide) (PES), is also a crystalline polymer with high melting point (210 °C), and like PMS, PES is insoluble in common organic solvents. The polythioethers with $z \geq 3$ are already soluble in common organic solvents. The conformational characteristics of both polyethers and polythioethers differ as a consequence of the larger dipole moment of C—S bonds (1.21 D) in comparison with that of C—O bonds (1.07 D). Moreover, the C—S bond length (1.821 A) is about 30% larger than the C—O bond length (1.43 Å), the CSC bond angle (100°) is nearly 10° smaller than the COC bond angle (110°), and the van der Waals radius of the sulfur atom is about 20% larger than that of the oxygen atom (2). The information at hand suggests that the lower the number of methylene groups in the repeating unit, the higher the difference between the physical properties of polyethers and poly(thioethers). Information on the conformational characteristics of PMS and PES was obtained from the critical interpretation of the dipolar correlation coefficient of molecular chains and model compounds with either CH_2S—CH_2—SCH_2 or SCH_2—CH_2S sequences in their structure. The principal conclusions obtained from the comparison of experimental and theoretical dipole moments are:

1. Gauche states about SCH_2—SCH_2 bonds have an energy about 1.2 kcal mol^{-1} below that of the corresponding trans states. Because the value of this energy calculated by semiempirical potential functions amounts to -0.36 kcal mol^{-1}, the indicated moiety presents a strong gauche effect. In this aspect the conformational characteristics of the SCH_2—SCH_2 and OCH_2—$OSCH_2$ sequences are similar (41).

2. Gauche states about SCH_2—CH_2S bonds, which give rise to first-order interactions between two sulfur atoms, have an energy about 0.4 kcal mol^{-1} above that of the alternative trans states (42, 43). As indicated before, similar states about OCH_2—CH_2O bonds in

Table IV. Experimental Value of g_{intra} for Several Polysulfides at 30 °C

Polymer	Repeating Unit	g_{intra}	Ref.
Poly(trimethylene sulfide)	$-S-(CH_2)_3-$	0.61	45
Poly(pentamethylene sulfide)	$-S-(CH_2)_5-$	0.76	44
Alternating copolymer of methylene sulfide and pentamethylene sulfide	$-S-CH_2-S-(CH_2)_5-$	0.26	41
Poly(propylene sulfide)	$-S-CH(CH_3)CH_2-$		
atactic		0.44	
isotactic		0.49	19
Poly(3,3-dimethylthiethane)	$-SCH_2C(CH_3)_2CH_2-$	0.62	46

PEO have an energy about 0.5 kcal mol^{-1} below that of the corresponding trans states.

3. Whereas gauche states about CH_2CH_2—OCH_2CH_2 bonds in polyoxides are strongly disfavored with respect to the trans states, these states about CH_2CH_2—SCH_2CH_2 bonds are slightly favored over the corresponding trans states (43).

4. Finally gauche states about CH_2CH_2—CH_2S bonds that produce first-order interactions between a sulfur atom and a methylene group have an energy about 0.6 kcal mol^{-1} above that of the alternative trans states (44). On the contrary, gauche states about CH_2CH_2—CH_2O bonds are slightly favored over the corresponding trans states.

The values of the dipolar correlation coefficient for several polysulfides are shown in Table IV. In general, polysulfides exhibit higher dipole moments than their polyoxide counterparts.

Polyesters and Polyamides

Linear aromatic polyesters, with repeating unit —$OOC(C_6H_4)COO$-$(CH_2)_z$—, have not been extensively studied with regard to their conformational-dependent properties. Only the second member of the series, poly-(ethylene terephthalate) (PET), has been characterized with respect to its spatial extension (1). The mean-square dipole moments of these polyesters were not measured largely because the high melting temperatures of the lower members of the series preclude the solubility of these polymers in nonpolar solvents. However, the introduction of oxygen atoms in the glycol residue decreases the melting point of the polyesters and improves their solubility. The dipolar correlation coefficients for several polyesters are

Table V. Experimental Values of g_{intra} for Polyesters at 30 °C

Polymer	Repeating Unit	g_{intra}	Ref.
Poly(diethylene glycol terephthalate)	$-OOCC_6H_4COO(CH_2)_2O(CH_2)_2-$	0.66	47
Poly(diethylene glycol isophthalate)	$-OOCC_6H_4COO(CH_2)_2O(CH_2)_2-$	0.70	48
Poly(diethylene glycol phthalate)	$-OOCC_6H_4COO(CH_2)_2O(CH_2)_2-$	0.64	49
Poly(propylene glycol terephthalate)	$-OOCC_6H_4COO(CH_2)CH(CH_3)-$	0.58	50
Poly(dipropylene glycol terephthalate)	$-OOCC_6H_4COOCH(CH_3)CH_2OCH(CH_3)-$	0.58	51
Poly(ethylene terephthalate)[a]	$-OOCC_6H_4COOCH(CH_2)_2-$	0.53	50
Poly(triethylene glycol terephthalate)	$-OOCC_6H_4COO(CH_2)_2O(CH_2)_2-O(CH_2)_2-$	0.68	52
Poly(ditrimethylene glycol terephthalate)	$-OOCC_6H_4COO(CH_2)_3O(CH_2)_3-$	0.81	53

[a] Calculated value.

shown in Table V. In polyesters based respectively on both terephthalic and isophthalic acids, coplanarity between the carbonyl and phenyl groups guarantees maximum overlapping of electrons of these groups; therefore, the rotational angles about C^{Ph}—CO bonds are restricted (1) to 0° and 180°. On the contrary, large repulsive intramolecular interactions between the two ester groups of the phthaloyl residue overcome the stabilizing effects of the coplanarity between the carbonyl and phenyl groups, and the rotational angles about C^{Ph}—CO bonds are $\pm 90°$ (49). The dipole vector associated with each ester group has a value of 1.89 D, and its direction makes an angle of 123° with the C^{Ph}—CO bond (54), whereas the dipole moment of the CH_2—O bond of the ether group has a value of 1.07 D.

In the analysis of the dipolar correlation coefficient for polyesters with ether functions in the glycol residue, use can be made of the conformational information collected for polyethers. As an example, the critical interpretation of the dipolar correlation coefficient of poly(ditrimethylene glycol terephthalate) (PDTT) will be described. With the exception of the O—CO bonds of the ester groups, which are restricted to trans states, and the virtual bond of the terephthaloyl residue in which the potential minima are located at 0° and 180°, the rotational angles of the remaining skeletal bonds of the repeating unit of PDTT (Figure 6) are assumed to be 0° and $\pm 120°$. The statistical weight matrices corresponding to the skeletal bonds of the repeating unit are (53)

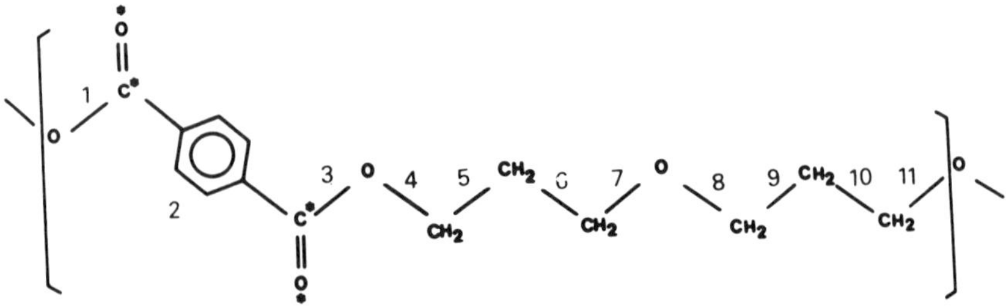

Figure 6. Repeating unit of PDTT in its planar all-trans conformation.

$$U_1 = (1); \qquad U_2 = (1 \; 1); \qquad U_3 = \begin{pmatrix} 1 \\ 1 \end{pmatrix}; \qquad U_4 = (1 \; \sigma_k \; \sigma_k)$$

$$U_5 = \begin{pmatrix} 1 & \sigma' & \sigma' \\ 1 & \sigma' & \sigma'\omega_k \\ 1 & \sigma'\omega_k & \sigma' \end{pmatrix}; \qquad U_6 = U_{10} = \begin{pmatrix} 1 & \sigma' & \sigma' \\ 1 & \sigma' & \sigma'\omega \\ 1 & \sigma'\omega & \sigma' \end{pmatrix};$$

$$U_7 = U_8 = \begin{pmatrix} 1 & \sigma'' & \sigma'' \\ 1 & \sigma'' & 0 \\ 1 & 0 & \sigma'' \end{pmatrix}; \qquad U_9 = \begin{pmatrix} 1 & \sigma' & \sigma' \\ 1 & \sigma' & 0 \\ 1 & 0 & \sigma' \end{pmatrix}; \qquad (30)$$

$$U_{11} = \begin{pmatrix} 1 & \sigma_k & \sigma_k \\ 1 & \sigma_k & \sigma_k\omega_k \\ 1 & \sigma_k\omega_k & \sigma_k \end{pmatrix}$$

The critical analysis of the unperturbed dimensions of PET (*1*) and both the dipole moments and the temperature coefficient of the unperturbed dimensions of poly(diethylene glycol terephthalate) (*47*) suggest that gauche states about O—CH$_2$ bonds of the ester residue have an energy 0.3–0.4 kcal mol^{-1} above that of the corresponding trans states. The rest of the conformational energies used to determine the statistical weight matrices were taken from the conformational analysis of polyethers. Specifically, the conformational energies (kcal mol^{-1}) used were: $E_{\sigma k} = 0.31$, $E_{\sigma'} = -0.15$, $E_{\sigma''} = 0.9$, $E_\omega = 0.45$, and $E_{\omega k} = 1.4$.

An inspection of the structure of PDTT reveals that the dipole moments of the chains should be very sensitive to the conformational energy of gauche states about CH$_2$—CH$_2$ bonds, $E_{\sigma'}$, as a consequence of the fact that the larger the value of this energy, the lower the fraction of all trans conformation of the glycol residue in which the dipoles are in almost anti-parallel direction. The changes on g_{intra} with $E_{\sigma'}$, depicted in Figure 7, show that the dipolar correlation coefficient decreases as the trans population about CH$_2$—CH$_2$ bonds increases. For example, g_{intra} increases from 0.70 to 1 kcal mol^{-1} when $E_{\sigma'}$ changes from 0.15 to -1 kcal mol^{-1}; as can be seen in Figure 7, agreement between theoretical and experimental re-

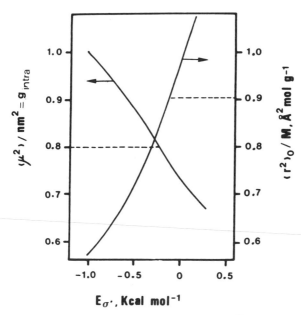

Figure 7. Dependence of both the molecular dimensions ($\langle \gamma^2 \rangle /M$) and the dipolar correlation coefficient g_{intra} ($\langle \mu^2 \rangle /nm^2$) for PDTT on the conformational energy E_σ, associated with gauche states about CH_2CH_2—CH_2O bond of the glycol residue. Horizontal lines represent the experimental values. (Reproduced from reference 53. Copyright 1988 American Chemical Society.)

sults is found for $E_{\sigma'} = -0.2$ kcal mol^{-1}. The statistics that describe the dipole moments of PDTT also give a good account of the unperturbed dimensions of these chains. The variation of $\langle r^2 \rangle_0 / M$ with $E_{\sigma'}$, where $\langle r^2 \rangle_0$ is the unperturbed mean-square end-to-end distance and M is the molecular weight, is also given in Figure 7. Concordance between theory and experiment is found for $E_{\sigma'} = -0.15$ kcal mol^{-1}, a value remarkably close to that found for this quantity in the analysis of the dipole moment of the chains (*53*). The conformational characteristics of cycloaliphatic and aliphatic polyesters in connection with their polarity have also been studied (*55*).

Studies on the polarity of polyamides in solution are scarce due to the difficulty of finding suitable solvents for the dielectric measurements and the ability of these polymers to form associates that can be of a different kind depending on the type of solvent, concentration, etc. The evaluation of the polarity of the chains requires knowledge of the modulus and orientation of the dipole corresponding to the amide group. The values of the experimental dipole moments, $\langle \mu^2 \rangle^{1/2}$, of *N*-methylacetamide, *N,N'*-dimethylalkoxamide, *N,N'*-dimethylmalonamide, *N,N'*-dimethylterephthalamide, and *N,N'*-dimethyl-*trans*-1,4-cyclohexanedicarboxamide amount to 4.22 ±

0.02, 1.27 ± 0.03, 3.28 ± 0.03, 4.01 ± 0.03, and 3.60 ± 0.05 D, respectively
(56). The critical interpretation of the dipole moments of these model
compounds permits the conclusion that the dipole moment of the amide
group is 3.99 D and it makes an angle $\tau = 115°$ with the CH_3—CO bond
(56–58).

The conformational characteristics of polyamides were recently investigated (59) on 1,5-dibenzamido-3-oxapentane (DEBA) and
1,8-dibenzamido-3,6-dioxaoctane (TEBA), model compounds of polyamides with repeating units [—$HNCOC_6H_4CONH(CH_2)_2O(CH_2)_2$—] and
[—$HNCOC_6H_4CONH(CH_2)_2O(CH_2)_2O(CH_2)_2$—], respectively. The
mean-square dipole moments for DEBA and TEBA lie in the ranges
24.6–24.5 D^2 and 25.4–26.0 D^2, respectively, at 30–60 °C. The dipole moments of these compounds were critically investigated by considering that
the NH—CO bond, like the O—CO bond in polyesters, is restricted to
trans states. Even though the polarity of these model compounds is very
sensitive to the modulus of the dipole associated with the amide groups, it
is rather insensitive to its orientation for values of 110–125°. An important
conclusion of this study is that gauche states about NH–CH_2 bonds of the
amide group are preferred over the alternative trans states, in sharp contrast
with CH_2—O bonds of the ester group in polyesters where the opposite
occurs (59).

Silicon-Based Polymers

Because of the large atomic polarization of these molecular chains, this
contribution to the total polarization has to be considered in the evaluation
of the dipole moment of silicon-based polymers. For the best known polymer of this family, poly(1,1-dimethyl siloxane) [—$Si(CH_3)_2O$—] (PDMS),
the molar atomic polarization (P_A) in cubic centimeters can be written as
(60)

$$P_A = 7.81 + 5.566x \tag{31}$$

where x is the degree of polymerization. In general these polymers exhibit
a low polarity (61–63), as indicated in Figure 8 where values of the dipolar
correlation coefficient for different fractions of PDMS are shown. The values of g_{intra} were obtained assuming that the dipole moment associated
with each skeletal bond is 0.60 D. The high preference for trans states of
the skeletal bonds may be held responsible for the relatively low polarity
exhibited by the chains. Actually, because of the large difference of the
skeletal bond angles of the repeating unit, displaying the chains in all trans
conformation gives a closed polygon that places the dipoles of the repeating
units in nearly antiparallel direction. Low polarity is also exhibited by sili-

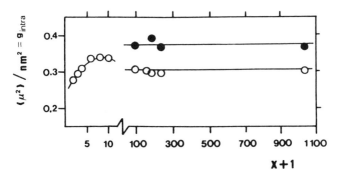

Figure 8. Experimental values for the dipolar correlation coefficient g_{intra} ($\langle\mu^2\rangle/nm^2$) for dimethyl siloxane chains as a function of the degree of polymerization. Filled circles refer to dielectric data obtained in cyclohexane at 25 °C. Open circles represent values of g obtained on the undiluted chains at the same temperature. (Reproduced from reference 63. Copyright 1974 American Chemical Society.)

con-based polymers with asymmetric centers in their structure such as poly-(methylphenyl siloxane) and copolymers of dimethylsiloxane–methylphenyl siloxane (*64*).

By schematically substituting the oxygen atom of the repeating unit of poly(dimethylsiloxane) by a methyl group, poly(1,1-dimethyl silmethylene) $[-Si(CH_3)_2-CH_2-]_x$ is obtained. Although trans states about the skeletal bonds of these molecules are more prevalent than in polysiloxanes, the dipolar correlation coefficient is somewhat larger than that of the latter polymers (*65*). The cause of this behavior lies in the near equivalence of the bond angles in the Si—C backbone that gives rise to long trans sequences that approximate a rectilinear zig-zag conformation in which the dipole moments associated with the repeating unit are in parallel direction and therefore have a much higher moment than the corresponding trans sequences in polysiloxanes. Low dipole correlation coefficient is also exhibited by poly(dimethyl silazane) polymers (*66*), which have a repeating unit $[-Si(CH_3)_2-NH-]$.

Polymers with Polar Side Groups

Polymers with polar side groups are an interesting kind of substances to study the relationship between structure and property in molecular chains. The dipolar correlation coefficient is in general dependent on the stereochemical structure. The polar side groups can be either rigidly attached to the backbone as occurs in poly(vinyl chloride), poly(vinyl bromide), poly(*p*-chlorostyrene), poly(vinyl carbazole), poly(*p*-bromostyrene), etc., or incorporated to flexible side groups as occurs in esters of acrylic and metacrylic acid. A wealth of information concerning the dipole moments in the first

Table VI. Values of g_{intra} for Some Vinyl and Acrylate Chains in the Vicinity of 30 °C

Polymer	Repeating Unit	g_{intra}	Ref.
Poly(vinyl chloride)	$-CH_2CHCl-$	0.66–0.72	67, 68
Poly(vinyl bromide)	$-CH_2CHBr-$	0.45–0.53	69, 70
Poly(p-chloro styrene)	$-CH_2CH(C_6H_5Cl)-$	0.53–0.80	18, 71–75
Poly(p-bromo styrene)	$-CH_2CH(C_6H_5Br)-$	2.20–2.44[a]	75
Poly(m-chloro styrene)	$-CH_2CH(C_6H_5Cl)-$	1.61[a]	73
Poly(vinyl carbazole)	$-CH_2CH(NC_{13}H_8)-$	0.41	76
Poly(methyl acrylate)	$-CH_2CH(COOCH_3)-$	0.68	81, 82
Poly(phenyl acrylate)	$-CH_2CH(COOC_6H_5)-$	0.62	83

[a] These values are given as $\langle \mu^2 \rangle / x$.

type of polymers can be found in the literature (67–79). Values of the dipolar correlation coefficient g_{intra} for some representative polymers are given in Table VI. A scheme of a polyvinyl chain showing meso and racemic configurations is shown in Figure 9. For polymers with planar side groups, the C—R bond is restricted to an orientation in which its plane is approximately perpendicular to the plane of the two skeletal bonds of the diad (e.g., bonds $i - 1$ and i in Figure 9) because steric repulsion by hydrogens of the adjoining methylene groups imposes this constraint. For the rotation in which the C—R bonds assume the position of the methylene in Figure 9, the planar R group impinges on one of the groups (R, CH_3, or H) pendant to the neighboring substituted carbons of the diad embracing the g bond and, consequently, this rotational state is suppressed. Therefore, an interesting feature of vinyl polymers with bulky side groups is that a 2 × 2 rotational states scheme instead of the 3 × 3 scheme is sufficient to describe their conformational properties. The statistical weight matrix for CH_2—C—CH_2 pair of bonds is conveniently expressed by

$$U' = \begin{pmatrix} 1 & 1 \\ 1 & 0 \end{pmatrix} \qquad (32)$$

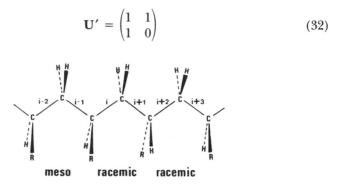

Figure 9. Segment of the vinyl chain in the planar all-trans conformation.

whereas the statistical weight matrices U_m'' and U_r'' for $C—CH_2—C$ bond pairs of meso and racemic diads, respectively, can be written as

$$
U_m'' = \begin{pmatrix} \omega'' & \dfrac{1}{\eta} \\[2mm] \dfrac{1}{\eta} & \dfrac{\omega}{\eta^2} \end{pmatrix}; \quad
U_r'' = \begin{pmatrix} 1 & \dfrac{\omega'}{\eta} \\[2mm] \dfrac{\omega'}{\eta} & \dfrac{1}{\eta^2} \end{pmatrix}
\tag{33}
$$

where the statistical weights have been normalized to unity for racemic tt, and the first-order parameter weights show the preference of trans over gauche. The second-order parameters ω, ω', and ω'' count for the repulsive effects between $CH_2\cdots CH_2$, $CH_2\cdots C_6H_5$, and $C_6H_5\cdots C_6H_5$, respectively. The dipole moments of poly(p-chlorostyrene) were theoretically studied by Saiz et al. (77), finding that sequences of preferred tt conformations in syndiotactic chains are preponderantly interrupted by gg conformations that place the dipole moments of the two sequences in nearly antiparallel direction so that g_{intra} should vanish in the limit $x \rightarrow \infty$. However, the occurrence of both tg and gt conformations and the fact that the dipole moments in tt conformations correlate favorably suggest that the larger the tt sequences the larger the dipole moment ratio of the chains. In isotactic chains tg and gt conformations are strongly favored over the alternative ones. Here a sequence gt of y_1 units can be followed by a sequence tg of y_2 units, the reverse combination being forbidden. Transition between tg and gt conformations requires the interposition of tt diads. The correlations between sequences are favorable and hence the polarity of isotactic chains is higher than that of syndiotactic chains.

The dipolar correlation coefficient depends on the stereochemical composition, and this dependence is strong for vinyl polymers with dipoles rigidly attached to the main chain. This effect can be seen in Table VII where values of g_{intra} for several polymers are shown. The results indicate that the influence of the tacticity of the chains on g_{intra} is rather small for chains in which the dipole entities are separated by three skeletal bonds as occurs in poly(propylene oxide) and poly(propylene sulfide) (80).

Among the polymers with polar groups incorporated to flexible side groups, the esters of polyacrylic and polyithaconic acids stand out. Theoretical and experimental studies on the mean-square dipole moments for these chains can be found in references 81–88 and 89 and 90, respectively. Here, the theoretical evaluation of the dipolar correlation coefficient of esters of poly(acrylic acid) will be emphasized. The contribution of each repeating unit to the polarity of molecular chains, such as poly(methyl acrylate), poly(cyclohexyl acrylate), and poly(phenyl acrylate), is the dipole associated with the ester group that forms an angle of $123°$ with the $C^\alpha—C^*O^*$ bond. The $C^\alpha—C^*O^*$ bond is restricted to two rotational states, $\tau = 0°$ and $180°$,

Table VII. Dependence of g_{intra} on the Tacticity for Several Polymers

Polymer	Repeating Unit	g_{intra}
Poly(vinyl chloride)	$-CH_2CHCl-$	
Isotactic		0.59
Atactic		1.0
Syndiotactic		4.0
Poly(p-chloro styrene)	$-CH_2CH(C_6H_4Cl)-$	
Isotactic		5.5
Atactic		0.65
Syndiotactic		0.92
Poly(propylene oxide)	$-CH(CH_3)CH_2O-$	
Isotactic		0.50
Atactic		0.45
Syndiotactic		0.42
Poly(propylene sulfide)	$-CH(CH_3)CH_2S-$	
Isotactic		0.41
Atactic		0.39

SOURCE: Data adapted from reference 80.

which correspond, respectively, to the conformations in which the carbonyl group is cis and trans to the methine CH bond. As in the vinyl chains, a 2 × 2 rotational states scheme would be enough to describe the statistics of the chains. However, to distinguish the two orientations of the ester group ($\tau = 0°$, $180°$), each of the two states of the skeletal bonds (t and g) are split into two. Consequently, the appropriate statistical weight matrices in the order (t, X = 0), (t, X = π), (g, X = 0), and (g, X = π) are (83–91)

$$U' = \begin{pmatrix} 1 & 0 & 1 & 0 \\ 0 & \rho & 0 & \rho \\ 1 & 0 & 0 & 0 \\ 0 & \rho & 0 & 0 \end{pmatrix} \tag{34}$$

for the CH_2—C^α—CH_2 bond and

$$U''_r = \begin{pmatrix} 1 & \gamma_1 & 0 & 0 \\ \gamma_1 & \gamma_2 & 0 & 0 \\ 0 & 0 & 0 & 0 \\ 0 & 0 & 0 & 0 \end{pmatrix}; \quad U''_m = \begin{pmatrix} 1 & \gamma & \beta & \beta \\ \gamma & 1 & \beta & \beta \\ \beta & \beta & 0 & 0 \\ \beta & \beta & 0 & 0 \end{pmatrix} \tag{35}$$

for racemic and meso configurations of the C^α—CH_2—C^α bond pair, respectively.

The theoretical dipolar correlation coefficient can be written as (1, 21)

$$g_{intra} = \frac{1}{Zx\mu_0^2} P_1 \left(\prod_{k=2}^{k=x-1} S'P_k \right) P_x \tag{36}$$

where Z is the rotational partition function, x is the degree of polymerization, and μ_0 is the dipole moment associated with the side group. Moreover, $\mathbf{S}' = \mathbf{U}' \otimes \mathbf{E}_5$, where the symbol $\otimes$ means direct product, $\mathbf{E}_5$ is the identity matrix of order 5, and $\mathbf{P}_k$ can be written as

$$
\mathbf{P}_k = \begin{pmatrix} \mathbf{M}'(t,\chi=0) & & & \\ & \mathbf{M}'(t,\chi=\pi) & & \\ & & \mathbf{M}'(g,\chi=0) & \\ & & & \mathbf{M}'(g,\chi=\pi) \end{pmatrix}_k \times (\mathbf{U}_k'' \otimes E_5)
$$

$$
\times \begin{pmatrix} \mathbf{M}''(t,\chi=0) & & & \\ & \mathbf{M}''(t,\chi=\pi) & & \\ & & \mathbf{M}''(g,\chi=0) & \\ & & & \mathbf{M}''(g,\chi=\pi) \end{pmatrix}_k
$$

(37)

$\mathbf{P}_1$ and $\mathbf{P}_x$ are, respectively, 1×20 row- and 20×1 column vectors that convert the matrix product of eq 36 into the scalar g_{intra}

$$
\mathbf{P}_1 = \text{row}(\mathbf{M}_{11},\mathbf{0},\mathbf{0},\mathbf{0}) \tag{38}
$$

$$
\mathbf{P}_x = \text{column}(\mathbf{M}_x(\chi = 0), \mathbf{M}_x(\chi = \pi), \mathbf{M}_x(\chi = 0), \mathbf{M}_x(\chi = \pi))
$$

where $\mathbf{M}_{11}$ and $\mathbf{M}_x$ represent the first row and last column of the matrix $\mathbf{M}$ defined by eq 16.

The evaluation of the polarity of the chains by the statistics described previously requires the determination of the components of the ester group in the reference frame of the C^α—CH_2 bonds of the repeating unit. For example, in the case of poly(cyclohexyl acrylate) the dipole moment of the ester group was taken to be that of the cyclohexyl isobutyrate whose experimental value amounts to 1.93 D. By assuming that the dipole moment forms an angle of 123° with the C^α—$C*O*$ bond, its components in the reference frame of the C^α—CH_2 bonds of the main chain are (84): $m_x = 1.141$ D, $m_y = -1.555$ D, $m_z = -0.027$ D for X = 0 and $m_x = -0.320$ D, $m_y = 0.513$ D, $m_z = 1.844$ D for $\chi = \pi$. The bond moment associated with the CH_2—C^α skeletal bond was assumed to be nil.

A common characteristic in the conformational analysis of these chains is that the dipolar correlation coefficient is in most cases extremely sensitive to the statistical weight parameters γ_1 and γ_2, which account for the relative orientation of two consecutive ester groups in racemic diads. This result is because for most acrylate polymers the fraction of racemic diads is predominant. For example, high values of γ_2 imply that the $t\pi,t\pi$ conformations, in which the dipoles of the two consecutive ester groups are nearly in antiparallel direction, are favored in racemic diads. For example, in increasing γ_2 from 0.44 to 3.20, the value of g_{intra} for poly(cyclohexyl acrylate) (84) drops from 1.71 ± 0.05 to 0.528 ± 0.005.

Some difficulties in the theoretical evaluation of g_{intra} arise when the flexibility of the side groups increase so that its dipole moment lies in a relatively wide range. For example, for poly(triethylene glycol acrylate) the dipole moment of the side group can oscillate between about 0.50 D and about 5 D, depending on the conformation (92). Methods were devised that permit the use of the statistics of less complex acrylic systems for the theoretical calculation of the dipolar correlation coefficient of these systems. In brief, the probability of the rotational states of each bond of the side group can be obtained by means of the expression

$$p(\zeta\eta) \;=\; Z^{-1} \left[\prod_{h=1}^{i-1} \mathbf{U}_h \right] \mathbf{U}'_i \left[\prod_{j=i+1}^{N} \mathbf{U}_j \right] \tag{39}$$

where Z is the rotational partition function, n is the number of bonds of the backbone of the side group, $\mathbf{U}'_i$ is a matrix in which all the elements are zero except those corresponding to the state, and $\mathbf{U}_1$ and $\mathbf{U}_n$ are, respectively, row and column vectors that convert the matrix product of eq 1 into a scalar. The variable contribution of the side groups to the polarity of poly(triethylene glycol acrylate) chains was determined by generating for each side group nine random numbers lying in the range 0 and 1 from which, and by using the probabilities obtained for the rotational states of each skeletal bond of the side group, the dipole moment of the conformation generated is calculated in the reference frame in which the x axis coincides with the C*O*—bond, whereas the y axis is in the plane defined by the C^αH—C*O*—O bonds. Finally, the contribution of these side groups in the reference frame of the C^α—CH_2 bond of the main chain is obtained for each of the two rotational states of the C^α—C*O* bonds and included in the generator matrices $\mathbf{M}$ of eq 37. The statistics that describe the polarity of poly(cyclohexyl acrylate), poly(phenyl acrylate), etc., also give a good account of the polarity of poly(triethylene glycol acrylate) and other acrylate polymers with long flexible side groups.

Dipolar Correlation Coefficients and Relaxation Strengths in Polymers in Bulk

An important issue in the study of the polarity of molecular chains is to predict the relaxation strength of polymers in the bulk as a function of their structure. This quantity is expressed by $\Delta\varepsilon = \varepsilon_r - \varepsilon_\infty$, where ε_r and ε_∞ are, respectively, the real part of the complex dielectric permittivity at zero and infinite frequencies obtained on polymers above the glass transition temperature. These values can be obtained from Cole–Cole plots. Whenever the global dipolar correlation coefficient, g, is known, $\Delta\varepsilon$ can

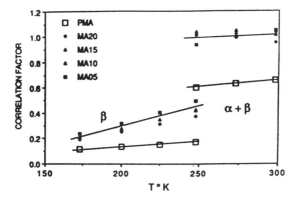

Figure 10. Correlation factors g from the Kirkwood–Onsager equation for both the β subglass process and the combined α + β processes for the MA—E system and for polymethyl acrylate.

be obtained by eq 5. However, the determination of *g*, given by eq 28, is a formidable task that defies any type of solution at hand. Because dipolar correlations between dipole entities sharply decrease as the distance between them increases, the approximation $g = g_{intra}$ is usually made. This approach was tested by Boyd and co-workers (*93, 94*) for copolymers of methyl acrylate–ethylene (MA–E) and vinyl acetate–ethylene (VA–E) and the results obtained, represented in Figures 10–12, show in most cases a good concordance between the intramolecular and global dipolar correla-

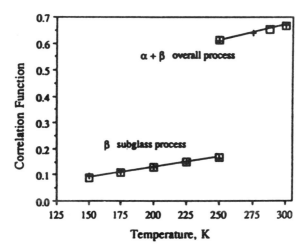

Figure 11. Comparison of calculated and experimental values of correlation coefficients for polymethyl acrylate. Crosses are experimental results; open squares are calculated. (Reproduced from reference 94. Copyright 1991 American Chemical Society.)

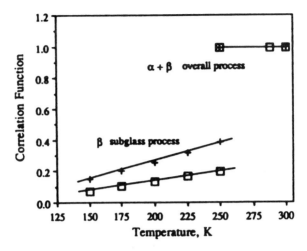

Figure 12. Comparison of calculated and experimental values of correlation coefficients for copolymers of MA/E containing 9 mol% acrylate units. Crosses are experimental results; open squares are calculated. (Reproduced with permission from reference 94. Copyright 1991 American Chemical Society.)

tion coefficients. However, this may not be the case for all polymers. Thus, the values of g for phenyl and chlorophenyl esters of polyacrylic acid, given in the third column of Table VIII, are in most cases significantly lower than those of g_{intra}, shown in the second column of the same table. These results suggest that intermolecular correlations may be important in some polymers.

In the glassy state, flexible polymers are supposed to have extreme difficulties in changing the conformations of the skeleton, whereas some rotational freedom still remains in the side groups. Therefore, the evaluation of the intramolecular dipolar correlation coefficient in the liquid state implies averaging μ^2 over conformational space that includes reorientations

Table VIII. Intramolecular g_{intra} and Global Dipolar Correlation Coefficient g, for Various Polymers

Polymer	g_{intra}	g
PPA	0.547	0.36 (83.5 °C)
POCPA	0.582	0.28 (83.6 °C)
PMCPA	0.570	0.40 (64.8 °C)
PPCPA	0.578	0.30 (92.2 °C)

NOTE: PPA is poly(phenyl acrylate), POCPA is poly(o-chlorophenyl acrylate), PMCPA is poly(m-chlorophenyl acrylate), and PPCPA is poly(p-chlorophenyl acrylate). Temperatures at which the values of ε_r and ε_g were obtained from Cole–Cole plots to determine g are indicated. Values for g_{intra} were determined at 30 °C.

of the entire chains over all spatial orientations, together with internal reorientations, even though the evaluation of the correlation factor, g_β, in the glassy region for asymmetric polymers with flexible side groups presumably only involves averaging of μ^2 over conformational space that includes reorientation of the side groups with respect to the frozen main chain. By using this approach, Boyd and Smith (94) developed a model, formally equivalent to that formerly described by Williams (95), that predicts that g is given by

$$g_\beta = \frac{1}{x \sum m_i^2} (\langle \mu^2 \rangle - \langle \mu \rangle \cdot \langle \mu \rangle) \tag{40}$$

where μ is the vector dipole moment associated with a chain in a given conformation. Whereas $\langle \mu^2 \rangle$ can be obtained by the procedures outlined previously (eq 36), the product $\langle \mu \rangle \cdot \langle \mu \rangle$ can be computed by means of the expression

$$\langle \mu \rangle \cdot \langle \mu \rangle = \frac{1}{N_t} \sum_{i=1}^{N_t} \left[\frac{1}{N_c} \sum_{j=1}^{N_c} \left[\left(\frac{1}{N_o} \sum_{k=1}^{N_o} \mu_{ijk}^x \right)^2 + \left(\frac{1}{N_o} \sum_{k=1}^{N_o} \mu_{ijk}^y \right)^2 + \left(\frac{1}{N_t} \sum_{k=1}^{N_o} \mu_{ijk}^z \right)^2 \right] \right] \tag{41}$$

where μ_{ijk} is the dipole moment of a chain having a configuration i (defined by a set of configurations for all the quaternary carbons of the skeleton), a conformation j (dcfined by a set of ϕ_i angles for the skeletal bonds), and orientation k (defined by the set of values of the rotational angles of the side groups). The way by which these calculations can be performed is described in detail elsewhere (89, 90).

The experimental values of g_β for the secondary relaxation can be obtained by means of the equation of Fröhlich (eq 5), where the values of ε_r and ε_∞ are directly obtained from Cole–Cole plots for the process. The results thus obtained for the dipolar correlation coefficient of MA–E and VA–E copolymers are compared with the theoretical results in Figures 10–12.

In the same way, an estimation of the relative number of dipolar entities participating in the relaxation can be estimated by means of eq 41. For poly(2-biphenyl acrylate) and poly(4-phenoxyphenyl acrylate) only about 3–5% of the side groups occupying the states of lowest energy participate in the conformational changes giving rise to the secondary β absorption (87, 88). In the case of poly(cyclohexyl acrylate) Boyd's model predicts that the β relaxation could be produced in this polymer by conformational changes taking place in about 9% of the side groups occupying the lowest energy states (84).

Static and Dynamic Correlation Coefficients

The phenomenological theories of dielectrics relate the complex dielectric permittivity to the macroscopic dipolar correlation or decay function $\phi(t)$ by means of the expression (96, 97)

$$\frac{\varepsilon^*(\omega) - \varepsilon_g}{\varepsilon_r - \varepsilon_g} = \Im\left(- \frac{d\phi(t)}{dt}\right) \tag{42}$$

where $\Im$ is the one side Fourier transform and the decay function is given by

$$\phi(t) = \frac{\langle\boldsymbol{\mu}(t)\cdot\boldsymbol{\mu}(0)\rangle}{\langle\boldsymbol{\mu}(0)^2\rangle} \tag{43}$$

For flexible polymers having a constant concentration of dipoles per unit of volume, the decay function can be written as

$$\phi(t) = \frac{g(t)}{g} = \frac{\langle\boldsymbol{\mu}_k(0)\cdot\boldsymbol{\mu}_k(t) + \sum_{k\neq k'}\langle\boldsymbol{\mu}_k(0)\cdot\boldsymbol{\mu}_k(t)\rangle}{\mu^2 + \sum_{k\neq k'}\langle\boldsymbol{\mu}_k(0)\cdot\boldsymbol{\mu}_{k'}(0)\rangle} \tag{44}$$

The normalized correlation function $\phi(t)$ can be evaluated from the experimental dielectric loss by using the relationship

$$\phi(t) = \frac{2}{\pi}\int_0^\infty \frac{\varepsilon''(\omega)}{\varepsilon_r - \varepsilon_u}\frac{\cos\omega t}{\omega}\,d\omega \tag{45}$$

Hence once $\phi(t)$ and g are known, the dynamic dipolar correlation coefficient $g(t)$ can approximately be determined by means of eq 44.

References

1. Flory, P. J. *Statistical Mechanics of Chain Molecules*; Wiley-Interscience: New York, 1969.
2. Riande, E.; Saiz, E. *Dipole Moments and Birefringence of Polymers*; Prentice Hall: Englewood Cliffs, NJ, 1992.
3. Suter, E. W.; Mattice, W. L. *Conformational Theory of Large Molecules: The Rotational Isomeric State in Macromolecular Systems*; Wiley-Interscience: New York, 1994.
4. Debye, P. *Phys. Z.* **1912**, *13*, 97; *Collected Papers: Polar Molecules*; Wiley: New York, 1954; p 173.
5. Onsager, L. *J. Am. Chem. Soc.* **1938**, *58*, 1486.
6. Kirkwood, J. G. *J. Chem. Phys.* **1939**, *7*, 911; *Ann. N.Y. Acad. Sci.* **1940**, *40*, 315; *Trans. Faraday Soc.* **1946**, *42A*, 7.

7. Fröhlich, H. *Trans. Faraday Soc.* **1948**, *44*, 238; *Theory of Dielectrics*; Oxford University Press: London, 1958.

8. Madden D.; Kivelson, P. *Ann. Rev. Phys. Chem.* **1984**, *35*, 75.

9. Guggenheim, E. A. *Trans. Faraday Soc.* **1949**, *45*, 714; **1951**, *47*, 573.

10. Smith, J. W. *Trans. Faraday Soc.* **1950**, *46*, 394.

11. Stockmayer, W. H. *Pure Appl. Chem.* **1967**, *15*, 539; Stockmayer, W. H.; Burke, J. J. *Macromolecules* **1969**, *2*, 647.

12. Marchal, J.; Benoit, H. *J. Chim. Phys. Chim. Biol.* **1955**, *52*, 818; *J. Polym. Sci.* **1957**, *23*, 223.

13. Nagai, K. L.; Ishikawa, T. *Polym. J.* **1971**, *2*, 416.

14. Mattice, W. L.; Carpenter, D. K. *Macromolecules* **1984**, *17*, 625.

15. Mattice, W. L.; Carpenter, D. K.; Barkley, M. D.; Kestner, N. R. *Macromolecules* **1985**, *18*, 2236.

16. Riande, E. *Makromol. Chem.* **1977**, *178*, 2001.

17. Liao, E. C. S.; Mark, J. E. *J. Chem. Phys.* **1973**, *59*, 3825.

18. Blasco, F.; Riande, E. *J. Polym. Sci: Polym. Phys. Ed.* **1983**, *21*, 835.

19. Riande, E.; Boileau, S.; Hemery, P.; Mark, J. E. *Macromolecules* **1979**, *12*, 702.

20. Riande, E.; Boileau, S.; Hemery, P.; Mark, J. E. *J. Chem. Phys.* **1979**, *71*, 4206.

21. Flory, P. J. *Macromolecules* **1974**, *7*, 381.

22. Hopfinger, A. J. *Conformational Properties of Macromolecules*; Academic: Orlando, FL, 1973; p 173.

23. Hopfinger, A. J. *Conformational Properties of Macromolecules*; Academic: Orlando, FL, 1973; p 45.

24. McCrum, N. G.; Read, B. E.; Williams, G. *Anelastic and Dielectric Effects in Polymeric Solids*; Wiley-Interscience: New York, 1967.

25. McClellan, A. L. *Tables of Experimental Dipole Moments*; Freeman and Co.: San Francisco, CA, 1963; Vol. I; Rahara Enterprises: El Cerrito, CA, 1974; Vol. II.

26. Abe, A.; Mark, J. E. *J. Am. Chem. Soc.* **1976**, *98*, 6468.

27. Porter, C. H.; Lawler, J. H. L.; Boyd, R. H. *Macromolecules* **1970**, *3*, 308.

28. Bak, K.; Elefante, G. E.; Mark, J. E. *J. Phys. Chem.* **1967**, *71*, 4007.

29. Riande, E. *J. Polym. Sci: Polym. Phys. Ed.* **1976**, *14*, 2231.

30. Mark, J. E.; Chiu, D. S. *J. Chem. Phys.* **1977**, *66*, 1901.

31. LeFevre, R. J. W.; Sundaram, K. M. *J. Chem. Soc., Perkin Trans.* **1972**, *2*, 2323.

32. Abe, A.; Hirano, T.; Tsuruta, T. *Macromolecules* **1979**, *12*, 1092.

33. Saiz, E.; Riande, E.; Guzmán, J.; de Abajo, J. *J. Chem. Phys.* **1980**, *73*, 958.

34. Garrido, L.; Garrido, E.; Guzmán, J. *Macromolecules* **1982**, *20*, 1805.

35. Riande, E.; de la Campa, J. G.; Guzmán, J.; de Abajo, J. *Macromolecules* **1984**, *17*, 1891.

36. Riande, E.; de la Campa, J. G.; Guzmán, J.; de Abajo, J. *Macromolecules* **1984**, *17*, 1431.

37. Riande, E.; Guzmán, J.; Garrido, L. *Macromolecules* **1984**, *17*, 1234.

38. Riande, E.; Mark, J. E. *Macromolecules* **1978**, *11*, 956.

39. Riande, E.; Mark, J. E. *J. Polym. Sci.: Polym. Phys. Ed.* **1979**, *17*, 2013.

40. Riande, E.; Mark, J. E. *Polymer* **1979**, *20*, 1188.

41. Welsh, W. J.; Mark, J. E.; Guzmán, J.; Riande, E. *Makromol. Chem.* **1982**, *183*, 2562.

42. Riande, E.; Guzmán, J. *Macromolecules* **1981**, *14*, 1234.

43. Abe, A. *Macromolecules* **1980**, *13*, 546.

44. Riande, E.; Guzmán, J.; Welsh, W. J.; Mark, J. E. *Makromol. Chem.* **1982**, *183*, 2555.

45. Guzmán, J.; Riande, E.; Welsh, W. J.; Mark, J. E. *Makromol. Chem.* **1982**, *183*, 2573.

46. Riande, E.; Guzmán, J.; Saiz, E.; de Abajo, J. *Macromolecules* **1981**, *14*, 608.
47. Riande, E. *J. Polym. Sci.: Polym. Phys. Ed.* **1977**, *15*, 1397; Riande, E.; Guzmán, J.; Lloreute, M. A. *Macromolecules* **1982**, *15*, 298.
48. Riande, E.; Guzmán, J.; de Abajo, J. *Makromol. Chem.* **1984**, *185*, 1943.
49. Riande, E.; de la Campa, J. G.; Schlereth, D. D.; de Abajo, J.; Guzmán, J. *Macromolecules* **1987**, *20*, 1641.
50. Riande, E.; Guzmán, J.; de la Campa, J. G.; de Abajo, J. *J. Polym. Sci.: Polym. Phys. Ed.* **1987**, *25*, 2403.
51. Díaz-Calleja, R.; Riande, E.; Guzmán, J. *Macromolecules* **1989**, *22*, 3654.
52. Riande, E.; Guzmán, J. *J. Polym. Sci.: Polym. Phys. Ed.* **1985**, *23*, 1235.
53. González, C.; Riande, E.; Bello, A.; Pereña, J. M. *Macromolecules* **1988**, *21*, 3230.
54. Saiz, E.; Hummel, J. P.; Flory, P. J.; Plavsic, M. *J. Phys. Chem.* **1981**, *85*, 3211.
55. Riande, E.; Guzmán, J.; de Abajo, J. *Macromolecules* **1989**, *22*, 4026; Riande, E.; Guzmán, J.; de la Campa, J. G.; de Abajo, J. *Macromolecules* **1985**, *18*, 2739; Riande, E.; Guzmán, J.; de la Campa, J. G. *Macromolecules* **1988**, *21*, 2128; Riande, E.; Guzmán, J.; Addabo, H. *Macromolecules* **1986**, *19*, 2567; Riande, E.; Guzmán, J. *J. Chem. Soc., Perkin Trans.* **1988**, *2*, 299 (1988).
56. Rodrigo, M. M.; Tarazona, M. P.; Saiz, E. *J. Phys. Chem.* **1986**, *90*, 2236.
57. Khanarian, G.; Msek, P.; Moore, W. J. *Bipolymers* **1981**, *20*, 1191.
58. Shipman, L. L.; Christoffersen, R. E. *J. Am. Chem. Soc.* **1973**, *95*, 1408.
59. de Abajo, J.; de la Campa, J. G.; Riande, E.; García, J. M.; Jimeno, M. L. *J. Phys. Chem.* **1993**, *97*, 8669.
60. Dasgupta, S.; Smyth, C. P. *J. Chem. Phys.* **1967**, *47*, 2911.
61. Sutton, C.; Mark, J. E. *J. Chem. Phys.* **1971**, *54*, 5011.
62. Beevers, M. S.; Mumby, S. J.; Carlson, S. J.; Semlyen, A. J. *Polymer* **1983**, *24*, 1565.
63. Mark, J. E. *Acc. Chem. Res.* **1974**, *7*, 218.
64. Salom, C.; Hernández-Fuentes, I. *Eur. Polym. J.* **1989**, *25*, 203; Salom, C.; Freire, J. J.; Hernández-Fuentes, I. *Polym. J.* **1988**, *20*, 1109.
65. Mark, J. E.; Ko, J. H. *Macromolecules* **1975**, *8*, 869.
66. Salom, C.; Riande, E.; Hernández-Fuentes, I.; Díaz-Calleja, R. *J. Polym. Sci.: Part B: Polym. Phys. Ed.* **1993**, *31*, 1591.
67. Le Fevre, R. J. W.; Sundaran, K. M. S. *J. Chem. Soc.* **1967**, 1494.
68. Blasco-Cantera, F.; Riande, E.; Almendro, J. P.; Saiz, E. *Macromolecules* **1981**, *14*, 138.
69. Le Fevre, R. J. W.; Sundaran, K. M. S. *J. Chem. Soc.* **1962**, 4003.
70. Saiz, E.; Riande, E.; Delgado, M. P.; Barrales-Rienda, J. M. *Macromolecules* **1982**, *15*, 1152.
71. Baysal, B.; Yu, H.; Stockmayer, W. H. *Dielectric Properties of Polymers;* Karasz, F. E., Ed.; Plenum: New York, 1972.
72. Smith, F. H.; Corrado, L. H.; Work, N. *Polym. Prep. Am. Chem. Soc. Div. Polym. Chem.* **1971**, *12*, 64.
73. Roig, A.; Hernández-Fuentes, I. *An. Quím.* **1974**, *70*, 668.
74. Yamaguchi, N.; Sato, M.; Ogawa, E.; Shima, M. *Polymer* **1981**, *22*, 1464.
75. Tonelli, A. E.; Belfiore, L. A. *Macromolecules* **1983**, *16*, 1740.
76. Salmerón, M.; Barrales-Rienda, J. M.; Riande, E.; Saiz, E. *Macromolecules* **1984**, *17*, 2728.
77. Saiz, E.; Mark, J. E.; Flory, P. J. *Macromolecules* **1977**, *10*, 967.
78. Yoon, D. Y.; Sundarajan, P. R.; Flory, P. J. *Macromolecules* **1975**, *8*, 776.
79. Sundarajan, P. R. *Macromolecules* **1982**, *13*, 512.
80. Riande, E.; García, M.; Mark, J. E. *J. Polym. Sci.: Polym. Phys. Ed.* **1981**, *19*, 1739.
81. Masegosa, R. M.; Hernández-Fuentes, I.; Ojalvo, E. A.; Saiz, E. *Macromolecules* **1979**, *12*, 862.

82. Ojalvo, E. A.; Saiz, E.; Masegosa, R. M.; Hernández-Fuentes, I. *Macromolecules* **1979**, *12*, 865.
83. Saiz, E.; Riande, E.; San Román J.; Madruga, E. L. *Macromolecules* **1990**, *23*, 785.
84. Diaz-Calleja, R.; Riande, E.; San Román, J. *Macromolecules* **1991**, *24*, 264.
85. Diaz-Calleja, R.; Riande, E.; San Román, J. *J. Phys. Chem.* **1992**, *93*, 931.
86. Diaz-Calleja, R.; Riande, E.; San Román, J. *Macromolecules* **1992**, *25*, 2875.
87. Diaz-Calleja, R.; Riande, E.; San Román, J. *J.Phys. Chem.* **1992**, *96*, 6843.
88. Diaz-Calleja, R.; Riande, E.; San Román, J. *Polymer* **1993**, *34*, 3456.
89. Díaz-Calleja, R.; Saiz, E.; Riande, E.; Gargallo, L.; Radic, D. *Macromolecules* **1993**, *26*, 3795.
90. Díaz-Calleja, R.; Saiz, E.; Riande, E.; Gargallo, L.; Radic, D. *J. Polym. Sci: Part B: Polym. Phys. Ed.* **1994**, *32*, 1069.
91. Yarim-Agaev, Y.; Plavsic, M.; Flory, P. J. *Polym. Prep. (ACS Div. Polym. Chem.)* **1983**, *24(1)*, 233.
92. Riande, E.; Guzmán, J. *Macromolecules* **1996**, *29*, 1728.
93. Buerger, D. E.; Boyd, R. H. *Macromolecules* **1989**, *22*, 2694, 2699.
94. Smith, G. D.; Boyd, R. M. *Macromolecules* **1991**, *24*, 2731.
95. Williams, G. *Adv. Polym. Sci.* **1979**, *33*, 159.
96. Williams, G. *Chem. Rev.* **1972**, *72*, 55.
97. Williams, G. *Chem. Soc. Rev.* **1978**, *7*, 89.

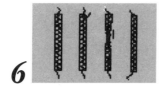

6

Unbiased Modeling
of Dielectric Dispersions

S. Havriliak Jr. and S. J. Havriliak

Dielectric relaxation parameters are generally obtained by using nonlinear regression techniques. Our review of the literature as well as reviews by others show that these parameters are obtained in a rather haphazard way. As a result of this action, considerable confusion exists in the literature about the details of dielectric relaxation mechanisms.

This chapter outlines rigorous statistical methods for representing dielectric relaxation data in terms of relaxation functions. Unbiased in the title is a statistical term and when applied to dielectric relaxation has the following components. First, the minimum in the nonlinear regression must be determined to a very high precision: that is, one figure in the eighth significant place. Minima are never reported in the literature. Second, all the alternative relaxation functions must be evaluated to determine which one has the lowest model standard error of estimate; model comparisons are seldom reported. Third, parameter confidence intervals must be reported to sort out the important contributions; this process has never been done.

The advantage of unbiased methods is given a set of data, anyone will come to the same results. Interestingly, papers always contain an experimental section, but they seldom if ever contain an analysis section.

Experimental dielectric relaxation methods cover the widest frequency range of any physical measurements, that is fully 17 decades of time. According to Jonscher (1), the decade range, from the age of the universe to the shortest wave length (high energy) radiation is about 40 decades. Furthermore he states "We are therefore in the uniquely favorable position of being able to cover experimentally up to two-thirds of the total logarithmic range of times and frequencies." This feat is truly a remarkable feat because it can be done with only three devices that operate (data acquisition and temperature control) automatically. Furthermore, the absolute accuracy of these measurements is for the real part about 0.5% and for the imaginary part about 2% depending on other influences such as temperature fluctuations during the measurement, lead problems connecting the device to the specimen holder (cell), sample preparations, and the like. Surprisingly, the total costs of these devices are less than they are for most physical devices with serious time-scale limitations such as NMR spectrometers, etc.

Another important feature of dielectric studies is that experimental dielectric relaxation data can be represented, at least for the past 50 years or so, by isothermal (empirical) relaxation functions with radian frequency as the experimental variable in terms of a few parameters. These parameters can often be represented in terms of simple temperature dependencies. Both the isothermal parameters and their dependence on temperature can be interpreted in terms of molecular theories. Molecular theories in this context means some well-defined property of the molecule such as dipole moment, polarizability, distribution of relaxation times, time-dependent correlation functions, and activation energy. Contrast this state of affairs to viscoelastic relaxation data that can only be made over a time range of a few decades and have accuracies of a few percent for the real part and about 10% for the imaginary part. These data are seldom if ever represented in terms of a relaxation function. A molecular theory is yet to be developed in the same sense that a molecular dielectric theory described previously exists.

The advantage of representing dielectric relaxation data in terms of a few parameters permits surveys of the literature to be in a form of parameter tables and their correlations with molecular parameters. The most comprehensive compilation of dielectric relaxation parameters is that of Böttcher and Bordewijk (2). They compiled the relaxation parameters of several thousand materials that met their relatively simple criteria for inclusion. These authors concluded:

> It is evident from the survey given in this section that experimental results concerning the dielectric behavior are often ambiguous, different types of relaxation behavior being reported for some kinds of systems by different authors, and sometimes even different types of relaxation behavior for the same system under the same conditions. Moreover, it is conceivable that in some cases the agreement between results of different authors is due to a common theoretical preference or to the convenience of the type of description chosen rather than to the similarity of the phenomena in question.

The objective of this work addresses these criticisms that are also our objectives. Several steps exist to the representation of dielectric relaxation data in terms of a few parameters that meet sound statistical or *unbiased* methods. First, all relaxation functions must be tested to determine which is the best one. The best function is always chosen in terms of the lowest standard model of estimate, which is analogous to the experimental standard deviation and defined in the similar way. Second, because the experimental data for all relaxation functions are often nonlinear with respect to their parameters, then nonlinear regression methods must be used. It is important that convergence criteria be specified and that a stable convergence is assured. Third, parameter confidence intervals should always be determined because the number of parameters is unimportant. What is important is how significant is the parameter. For example if a parameter and its confidence interval are determined and the parameter includes 0 in the 50% confidence interval, then that parameter is hardly significant. Parameter confidence intervals for linear regressions are trivial estimations. In today's spread sheets, parameter confidence intervals are usually cited, even in linear multiparameter regressions. Parameter confidence intervals for the nonlinear case are calculated by assuming that at convergence the linear equations apply (*3*). The nonlinear case is troublesome but with computational capabilities it is trivial once the programs are written.

These three steps

- defining convergence criteria
- testing the alternatives
- determining the parameter confidence intervals

form the basis of *unbiased* modeling of dielectric dispersions. These statistical steps have been known and available for many years. We have reviewed all the papers cited by Böttcher and Bordewijk plus another several hundred since that review. There are only a few cases where these procedures have been followed. Omitting this kind of information is the equivalent of reporting dielectric experimental data without specifying the experimental equipment and sample preparation used to obtain the data. A paper without this experimental information would never be accepted in any journal, yet papers are routinely accepted without this three-step regression statistical information.

Relaxation Functions

Havriliak–Negami and Related Functions

Dielectric relaxation data of polymers can be represented in terms of the Havriliak–Negami (HN) function (*4, 5*) defined by equation 1:

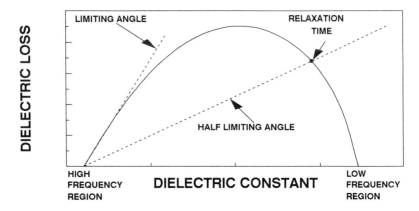

Figure 1. Complex plane plot for an arbitrary set of parameters indicating the graphical definition of the parameters.

$$\varepsilon^*(\omega) = \varepsilon_\infty + \frac{\varepsilon_0 - \varepsilon_\infty}{\{1 + (i\omega\tau_0)^\alpha\}^\beta} \tag{1}$$

In this expression, $\varepsilon^*(\omega)$ is the complex dielectric constant (or complex relative permittivity) measured at radian frequency $\omega = 2\pi f$, where f is in hertz and i is $\sqrt{-1}$. (In this work we use the term dielectric constant, meaning that it is an intensive property of the material and not independent of temperature or frequency. The reader may wish to substitute the term relative permittivity in its place.) The real part of this expression is denoted by $\varepsilon'(\omega)$, whereas the imaginary part is denoted by $\varepsilon''(\omega)$. These parts are related to the complex dielectric constant through the expression $\varepsilon^*(\omega) = \varepsilon'(\omega) - \varepsilon''(\omega)$. The real and imaginary parts are extracted from eq 1 by the successive application of DeMoiver's theorem. The results are

$$\varepsilon'(\omega) - \varepsilon_\infty = r^{-\beta/2}(\varepsilon_0 - \varepsilon_\infty) \cos \beta\Theta = \Delta\varepsilon r^{-\beta/2} \cos \beta\Theta \tag{2}$$

$$\varepsilon''(\omega) = r^{-\beta/2}(\varepsilon_0 - \varepsilon_\infty) \sin \beta\Theta = \Delta\varepsilon r^{-\beta/2} \cos \beta\Theta \tag{3}$$

$$r = \left\{1 + (\omega\tau_0)^\alpha \sin \frac{\alpha\pi}{2}\right\}^2 + \left\{(\omega\tau_0)^\alpha \cos \frac{\alpha\pi}{2}\right\}^2 \tag{4}$$

$$\Theta = \arctan\left\{\frac{(\omega\tau_0)^\alpha \cos \dfrac{\alpha\pi}{2}}{1 + (\omega\tau_0)^\alpha \sin \dfrac{\alpha\pi}{2}}\right\} \tag{5}$$

It is instructive to calculate $\varepsilon'(\omega)$ and $\varepsilon''(\omega)$ for an arbitrary set of the five parameters on the right-hand side of eq 1; the results are represented in a complex plane plot in Figure 1. A complex plane plot is constructed

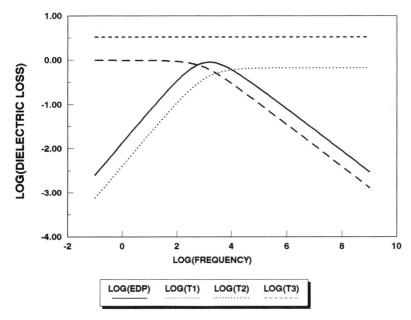

Figure 2. Plot of log(ε″(ω)) and its components T1, T2, and T3 with log(ω).

by plotting the $(-)$imaginary part against its real part for each frequency of measurement. The result is that frequency is a running variable along the locus with the low frequency end to the right and high frequency end to the left. It is relatively straightforward (4, 5) to show that ε_0 is the static or equilibrium dielectric constant (6) obtained at low frequencies, ε_∞ is the instantaneous dielectric constant obtained at very high frequencies, and τ_0 is the relaxation time and in general is not the time (frequency) of maximum loss. Rather it is the point that the high frequency angle bisector line intersects with the complex plane locus. The significance of the α and β parameters is most evident when log(loss) is plotted against log(-frequency) (*see* the heavy line in Figure 2). Taking the logarithm of eq 3 yields the sum of three terms (log(T1), log(T2), and log(T3)), which are also plotted in Figure 2. In such plots, the limiting low frequency slope yields α whereas the limiting high frequency slope yields $\alpha\beta$. Changing α changes both slopes equally. Changing β changes only the high frequency slope. Hence the interpretation that β represents the width of the relaxation process whereas β represents its skewness. The related functions are the Cole–Cole (CC) (7) function, which is obtained when $\beta = 1$ in eq 1, and the Davidson–Cole (DC) (8) equation, when $\alpha = 1$ in eq 1.

  An incorrect statement is often made in the literature about α and β limits: that is, $0 \le \alpha, \beta \le 1$. The falseness of this assertion is readily apparent from the work of Jonscher (9). He proposed a method of data representa-

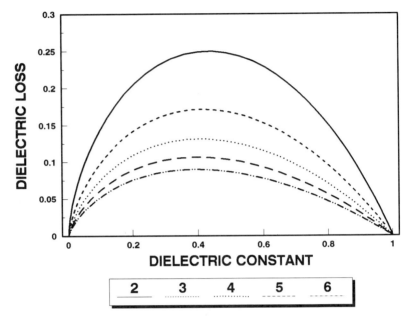

Figure 3. Complex plane plots for the case of αβ = 1.

tion based on recording the low (m) and high $(1 - n)$ frequency slopes in a log(loss) versus log(frequency) plot. Experimentally, he found (9) that the range of materials in such plots is $0 \leq m, 1 - n \leq 1$. This result translates to $0 \leq \alpha, \alpha\beta \leq 1$. In as much as α is always ≤ 1 and the product $\alpha\beta$ is always ≤ 1, then β can be >1. Inserting arbitrary values of α and β that meet the condition $\alpha\beta = 1$ and $\beta > 1$ into eq 1 and calculating $\varepsilon^*(\omega)$ over an extensive frequency range shows that the skewness either in a complex plane plot (Figure 3) or in a log(loss) versus log(frequency) plot is reversed (10).

The distribution of relaxation times can be obtained from eq 1 and the corresponding integral equation relating it to a distribution of relaxation times (5). The distribution function, $F(y)$, is given by

$$F(y) = \left(\frac{1}{\pi}\right) y^{\alpha\beta} (\sin \beta\theta)(y^{2\alpha} + 2y^\alpha \cos \pi\alpha + 1)^{-\beta/2} \qquad (6)$$

In this expression $y = \tau/\tau_0$ and

$$\Theta = \arctan \left(\frac{\sin \pi\alpha}{y^\alpha + \cos \pi\alpha}\right) \qquad (7)$$

Results of substituting the α,β parameters used in Figure 3 yields the results shown in Figure 4. The distribution of relaxation times calculated

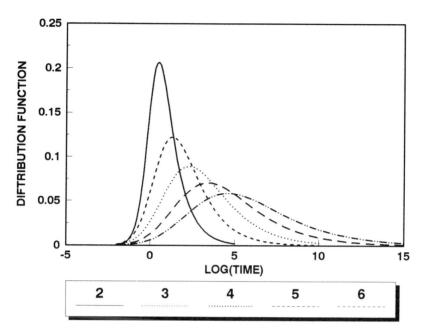

Figure 4. Distribution of relaxation times calculated from eq 6 and the β parameters listed in the legend.

from such parameter values is physically reasonable: that is, the distribution function is single valued and positive, although the skewness is reversed.

Gaussian Function

Wagner (*11*) proposed that the distribution of relaxation times be given by an equation of the form

$$F(y) = \left(\frac{1}{\sigma\sqrt{2\pi}}\right) \exp\left\{\left(\frac{-1}{2}\right)\left(\frac{y}{\sigma}\right)^2\right\} \tag{8}$$

In this expression σ is known as the standard deviation and indicates the breadth of the dispersion. For the case $1/\sigma \to 0$, the distribution reduces to a single relaxation time. From the form of this expression it is obvious that the distribution function hence $\varepsilon''(\omega)$ is symmetrical about the central or relaxation time. Böttcher and Bordewijk (*12*) discussed the relationship between this function and other symmetrical relaxation functions. Unfortunately for the experimenter the discussions are always carried out at very low or very high frequencies relative to the relaxation times. In contrast, the experimenter's data are usually analyzed in the relaxation region where

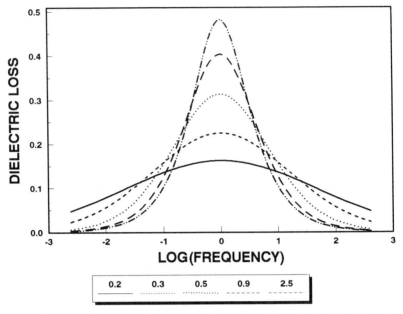

Figure 5. Plot of $\varepsilon''(\omega)$ *versus log(frequency) for various values of* σ *listed in the legend.*

these discussions are not helpful. Tables of $\varepsilon^*(\omega)$ for an $\ln(\tau/\tau_0)$ range of -6 to 6 and a parameter σ range of 0.4 to 5 are available from tables developed by Lichter and McDuffie (*13*). Plots of $\varepsilon''(\omega)$ against log (frequency) are given in Figure 5, whereas in Figure 6, $\log(\varepsilon''(\omega))$ is plotted against log(frequency). Figure 6 clearly shows that the very high and low frequency slopes become parallel with a slope of 1: that is, they become Debye like.

Fuoss–Kirkwood Function

Another symmetric distribution function was proposed by the Fuoss and Kirkwood function (*14*). They noted that $\varepsilon''(\omega)$ for the Debye relaxation process could be arranged to yield

$$\frac{\varepsilon''(\omega)}{\varepsilon''_{max}} = \text{sech}(\ln\omega\tau_0) \tag{9}$$

where ε''_{max} is the dielectric loss maximum. The left-hand side will be referred to as the scaled dielectric loss. Fuoss and Kirkwood introduced a parameter m into eq 9 to yield:

$$\frac{\varepsilon''(\omega)}{\varepsilon''_{max}} = \text{sech}[m(\ln\omega\tau_0)] \tag{10}$$

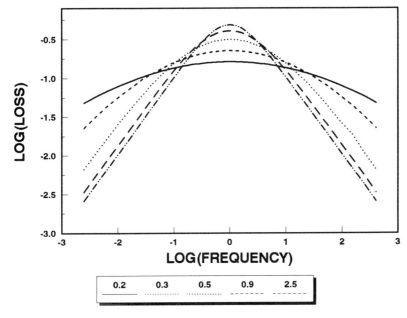

Figure 6. Plot of log($\varepsilon''(\omega)$) versus log(frequency) for the same values of the Gaussian σ indicated in Figure 5.

where $0 < m \leq 1$. The dependence of $\varepsilon''(\omega)$ on the Fuoss–Kirkwood m parameter with log(frequency) is shown in Figure 7. Böttcher and Borde-wijk (*12*) have shown that the frequency dependence of the loss at very high or low frequencies relative to the relaxation time converges to the CC function: that is, $\beta = 1$ in eq 1. Unfortunately for the experimenter such discussions are always carried out at very low or very high frequencies rela-tive to the relaxation times. In contrast, the experimenter's data are always obtained in the relaxation region where these discussions are not helpful. This result is evident in Figure 8 where the log($\varepsilon''(\omega)$) is plotted against log(frequency). We see that the slopes all depend on the Fuoss–Kirkwood m parameter. This behavior is entirely different from the behavior shown in Figure 6 for the Gaussian function.

Jonscher Function

Equation 10 can be rearranged to read

$$\varepsilon''(\omega) = 2\varepsilon''_{max} \left\{ \frac{(\omega\tau_0)^m}{1 + (\omega\tau_0)^{2m}} \right\} \tag{11}$$

It can be shown that ε''_{max} is given by

$$\varepsilon''_{max} = \frac{\varepsilon_0 - \varepsilon_\infty}{2} m \tag{12}$$

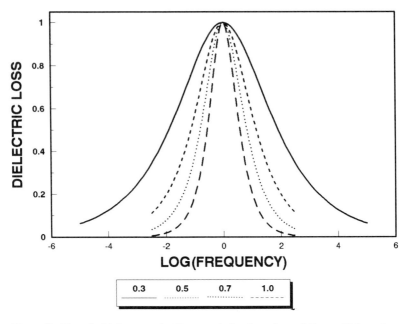

Figure 7. Plot of $\varepsilon''(\omega)$ versus log(frequency) for the values of Fuoss–Kirkwood m indicated in the legend.

Equation 10 then becomes

$$\varepsilon''(\omega) = \frac{\varepsilon_0 - \varepsilon_i}{2}\, m\, \text{sech}[m(\ln\omega_0)] \tag{13}$$

$$\varepsilon''(\omega) = \frac{\varepsilon''_{\max}}{(\omega/\omega_0)^{-m} + (\omega/\omega_0)^{m}} \tag{14}$$

Jonscher (*15*) suggested an expression for $\varepsilon''(\omega)$ that is a generalization of equation 14:

$$\varepsilon''(\omega) = \frac{\varepsilon''_{\max}}{(\omega/\omega_0)^{-m} + (\omega/\omega_0)^{1-n}} \tag{15}$$

The parameters m and $1 - n$ can be determined from a $\log\varepsilon''(\omega)$ versus log(frequency) plot. This plot is linear at high frequencies because, as $\omega \to 0$,

$$\varepsilon''(\omega) = A\left(\frac{\omega}{\omega_1}\right)^{m} \tag{16}$$

At low frequencies, as $\omega \to \infty$ the plot becomes

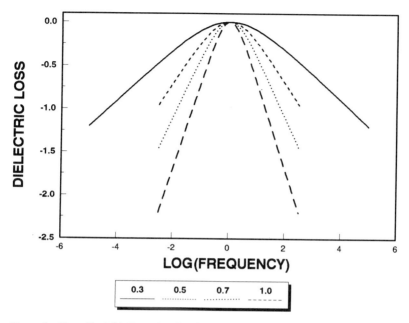

Figure 8. Plot of $\log(\varepsilon''(\omega))$ against log(frequency) for the same values of Fuoss–Kirkwood m indicated in Figure 7.

$$\varepsilon''(\omega) = A\left(\frac{\omega}{\omega_2}\right)^{n-1} \tag{17}$$

From these equations we determine that the high frequency slope is given by $n - 1$ and the low frequency intercept is given by m, where $0 < m \leq 1$, $0 \leq 1 - n < 1$, which are the Debye limits. These are the same limits discussed earlier for the HN function, which are $0 \leq \alpha$, $\alpha\beta \leq 1$. These slopes are related to the α,β parameters as shown in Figure 2.

Stretched Exponential or Kohlrausch–Williams–Watts Function

So far, all of the relaxation functions are explicit in the frequency domain. We now consider a function explicit in the time domain, i.e. the so-called stretched exponential, or Kohlrausch–Williams–Watts (KWW) function (*16–18*), given in the normalized form by eq 18

$$\varepsilon_n(t) = 1 - \exp[-(t/\tau_0)^k] \tag{18}$$

In this expression k is the KWW parameter.

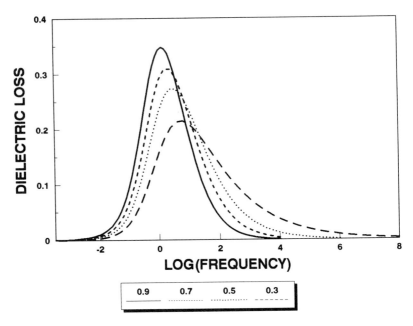

Figure 9. Plot of $\varepsilon''(\omega)$ versus log(frequency) for the values of KWW k indicated in the legend.

Koizumi and Kita (*19*) transformed $\varepsilon_n(t)$ to $\varepsilon_n^*(\omega)$ calculated over a wide time range from the stretched exponential KWW function for given values of the k parameter from 0.29 to 1.0 incremented by 0.01. Dishon, Weiss, and Bendler (*20*) also treated the same problem but did not mention the results of Koizumi and Kita. Their procedure was similar and care was given to ensure convergence. They stated that the calculations were carried in double precision and the results reported to six places. They covered a similar parameter and time range as did Koizumi and Kita. In both cases the log(frequency) range is ± 3, which is not long enough to define the frequency dependence for low k values. For this reason we use the results from the extended Schwarzl method (*21*) from which $\varepsilon''(\omega)$ can be calculated over an extensive range as shown in Figure 9 for various KWW k values. These results show that the breadth and skewness increase with decreasing k value. A plot of $\log(\varepsilon''(\omega))$ with log(frequency) is shown in Figure 10. We see from that figure the low frequency slopes are equal and have a value of 1, whereas the high frequency slopes depend on k value. In other words the high and low frequency behavior of the KWW function is the same as the DC function as pointed out by Böttcher and Bordewijk (*12*).

Alvarez, Algeria, and Colmenero (*22*) calculated numerically the distribution of relaxation times for given values of k in the range of 0.1 to 1.0 and then fitted the numerical results to eq 6 by using an unspecified nonlinear

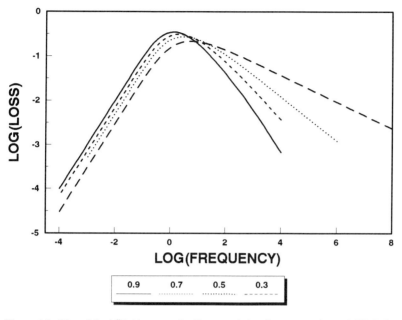

Figure 10. Plot of log(ε″(ω)) versus log(frequency) for the same values of KWW k indicated in Figure 9.

regression routine. Plots of α against β for different *k* values are given in Figure 11. In addition the results from directly fitting the transformed results by using the methods in the next section are also given there. In other words for a given KWW *k* parameter there exists a specific α,β pair. The converse is not true because CC (set β = 1 in eq 1), Fuoss–Kirkwood, and Gaussian systems that are symmetrical functions cannot be represented by the unsymmetrical KWW function. This point has been discussed by Alvarez, Alegria, and Colmenero (*23*).

Unbiased Regression Methods

In this section we assume that the proper analytical function to represent the data is eq 1. The method outlined here is identical for the other relaxation functions. In a later section we examine surveys on relaxation data that support the use of eq 1. For the parameters of eq 1 to be meaningful, they should be determined by rigorous statistical techniques (*24–27*). The usual procedure is to evaluate the five parameters at constant temperature and then determine the temperature dependence of the parameters. For a typical experimental case at constant temperature there would be about 20 experimental frequencies, covering five decades. Because there are five

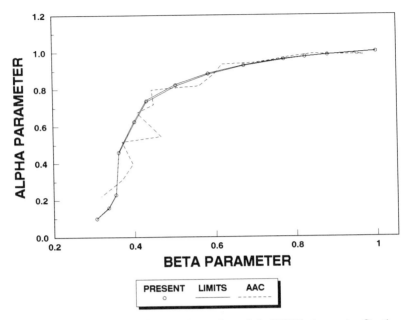

Figure 11. Plot of α versus β for various values of the KWW k parameter. Starting from the lower left hand side, the circle k values are 0.04, 0.07, 0.1, 0.2, 0.3, 0.4, 0.5, 0.6, 0.7, 0.8, 0.85, 0.9, and 1.0. The diameter of the circles represents the parameter confidence interval.

parameters in eq 1, this amount would lead to an experiment-to-parameter ratio (R) of 4. This R could be enhanced considerably by pooling (combining) the experimental data and proposing temperature dependencies for the five parameters. For the simple case of 10 temperatures leading to 200 experimental points and increasing the number of parameters from five to 10 leads to an R of 20, a significant improvement. Even for the more complicated and unusual case of quadratic temperature dependencies requiring 20 parameters, the ratio is 10. Another benefit of pooled data sets is that data at temperatures that could not be regressed because they were too far to one side or the other of the relaxation time can be added to the data set and still maintain a balanced data set.

To proceed further we need to propose a set of general model equations representing the temperature dependence for the five parameters. The general model equations used in this work are given by eq 19–23. All the parameters are quadratic functions of temperature except for $\ln(f_0)$. The assumed form for $\ln(f_0)$ has been discussed (28) from the point of view of Bueche's jumping model (29). These equations also include a centering temperature (T_0) in the general model equations. This temperature is chosen to be in the center of the experimental temperature range and has the effect of reducing the size of intercepts of the general model eqs 19–23.

$$\varepsilon_0 = I_1 + C_1(T - T_0) + D_1(T - T_0)^2 \tag{19}$$

$$\varepsilon_\infty = I_2 + C_2(T - T_0) + D_2(T - T_0)^2 \tag{20}$$

$$\ln f_0 = I_3 + C_3\left(\frac{1}{K} - \frac{1}{K_0}\right) + D_3(T - T_0)^2 \tag{21}$$

$$\alpha = I_4 + C_4(T - T_0) + D_4(T - T_0)^2 \tag{22}$$

$$\beta = I_5 + C_5(T - T_0) + D_5(T - T_0)^2 \tag{23}$$

The procedure used to determine which parameters must be retained in the polymer model equations is fully discussed in references 26 and 27. Briefly, the minimum number of parameters required for convergence is first determined by using nonlinear regression techniques and the residual sums of squares (rss) noted. Each remaining term in the general model equations is then introduced one at a time into the regression and its effect on rss noted. The term having the greatest reduction in rss is retained in the polymer model equations and the regression cycle repeated. A plot of log(rss) versus the number of parameters shows a break at some number of parameters and then generally remains independent of number of parameters. This level of rss can be referred to as imbedded error and includes experimental error as well as model breakdown errors. Parameters that have no effect on rss are not retained in the polymer model equation. In addition the confidence intervals for each parameter are determined. The final set of polymer model equations used to represent the dielectric relaxation data of the polymer must meet two criteria. First, there must be a significant reduction in rss. Second, except for α and β, zero must not be in the parameter 95% confidence interval. In the case that α,β has 1 in the 95% confidence interval, the parameter is retained and the reader may assume that the process is of the CC or DC type.

Although nonlinear regression can be carried out in a number of different ways, the software chosen here is PROC NLIN available through SAS (*30*). The convergence criterion for this software is given by

$$\frac{d_{i-1} - d_i}{d_i + 10^{-6}} < c \tag{24}$$

where $c = 10^{-8}$. This criterion means that the minimum has been determined to the eighth significant figure. In addition this software reports the parameter confidence intervals, determined at convergence. Initially the regressions were carried out on an IBM mainframe. However, recent advances in personal computer technology made it possible to carry out these calculations on computers of modest cost: that is 25 MHz clock speed with 3 Mbyte of random access memory. One step in the stepwise procedure is generally carried out in less than 2 min.

Examples of Unbiased Analyses

In this section we consider several cases that illustrate the nature of the problems associated with parameter estimation. In two of these cases we assume that the data are given by one of the relaxation functions, whereas the third case is actual experimental data. One of the problems with analyzing experimental data is that although the frequency range may be extensive there is seldom only one relaxation process. In addition measurements are made at some temperature other than room temperature so that there is the effect of temperature fluctuations on the relaxation times, which are known to be very sensitive to temperatures.

Gaussian Relaxation Function

For the present purpose, we assume that the relaxation data are of the Gaussian form and we shall try to represent the data in terms of the HN function. The specific case of $\sigma = 0.5$ shown in Figure 5 is used for analysis. The results are given in Figure 12, where the parameters are also listed.

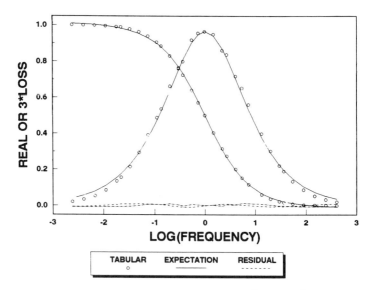

Figure 12. Plot of the tabular and expectation values of $\varepsilon'(\omega)$ and $\varepsilon''(\omega)$ as a function of log(frequency) for the Gaussian parameter $\sigma = 0.5$. The parameters are $\alpha = 0.703(0.009)$ and $\beta = 1.05(0.05)$ with $\sigma' = 0.4\%$ and $\sigma'' = 3\%$.

The results of the analysis are that this process can be represented by an HN function with β = 1.05 (0.05): that is, the CC function. Unless the investigator made a log–log plot of the data as shown in Figure 6 to determine the slopes this misrepresentation would have been missed. The converse is equally true, representing a symmetrical relaxation process: that is, a CC or Fuoss–Kirkwood in terms of a Gaussian function.

KWW Function

Consider the KWW case for k = 0.5 shown in Figure 9. Once again we assume that the experimental data are in fact a KWW type and we try to represent the data in terms of the HN function. Data in Figure 9 were regressed using the *unbiased* techniques described in the previous section. The results of this regression are listed in Figure 13 and the parameters in the caption. Once again the results appear to be acceptable. Unless one constructs a log–log plot the systematic deviations are not apparent and the limiting frequency behavior is not observed. This method was used to evaluate the α,β for the range of k values shown in Figure 11.

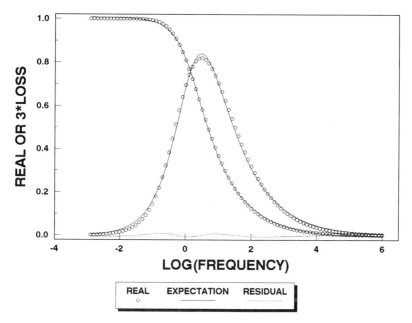

Figure 13. Plot of the transformed and expectation values of $\varepsilon'(\omega)$ and $3\varepsilon''(\omega)$ as a function of log(frequency) for the KWW k = 0.5. The parameters are α = 0.818(0.005) and β = 0.513(0.007) with $\sigma' = 0.3\%$ and $\sigma'' = 5\%$.

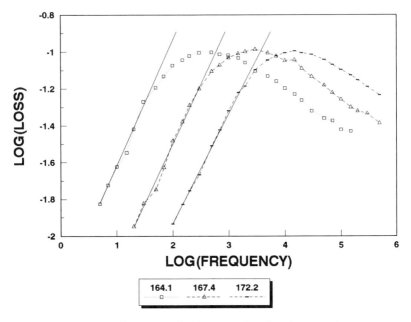

Figure 14. Plot of log(ε″(ω)) versus log(frequency) for polycarbonate at the temperatures listed in the legend.

Polycarbonate

The dielectric relaxation data for polycarbonate (*31*) in the α-process region are shown in Figure 14, as log(ε″(ω)) − log(frequency) plots at three temperatures. The high and low frequency slopes are listed in Table I. On the basis of the discussions in the previous sections we know that this relaxation data cannot be represented by the KWW, DC, CC functions nor any of the other symmetrical ones such as the Gaussian or Kirkwood–Fuoss functions. In other words a simple log(ε″(ω)) − log(frequency) rules out many of the alternatives.

We can extend the analysis to all seven temperatures in the α-process

Table I. High and Low Frequency Slopes for Polycarbonate at Various Temperatures

Temp.	Low Frequency	σ	High Frequency	σ
164.1	0.69	0.04	−0.215	0.008
167.4	0.65	0.04	−0.212	0.008
172.2	0.60	0.01	−0.193	0.004

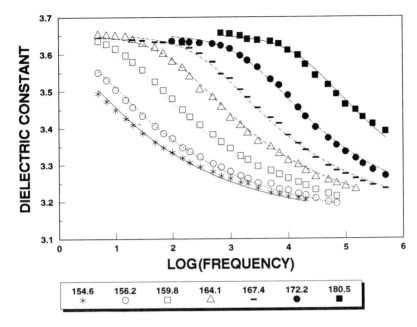

Figure 15. Plot of the experimental and expectation values of $\varepsilon'(\omega)$ for polycarbonate at the temperatures listed in the legend.

region shown in Figure 15 by combining or pooling all seven temperatures into a single data set and determine what parameters in the general model equations meet the conditions of significant ssr reduction and which parameters do not include 0 or 1 as the case may be, in their confidence intervals. The simplest regression model for the HN parameters and their dependence on temperature is a five-parameter model. The model is outlined in Table II as test number 1, where the ssr is listed. Setting I_5 (β) = 1, and

Table II. Summary of Model Building Steps To Determine Model To Represent Dielectric Relaxation Data of Polycarbonate

Test No.	Comments	ssr ($\times$ 10^5)
1	Use I_1, I_2, I_3, C_3, and I_5 (β), set I_4 (α) = 1, and all other parameters set to 0.	4.13
2	Same as test 1 but set I_5 = 1, use I_4 as a variable of regression, and all other parameters set to 0.	4.77
3	Use I_1, I_2, I_3, C_3, I_4 (α), and I_5 (β) as variables of regression and all other parameters set to 0.	3.67
4	Introduce C_2 as a variable of regression into test 3.	3.27
5	Test 4 plus any other parameter in the general model eqs 9–13 does not significantly reduce ssr and in addition includes 0 in their 95% confidence interval.	—

Table III. Parameters and Limits for Representing Polycarbonate as a Function of Temperatures in Terms of Model Eqs 19–23 and Setting $T_0 = 164.1$

Parameter	Limit	Parameter	Limit	Parameter	Limit
I_1	3.651	I_3	6.63	I_5	0.27
σ	0.002	σ	0.08	σ	0.02
C_1	—	C_3	-79.9	C_5	—
σ	—	σ	0.06	σ	—
I_2	3.13	I_4	0.79	σ' (%)	0.5
σ	0.01	σ	0.03	σ''	13
C_2	-0.002	C_4	—		
σ	0.0003	σ	—		

permitting I_4 (α) to be a regression variable is test number 2 in Table II. Finally, permitting α and β to be regression variables leads to case 3. These results show that once again α and β must be included in the polymer model equations. In a similar way, all the other parameters in the model eqs 19–23 were tested and only B_i was found to be significant. The result of this stepwise regression procedure is a seven-parameter model, and the parameters are listed in Table III. We see that α and β are different from 1 by 7 and 35 σ, respectively. Also, the number of experiments is 171 and the number of parameters is 7, so that the ratio is significant. A plot of the experimental and expected values for $\varepsilon'(\omega)$ is given in Figure 15.

Examples of Biased Analysis

Normally, one does not discuss poor results. In the present work, such results are those that come from analyzing data with built-in prejudices by not examining the alternatives. As a result, some of the conclusions reported in the literature about the relaxation parameters are not only *biased* but wrong. Commenting on such results is important because otherwise readers will assume that the reported results, hence their conclusions, are statistically valid. As indicated earlier, Böttcher and Bordewijk were concerned about this state of affairs. In this section we consider a number of mistakes that have been made in the literature.

Bergmann Analysis

Some workers analyzed relaxation data by using the method of Bergmann (*32*). This method notes that $\varepsilon''(\omega)$ for two Debye processes can be written as

$$\frac{\varepsilon''(\omega)}{\varepsilon_0 - \varepsilon_\infty} = \frac{C_1 \omega t_1}{1 + \omega^2 t_1^2} + \frac{C_2 \omega t_2}{1 + \omega^2 t_2^2} \tag{25}$$

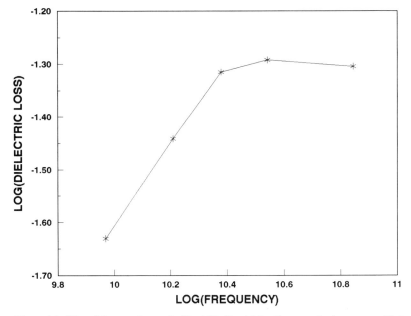

Figure 16. Plot of the experimental of log($\varepsilon''(\omega)$) with log(frequency) of an unspecified material. The low frequency slope is $\alpha = 0.79$.

where C_1 and C_2 are the relative weights for each process and $C_1 + C_2 = 1$. Equation 25 is a four-parameter model and requires just as many parameters as does the HN and one more than for the CC and DC equations. Furthermore, the proponents of the Bergmann method never compared the results of eq 25 with results from the C–C or the C–D models to determine the best method of data representation.

Consider the case for the data in Figure 16. The six experimental quantities in Figure 16 were used to estimate the parameters of eq 25. This estimation is done by plotting the data as $\varepsilon''(\omega)$ against $\varepsilon''(\omega)\log(\omega)$, which should be a straight line for a single Debye process. In other words deviations from linear Debye behavior were used to estimate the parameters of a second or even third relaxation time, hence process (Figure 17). These data were interpreted in terms of four relaxation processes. It is also obvious that the relaxation times are predetermined by the arbitrary spacing of the experimental frequencies.

This analytical procedure was criticized by Copeland and Denney (*33*), who argued against the arbitrary resolution of relaxation data into a small number of Debye processes, particularly when only one loss maximum is observed. Dannhauser (*34*) made somewhat more rigorous arguments about the dangers of assuming two closely spaced relaxation processes.

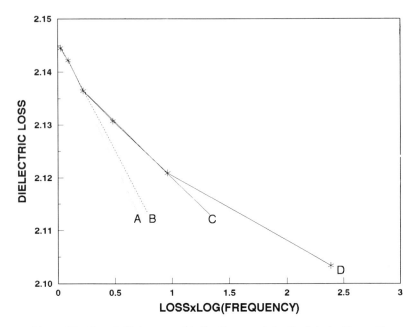

Figure 17. Plot of ε′(ω) versus ε′(ω)log(frequency) for the data in Figure 16.

Overlapping Dispersions

Such arbitrary resolution of relaxation data into two Debye processes is not limited to cases with a limited number of observations. Consider the case shown in Figure 18, which is a mixture of two polar liquids. In this case each component could be represented by a single or Debye process. However, the mixture gave the results shown in Figure 18. The authors tried to resolve this data into two Debye processes. Attempts by us to resolve this data into two processes by using the *unbiased* methods of the previous section failed because two loss peaks are not observed. However, the data could be represented in terms of the HN function (lines in Figure 18), with $\alpha = 1.00 \pm 0.02$ and $\beta = 0.75 \pm 0.05$. In other words, the relaxation process could have been represented in terms of a DC function, which requires fewer parameters than two Debye processes. This representation leads to an entirely different relaxation mechanism and one based on a distribution of relaxation times due to a distribution of local structures.

Conclusions

We would like to express an opinion based on reading thousands of papers, some of which were published in the most prestigious journals. With the

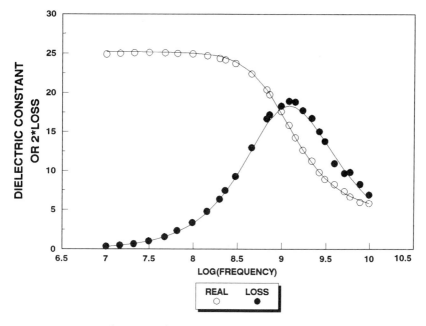

Figure 18. Plot of $\varepsilon'(\omega)$ and $2\varepsilon'(\omega)$ against log(frequency) for an unnamed material.

advent of minicomputers and later personal computers, many investigators choose to write regression routines; we do not recommend such an undertaking. This conclusion is based on the observation that in the papers that we have read, convergence criteria are seldom if ever specified, and none of the papers reports parameter confidence intervals. Should these investigators choose to write their own routines, then we recommend they consult standard statistical sources such as Draper and Smith (*3*). Also, many prepackaged nonlinear regression routines are also limited because they do not report parameter confidence intervals. We also know of regression routines that limit $\beta \leq 1$.

We recommend to these users of *canned* routines that they construct trial data sets with and without various levels of noise and determine if the routines yield the same parameters that were used to construct the trial data. This procedure is similar in concept to the one experimenters use when they build their own dielectric devices; that is, they construct lumped circuits with known precision components and determine if their devices yield the known values of the standards. Testing of canned routines is not any different.

The results in the previous sections show the importance of examining $\log(\varepsilon''(\omega))$ versus $\log$(frequency) data. Jonscher reviewed such plots for 100 materials about 15 years ago, and his results are reproduced here in Figure

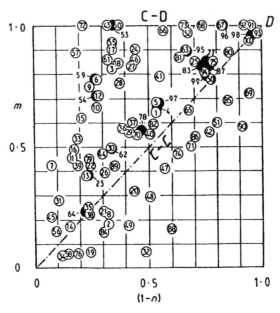

Figure 19. Plot of m *versus 1 −* n *for 100 different materials taken from the work of Jonscher. In this Figure D represents the Debye condition, C-D represents the Cole–Davidson, whereas C-C represents the Cole–Cole condition. (Reproduced with permission from reference 36. Copyright 1971 American Institute of Physics.)*

19 (*35*). The numbers next to the circles identify the materials in the original reference. The data are spread throughout the entire plane. We see no clustering around the coordinates 1, 1 (D in Figure 19), a result indicating that few if any processes are Debye-like. The Gaussian function is also ruled out. There are 12 data points along the DC line that indicate that only a maximum of 12 dispersions out of 100 are of the KWW or DC types. Only a few data points fall on the CC line and indicate a maximum number for the CC or Fuoss–Kirkwood types. On the basis of these results, the HN function is the only function that represents most if not all of the data.

We published a second important set of results in references 11 and 26 and reproduced the results in Figure 20. In Figure 20 alpha refers to the α-process in polymers. Beta signifies the β-process in polymers. Mashimo represents an α,β pair determined from a composite loss curve of polymer solutions (*36*). Log(β) was chosen for the *x*-axis because it has a considerable range. The line labeled *limit* in Figure 20 represents the condition αβ = 1. Solution represents those measurements on dilute polymer solutions. Considerable overlap exists between the α- and β-processes. Also, a surprising number of α-processes have parameters that are similar to those

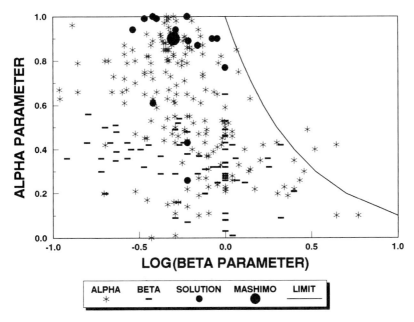

Figure 20. Plot of α versus log(β) for the classes of materials listed in the legend. See text for definition of the terms in the legend.

of dilute solutions. The results support those of Jonscher. Little evidence exists to support the Debye, CC, DC, KWW, or Gaussian functions.

References

1. Jonscher, A. K. *Dielectric Relaxation in Solids;* Chelsea Dielectric: London, 1983; p 5, Figure 1.1.
2. Böttcher, C. J. F.; Bordewijk, P. *Theory of Electric Polarization: Vol. II. Dielectrics in Time-Dependent Fields;* Elsevier: New York, 1978; pp 116–117.
3. Draper, N. R.; Smith, H. *Applied Regression Analysis,* 2nd ed.; Wiley: New York, 1981.
4. Havriliak, S., Jr.; Negami, S. *J. Polym. Sci. C.* **1966,** *14,* 99.
5. Havriliak, S. Jr.; Negami, S. *Polymer,* **1967,** *8,* 161.
6. In this work we use the term dielectric constant, meaning that it is an intensive property of the material and not independent of temperature or frequency. The reader may wish to substitute the term relative permittivity in its place.
7. Cole, K. S.; Cole, R. H. *J. Chem. Phys.* **1941,** *9,* 341.
8. Davidson, D. W.; Cole, R. H. *J. Chem. Phys.* **1950,** *18,* 1417.
9. Jonscher, A. K. *Dielectric Relaxation in Solids;* Chelsea Dielectric: London, 1983; p 199.
10. Havriliak, S., Jr.; Havriliak, S. J. *J. Non-Crystal Solids* **1994,** *172–174,* 297.
11. Wagner, K. W. *Ann. Phys. I,* **1913,** *40(4),* 817.
12. Böttcher, C. J. F.; Bordewijk, P. *Theory of Electric Polarization: Vol. II. Dielectrics in Time-Dependent Fields.;* Elsevier: New York, 1978; pp 83–85.

13. Lichter, J. J.: McDuffie, G. E. *Tables of the Davidson–Cole and The Log Gaussian Distribution Functions of Relaxation Times;* Report No. NOLTR 64–170, U.S. Naval Ordinace Laboratory: White Oak, MD, 1964.
14. Fuoss, R. M.; Kirkwood, J. G. *J. Am. Chem. Soc.* **1941,** *63,* 385.
15. Jonscher, A. K. *Dielectric Relaxation in Solids;* Chelsea Dielectric: London, 1983; p 199.
16. Kohlrausch, R. *Pogg. Ann. Phys. Chem.* **1854,** *91,* 179.
17. Kohlrausch, F. *Pogg. Ann. Phys. Chem.* **1963,** *119,* 337.
18. Williams, W.; Watts, D. G. *Trans. Faraday Soc.,* **1970,** *66,* 80.
19. Koizumi, N.; Kita, Y. *Bull. Inst. Chem. Res.* **1978,** *56(6),* 300, (Kyoto University).
20. Dishon, M.; Weiss, G. H.; Bendler, J. T. *J. Res. Natl. Bur. Stand.* **1985,** *90(1),* 27.
21. Havriliak, S., Jr.; Havriliak, S. J. *Polymer* **1995,** *36(14),* 2675.
22. Alvarez, F.; Alggria, A.; Colmenero, J. *Phys. Rev. B.,* **1991,** *44(11),* 7306.
23. Alvarez, F.; Alggria, A.; Colmenero, J. *Phys. Rev. B.* **1993,** *47(1),* 125.
24. Havriliak, S., Jr.; Watts, D. J. *Design, Data, and Analysis, by some Friends of Cutbert Daniels;* Mallows, C., Ed.; Wiley: New York, 1986.
25. Havriliak, S., Jr.; Watts, D. G. *Polymer,* **1986,** *27,* 1509.
26. Havriliak, S., Jr.; Havriliak, S. J. *J. Mol. Liq.* **1993,** *56,* 49.
27. Havriliak, S., Jr.; Havriliak, S. J. *Dielectric and Mechanical Relaxation in Materials: Analysis and Application to Polymers, Their Solutions and Other Systems;* Hanser Verlag: New York, 1996.
28. Havriliak, S., Jr. *Colloid Polym. Sci.* **1990,** *268,* 268.
29. Bueche, F. *J. Chem. Phys.* **1954,** *21(10),* 1850.
30. SAS®, SAS®Institute, SAS® Circle, Box 8000, Cary, NC, 27512–8000.
31. The data in Figure 14 of reference 5 are only a single temperature of a much more complete data set.
32. Bergmann, Ph.D. thesis, Freiburg/Bresgau, West Germany, 1957.
33. Copeland, T. G.; Denney, D. J. *J. Phys. Chem.* **1976,** *80(2),* 210.
34. Dannhauser, W. *J. Chem. Phys.* **1971,** *55,* 629.
35. Jonscher, A. K. *J. Chem. Phys.* **1971,** *55,* 199, Figure 5.27.
36. Mashimo, S.; Chiba, A. *Polym. J.,* **1973,** *5,* 41.

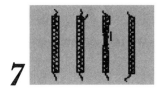

7

Application High-Frequency Dielectric Measurements of Polymers

Satoru Mashimo†

Dielectric measurements in the megahertz to gigahertz region are very important for investigating molecular dynamics of polymers, especially in solution. In general polar polymers with a dipole component perpendicular to chain contour in organic solvents show a dielectric dispersion in this region. The relaxation is well interpreted by employing the Kohlrausch–Williams–Watts (KWW) function as a dielectric response function. However, a definite deviation of experimental absorption curve from that calculated by the function is commonly seen on the high-frequency side of the curve. This deviation may be explained by the mode-coupling theory. In the case of nonelectrolyte solutions, chain dynamics investigated by dielectric studies are interpreted as the same mechanism as that in organic solvent. Examples are shown for poly(vinyl pyrrolidone) and poly(vinyl methyl ether). In the case of polyelectrolyte solutions, the chain dynamics are explained in the same way as those in nonelectrolyte solutions. This result is shown for poly(glutamic acid) in aqueous solution. Aqueous solution of biopolymer is usually an electrolyte solution, and the biopolymer generally has a configurational structure. As examples, DNA usually takes a double helix of B-type, and collagen usually takes a triple helix. Proteins such as albumin and lysozyme take globule structures.

† Deceased.

In these biopolymers in aqueous solution, a dielectric relaxation peak due to
bound water was found. These findings are the first of bound water irrespec-
tive of bulk water.

Dielectric behavior of a polymer at a frequency region higher than
10 MHz has not been examined satisfactorily. In the case of amorphous
polymers, a frequency region lower than 1 MHz was used exclusively for the
dielectric study on the chain motion in connection with a glass transition
phenomenon. The motion at the glass transition temperature shows a dielectric absorption peak at about 1 mHz (*1*). Therefore, dielectric measurements in the high-frequency region were not thought of as important for
the study of the chain motion. Recently, the dielectric relaxation due to
the chain motion was explained by the Kohlrausch–Williams–Watts (KWW)
function (*2, 3*). A comparison of the experimental complex permittivity
with that calculated by the KWW function has been carried out extensively
for many amorphous polymers. Agreement between these values is considerably good in a frequency range $f/f_m < 10^2$, where f_m is a frequency of
the absorption peak (*3*). However, disagreement between them is definite
in the frequency range $f/f_m > 10^2$. Experimental dispersion or absorption
is always larger than that calculated in the frequency range (*3*).

In the case of polymer solution, the absorption or the dispersion caused
by the chain motion can be explained well by the KWW function (*4*). In
this case too, agreement is almost complete in the frequency range f/f_m
$< 10^2$. The fact that the dielectric behavior of polymer in solution can be
explained well by the coupling model (*5*) means that correlation of the
chain motions along the same chain is caused by the chain connectivity.
Even in dilute solution the dielectric response function is of the KWW
function type (*4*).

In general polar polymers in dilute solution of nonpolar or weakly polar
solvent show an absorption peak at a frequency between 10 MHz and 1 GHz,
if they have a dipolar component perpendicular to their chain contour (*6*).
Frequency of the peak does not depend on molecular weight of the polymer
(*6, 7*). Relaxation time is approximately proportional to the solvent viscosity, and apparent activation energy for the relaxation process is a sum of
that of the solvent viscosity and a constant (*7*). The constant takes a value
of 12–16 kJ/mol, which is reasonable for the barrier height for rotation
around a chain bond. Frequency dependence of the dispersion and absorption curves is well explained by the KWW function (*4*).

On the other hand the chain motion in polar solvent such as water has
not been sufficiently clarified. The solution is often a polyelectrolyte solution and has a big dc conductivity. Further, water has a big static dielectric
constant and a big dispersion around 10 GHz. These characteristics prevent
an accurate measurement of dielectric relaxation due to the motions of

polar polymer. However, it is very important to know dielectric behavior of the aqueous polymer solutions, because the solutions of biopolymer such as DNA or protein are always electrolyte solutions and their dynamical features are, in many cases, related to their biological functions. Even in the case of nonelectrolyte aqueous solution, dielectric behavior of polymer is very interesting whether it can be interpreted in the same manner as that of the polar polymer in nonpolar solvent.

In a frequency region covering a mega- to gigahertz region, an accurate dielectric measurement on the aqueous solution of polymer by the conventional methods using impedance bridges has been thought impossible. The method usually does not reach the megahertz region. Further measurement of standing wave in a wave guide is considerably complicated and gives no precision data for dilute polymer solution.

In 1980, time-domain reflectometry (TDR) was shown to have possibility for measuring dielectric properties of polymers in dilute solution in the mega- to gigahertz region (*8*). Ten years later, the aqueous polymer solution including electrolyte solution can be measured with sufficient accuracy by TDR (*9*). Development of TDR in these 20 years is remarkable. A frequency range from 100 kHz to 30 GHz can be covered by TDR, and dielectric properties of aqueous polymer solutions are going to be clarified quickly. In this chapter the TDR method is described in detail as a new experimental technique of dielectric measurement, and dielectric properties of aqueous polymer solutions measured first by TDR are shown for further discussions.

Time-Domain Reflectometry

When an electric wave passing through a coaxial line is reflected at the top of the line, where a dielectric sample cell is attached, the reflective coefficient is given as a function of complex permittivity of the sample, an electric cell length, and a geometric cell length. When a step-pulse voltage is used as an incident pulse passing through the coaxial line and its reflected wave form is measured in time domain, a ratio of Fourier transforms of the reflected wave and the incident pulse is given as a function of complex permittivity of the sample. The permittivity ε^* of the sample is given by (*10*)

$$\varepsilon^* = \frac{c}{j\omega\gamma d} \frac{v_0(j\omega) - r(j\omega)}{v_0(j\omega) + r(j\omega)} \, x \cot x \qquad (1)$$

where

$$x = \frac{\omega d}{c} \varepsilon^{*1/2} \qquad (2)$$

Here c is the speed of propagation in vacuo, γd is the electric length of the sample cell, d is the geometric cell length, ω is the angular frequency, j is the imaginary unit, $v_0(j\omega)$ is the Fourier transform of the incident pulse $V_0(t)$, and $r(j\omega)$ is that of the reflected wave form $R(t)$. The term $x \cot x$ accounts for multiple reflections in the sample section.

If two measurements of the sample (x) with unknown permittivity ε_x^* and the reference sample (s) with known permittivity ε_s^* are performed, the permittivity ε_x^* is given by (11)

$$\varepsilon^* = \varepsilon_s^* \frac{1 + \{(cf_s)/[j\omega(\gamma d)\varepsilon_s^*]\}\rho}{1 + \{[j\omega(\gamma d)]\varepsilon_s^*/cf_s\}\rho} \frac{f_x}{f_s} \tag{3}$$

where

$$\rho = \frac{\displaystyle\int_0^\infty [R_s(t) - R_x(t)]e^{-j\omega t}\,dt}{\displaystyle\frac{2}{j\omega}\int_0^\infty \left[\frac{d}{dt}R_s(t)\right]e^{-j\omega t}\,dt - \int_0^\infty [R_s(t) - R_x(t)]e^{-j\omega t}\,dt} \tag{4}$$

and

$$f = x \cot x \tag{5}$$

$R_s(t)$ is the reflected wave form from the reference sample, and $R_x(t)$ is that from the unknown sample. Equation 3 is derived easily, because both ε_x^* and ε_s^* can be written by eq 1.

In the measurement of electrolyte solution, we chose water with sodium chloride as the reference sample, and we adjusted concentration of sodium chloride to have the same dc conductivity of the unknown sample. The permittivity of water with sodium chloride is measured before the measurement of the unknown sample. The permittivity of water with sodium chloride is nearly the same as that of pure water except for the dc conductivity if the concentration of the sodium chloride is low. Therefore, for the measurement of electrolyte solution, we use $\varepsilon_s^*(\omega) - \sigma/j\omega$ instead of $\varepsilon_s^*(j\omega)$ in eq 3, where σ is the adjusted dc conductivity of reference sample.

At first, a time window within which $R_s - R_x$ reaches zero is employed to measure the total wave form of $R_s - R_x$. The time t_1 is chosen so that $R_s - R_x$ reaches zero at $t = t_1$. Then, we get

$$\rho = \frac{\displaystyle\int_0^{t_1} [R_s(t) - R_x(t)]e^{-j\omega t}\,dt}{\displaystyle 2\int_0^{t_1} [R_s(t)]e^{-j\omega t}\,dt - \int_0^{t_1} [R_s(t) - R_x(t)]e^{-j\omega t}\,dt} \tag{6}$$

There are 500 points of time intervals in the time window of the digitizing oscilloscope (HP54121 T). Therefore, the limit of the highest frequency f_m obtained from the Fourier transform is determined by the time interval of Δt as $f_m = 1/(2\Delta t)$. Then, the measurements of R_s and R_x in shorter time ranges are required to obtain more precise data in a higher frequency region. If $R_s - R_x$ does not drop to zero at the longest time, t_2, in the time window employed, the value $R_s - R_x$ at $t > t_2$ is assumed to be constant, which is determined as the average of the last 20 or 30 points of $R_s - R_x$. We connect two complex permittivities measured using two time windows in the overlapping area, from which dc conductivity is already subtracted. They are usually in good agreement with each other in the area. In the case where the value of $R_s - R_x$ in the vicinity of $t = t_2$ decreases apparently, this procedure sometimes does not work well. In such a case, an exponential tail of $A\exp(-t/\tau_a)$ is added to $R_s - R_x$. The values of A and τ_a are determined by a least-square fitting procedure using the last 50 points. The procedure always works well, and the agreement of two complex permittivities in the overlapping area is satisfactory. However, if $R_s - R_x$ does not change much, the procedure with a constant value is much easier. In this measurement, we employed four kinds of cells made of gold having $\gamma d = 4.3, 1.9, 0.38,$ and 0.1 mm and $d = 2.1, 1.5, 0.3,$ and 0.01, respectively.

A fast measurement can be done easily even in a wide frequency range from 100 kHz to 1 GHz by employing the TDR method. An experimental error both for the relaxation strength and the relaxation time of the relaxation process is within $\pm 3\%$, and it is unavoidable because of a slight jitter and the signal-to-noise ratio of the signals R_s or R_x. In the case where the dispersion concerned is close to the dispersion of the solvent, the error is reduced to within $\pm 1\%$ in a frequency range higher than 1 GHz. Nevertheless, the TDR method is very useful for measuring the multiple relaxation processes in the frequency region from 100 kHz to 20 GHz and extracting each relaxation, because the dispersion and absorption curves can be obtained continuously in the frequency domain.

Dielectric Relaxation Due To Micro-Brownian Motion Described by KWW Function

Micro-Brownian motion of the polymer chain is one of the most important subjects in polymer physics. Relaxation phenomena such as viscoelastic relaxation, dielectric relaxation, and NMR relaxation are dominated more or less by such motions. Recent studies on the micro-Brownian motion showed that the response function of the relaxation is of the KWW type for a variety of polymers.

Dielectric relaxation measurement is a reliable method of investigating the motion of the polymer if the polymer is polar. In the case of amorphous polymers, the complex permittivity ε^* can be expressed over a frequency range of $f/f_m < 10^2$, where f_m is the frequency of the maximum dielectric loss, by the KWW function $\Psi_K(t)$ as (2)

$$\frac{\varepsilon^* - \varepsilon_\infty}{\varepsilon_0 - \varepsilon_\infty} = \int_0^\infty \exp(-j\omega t)\left[\frac{-d\varphi_K(t)}{dt}\,dt\right]dt \qquad (7)$$

and

$$\varphi_K(t) = \exp[-(t/\tau)^{\beta_K}] \qquad 0 < \beta_K < 1 \qquad (8)$$

where ε_0 and ε_∞ are static and limiting high-frequency dielectric constants, respectively; τ is a relaxation time; and β_K is a coupling parameter.

As an example, dielectric dispersion and absorption curves of poly(vinyl acetate) (PVAc) (1) are shown in Figure 1. Dispersion and absorption curves at frequencies lower than f_m are well explained by eqs 7 and 8. However in a high-frequency region of $f > 10^2 f_m$ both dispersion and absorption curves deviate definitely from the calculated ones.

The coupling model suggests the deviation of dispersion and absorption curves from those calculated by the KWW function. According to the model (5), the response function $\Psi_K(t)$ is given as

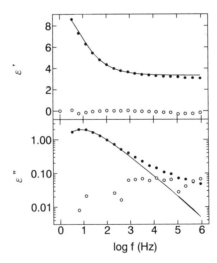

Figure 1. Comparison of dielectric dispersion and absorption curves obtained experimentally at 53 °C with those calculated by the KWW function. Closed circles are obtained experimentally for the complex permittivity. Solid lines indicate the calculated permittivity from eqs 7 and 8. Open circles are the difference between them.

$$\varphi_K(t) = \exp\left[-\left(\frac{t}{\tau^*}\right)^{1-n} \right] \qquad t \gg t_c \tag{9a}$$

and

$$\varphi_K(t) = \exp\left(-\frac{t}{\tau_0} \right) \qquad t \ll t_c \tag{9b}$$

where

$$\tau^* = [(1 - n)\omega_c^n \tau_0]^{1/1-n} \tag{10}$$

and n is a coupling constant related to β_K as

$$1 - n = \beta_K \tag{11}$$

Usually t_c is of the order of 10 ps. Therefore, the complex permittivity calculated by the coupling model on the high-frequency side deviates sometimes from that predicted by the KWW function and shows a behavior of the Debye type. However, the deviation obtained experimentally cannot be explained definitely by the model. Its imaginary part may be a constant on the high-frequency side or it may be due to the relaxation with an extremely wide distribution of relaxation times.

In Figure 2, dielectric dispersion and absorption curves for concen-

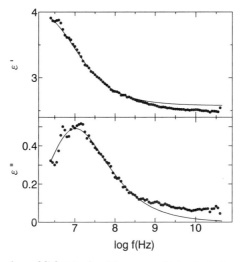

Figure 2. Comparison of dielectric absorption curve obtained experimentally for solution of PVAc (40 wt%) in benzene at 6 °C with that calculated by the KWW function.

trated PVAc (40 wt%) in benzene measured by the TDR method are shown to indicate clearly that the absorption observed deviates from that calculated by eq 7 on the high-frequency side in this case, too. This feature may be universal for all polymers.

In the case of dilute polymer solution in organic solvent, the relaxation time observed is completely independent of the molecular weight of the polymer, if the polymer is atactic and each repeat unit has a dipolar component perpendicular to the chain contour (6, 7). The relaxation is of the KWW type. Furthermore, a dielectric study on (4-chlorostyrene, 4-methylstyrene) copolymers in benzene indicated that even the dielectric response function of isolated dipoles in the chain is expressed as the KWW function (4).

The deviations of dispersion and absorption from those calculated by the KWW function are explained by the mode-coupling theory (12). However, further data are required to discuss this problem.

Micro-Brownian Motion in Aqueous Solution

In aqueous polymer solution the motion of polymer chain has not been observed. A clear evidence of the observation has not been reported yet. Polyethylene glycol (PEG) shows a dielectric relaxation process in aqueous solution in the gigahertz region (9). The absorption peak shifts to high frequency as the polymer concentration decreases and that extrapolated to zero concentration agrees with that of pure water. However, PEG shows a relaxation peak at 5 GHz and 25 °C in dilute benzene solution, which locates close to that of pure water. These results imply that the relaxation observed in aqueous solution is caused by orientation of water molecules, supplemented by motions of the polymer chain.

Poly(vinyl pyrrolidone) (PVP) shows two relaxation peaks in aqueous solution (9, 13). Dielectric dispersion and absorption curves for PVP with a viscosity average molecular weight of 4.0×10^4 in aqueous solution are shown in Figures 3 and 4. The peak observed at the high frequency coincides with that of pure water if the solution is dilute.

The relaxation of the PVP aqueous solution is thus represented as a sum of two relaxations

$$\varepsilon^* - \varepsilon_\infty = \varepsilon_h^*(\omega) + \varepsilon_l^*(\omega) \tag{12}$$

In the case of low polymer concentration, $\varepsilon_h^*(\omega)$ for the high-frequency process can be explained by the Cole–Cole representation

$$\varepsilon_h^* = \frac{\Delta\varepsilon_h}{1 + (j\omega\tau_h)^{\beta_h}} \tag{13}$$

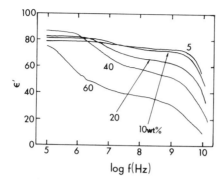

Figure 3. Dielectric dispersion curves for PVP aqueous solutions with various concentrations at 25 °C. (Reproduced with permission from reference 13. Copyright 1992 American Institute of Physics.)

where $\Delta\varepsilon_h$ is the relaxation strength, τ_h is the relaxation time, and β_h is the Cole–Cole parameter. If the permittivity $\varepsilon_h^*(\omega)$ given by eq 13 and ε_∞ are subtracted from the total permittivity $\varepsilon^*(\omega)$, the remaining permittivity $\varepsilon_l^*(\omega)$ for the low-frequency process can be fitted to those calculated by eqs 7 and 8, as shown in Figure 5. The low-frequency relaxation is of the KWW type. For the solution with high polymer concentration, the dispersion and absorption curves are similarly explained by the sum of two types of relaxations. The absorption and dispersion curves for the low-frequency process are fitted to those calculated by eqs 7 and 8 (Figure 5).

Relaxation caused by the micro-Brownian motion is commonly explained by employing the KWW function as a response function. The relaxation peak for dilute solution is found at a frequency between 10 and 100

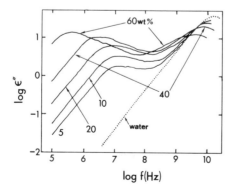

Figure 4. Dielectric absorption curves for PVP aqueous solutions with various polymer concentrations at 25 °C. (Reproduced with permission from reference 13. Copyright 1992 American Institute of Physics.)

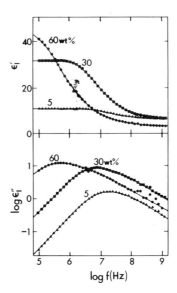

Figure 5. Dispersion and absorption curves for the low-frequency process of PVP aqueous solutions with c = 5, 30, and 60 wt%, respectively. The solid curves are calculated by the KWW function. (Reproduced with permission from reference 13. Copyright 1992 American Institute of Physics.)

MHz (6, 7), where the peak of PVP solution with low polymer concentration locates.

The relaxation time for the motion depends on the concentration and increases with increasing concentration. In the case of poly(vinyl acetate) near the glass transition temperature T_g, the relaxation time for the toluene solution with 68 wt% polymer concentration is about 30 times greater than that for the dilute solution (14). This trend is quite similar to the present case of PVP, where τ_l for a 60 wt% solution is 30 times longer than that for the dilute solution. Similar behavior is observed for a variety of polymers, such as polystyrene (15) and poly(p-chlorostyrene) (6). If the polymer–diluent system is a uniform mixture, concentration dependence of the relaxation time can be explained by the free-volume theory. In the case of PVP solution, dependence of τ_l on the polymer concentration is described by an equation of the free-volume type (16)

$$\log \tau_l = -9.77 - \frac{1.35}{C_d - 2.07} \qquad \text{at 25 °C} \qquad (14)$$

where C_d is the concentration expressed by g/mL. The value of $\log \tau_l$ extrapolated to pure polymer ($C_d = 1.24$ g/mL) is 9.5, which is reasonable for $T < T_g$. The T_g of PVP is 50–60 °C (17). This result offers evidence that the relaxation caused by the micro-Brownian motion of polymer chain is related to the glass transition phenomenon. The relaxation strength $\Delta\varepsilon_1$ depends on the polymer concentration. If the dipole moment $\langle\mu^2\rangle^{1/2}$ of a repeat unit is calculated for dilute solution by using Debye formula

$$\langle\mu^2\rangle = \frac{9kTM}{4\pi N_A C_d} \frac{3\Delta\varepsilon_1}{(\varepsilon_\infty + \Delta\varepsilon_h + 2)^2} \tag{15}$$

where M is molecular weight of the monomer unit, k is the Boltzmann constant, and N_A is the Avogadro's number, a value of 0.52 D is obtained for the perpendicular component of the dipole moment of the repeat unit. This value compares quite reasonably with the calculated component of 0.7 D, which is obtained by taking the dipole moment for 2.3 D of carbonyl group into account.

The coupling parameter β_1 is given as a linear function of C_d,

$$\beta_1 = 0.662 - 0.182C_d \tag{16}$$

Ngai and co-workers (5) have shown that $n = 1 - \beta_1$ is a parameter representing a degree of coupling strength brought from the molecular environment. Inter- and intrachain interactions increase with the increase of polymer concentration. Accordingly, β_1 decreases with C_d.

Aqueous solution of poly(vinyl methyl ether) also shows two relaxation peaks (*18*). One observed around 10 GHz is due to orientation of water molecules and the other observed in a frequency region of 10–100 MHz is caused by the micro-Brownian motion of the polymer chain. The solution has a lower critical solution point (LCST) (*19*). This characteristic means that intra- and interchain interactions increase as the temperature increases and a localized condensation occurs in the vicinity of LCST. In Figure 6, the coupling parameter β_1 is plotted against temperature. It decreases with

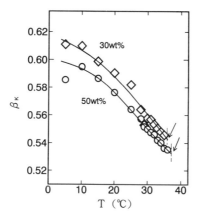

Figure 6. Temperature dependence of the coupling parameter β_k for poly(vinyl methyl ether) in aqueous solutions. Phase-separation temperature is indicated by an arrow.

the increase of temperature. This result offers evidence of the validity of the coupling model.

When an aqueous polymer solution is nonelectrolytic and the polymer has the perpendicular component, dielectric relaxation due to the micro-Brownian motion of the polymer chain is well described by the KWW response function. On the other hand, dielectric behavior of electrolyte polymer solution is still not clear, because the solution always has a big dc conductivity (20). Therefore it is thought to be difficult to extract a small relaxation peak due to the chain motion from the total absorption observed. However, the TDR method is very useful for such an electrolyte solution.

Molecular Motions of Polyelectrolyte Solutions

Poly(glutamic acid) (PGA) is one of the good analogs to investigate molecular dynamics of real proteins. It exhibits a helix-coil transition in aqueous solution around pH = 5.20. If the pH value is less than 5, it prefers the helix structure. The transition could be seen in dielectric property of the solution (21).

The repeat unit of PGA has two dipole components: one is perpendicular to the chain contour and the other is parallel to it. Dielectric relaxation due to the perpendicular component reflects the micro-Brownian motion of the chain, and that due to the parallel component reflects the overall rotation of the polymer. The relaxation peak caused by the perpendicular component is usually located at a frequency around 100 MHz. If PGA takes a helix structure, hydrogen bonds between oxygen and protons bring a fairly big parallel component and the resultant dipole moment is proportional to the end-to-end vector of the polymer. Therefore, dielectric relaxation strength and relaxation time depend on the molecular weight of the polymer. If PGA takes a randomly coiled structure, the relaxation caused by the parallel component has a small relaxation strength and a short relaxation time.

Relaxation peak of PGA in aqueous solution around 100 MHz was first observed by the use of TDR (21). After this observation, it is suggested that the peak has a strong possibility of being due to truncation error of the Fourier transform or evaluation of the dc conductivity of the electrolyte solution (22). However, a high-frequency impedance measurement on the PGA aqueous solution has exhibited the existence of the relaxation peak between 10 and 100 MHz (23).

Once it was thought that the TDR method is inadequate to measure the electrolyte materials such as PGA aqueous solution. It was believed almost impossible to evaluate the contribution of the dc conductivity to the complex permittivity, especially in the low-frequency region. However, it was shown recently that if a standard sample with known permittivity and

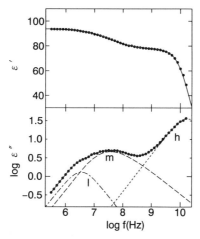

Figure 7. Frequency dependence of dielectric dispersion and absorption for a PGA aqueous solution (5 mg/cm³, pH = 6.8) at 25 °C. ● are obtained experimentally. The solid curve is calculated by eq 18. Three relaxation peaks denoted by l, m, and h can be seen definitely for the absorption curve at about 4 MHz, 30 MHz, and 18 GHz, respectively.

dc conductivity is employed as a reference and its dc conductivity is adjusted to be nearly the same as that of the unknown sample, the complex permittivity of the unknown sample can be obtained as a function of the ratio of Fourier transforms of two reflected waves from the known and the unknown samples.

The dielectric absorption curves of PGA, the viscosity-average molecular weight of which is 5.1×10^4, in aqueous solution with pH = 6.77 show three relaxation peaks (Figure 7) (24). On the other hand for the solution with the pH value less than 4.75, only two relaxation peaks could be observed and the peak observed at a frequency between 10 and 100 MHz for the solution with the pH value greater than 5 disappeared. The high-frequency process, the peak of which is located at 18 GHz, can be attributed to the bulk water process, judging from the relaxation time and relaxation strength. This result is because the solution is dilute and the region of the pH value is not so wide that the relaxation time and relaxation strength do not change with the pH value.

The low-frequency process can be explained sufficiently by the Debye relaxation, even though the relaxation time and the relaxation strength depend on the pH value. If the complex permittivity is described by the Cole–Davidson representation

$$\varepsilon_l^* = \frac{\Delta \varepsilon_l}{(1 + j\omega \tau_l)^{\alpha_l}} \tag{17}$$

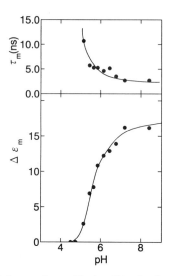

Figure 8. Variation of $\Delta \varepsilon_m$ and τ_m with the pH value for a PGA aqueous solution (5 mg/cm^3) at 25 °C.

the solution with the pH value greater than 5 takes a value of unity for α_l. Only a slight deviation from unity is observed for the solution with the pH value less than 5. Even in this case, α_l takes a value about 0.95, close to unity.

The intermediate process can be thus extracted in the manner described by

$$\varepsilon_m^* = \varepsilon^* - \left[\varepsilon_\infty + \frac{\Delta \varepsilon_h}{1 + j\omega\tau_h} + \frac{\Delta \varepsilon_l}{(1 + j\omega\tau_l)^{\alpha_l}} \right] \quad (18)$$

The complex permittivity of the intermediate process m is well explained by the KWW function. The relaxation time τ_m and the relaxation strength $\Delta \varepsilon_m$ for the process are plotted against the pH value in Figure 8. An abrupt change is seen in $\Delta \varepsilon_m$ at pH ~ 5.5.

In the case of the low-frequency process, $\Delta \varepsilon_l$ changes little with the pH value if the value is greater than 6.5, but steeply changes in the vicinity of pH = 5.5. Similarly, τ_l changes with the pH value (Figure 9).

To trace the conformational change, the molar ellipticity at 222 nm $[\theta]_{222}$ was measured by circular dichroism (CD) spectrophotometer (J-600) and the fraction of the coil in PGA was estimated as

$$f_c = \frac{[\theta]_{222} - [\theta]_{222}^h}{[\theta]_{222}^c - [\theta]_{222}^h} \quad (19)$$

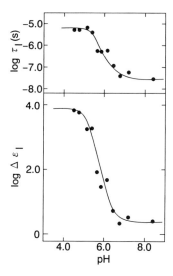

Figure 9.   Variation of $\Delta\varepsilon_l$ and τ_l with the pH value for a PGA aqueous solution (5 mg/cm^3) at 25 °C.

where $[\theta]_{222}^c$ and $[\theta]_{222}^h$ are the limiting molar ellipticity of random coil and that of helix, respectively. In Figure 10, plots of $\Delta\varepsilon_m$ and τ_m against f_c are shown, and those of $\Delta\varepsilon_l$ and τ_l are shown in Figure 11.

The total dipole moment $\overrightarrow{M}$ of PGA with degree of polymerization n is written as

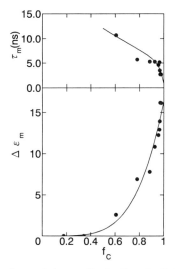

Figure 10.   Dependence of τ_m and $\Delta\varepsilon_m$ on the fraction of coiled part in a PGA aqueous solution at 25 °C. The solid curve for $\Delta\varepsilon_m$ is calculated by eq 34.

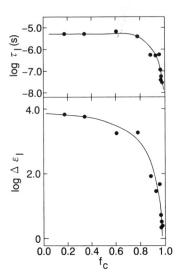

Figure 11. Dependence of $\log \tau_l$ and $\log \Delta \varepsilon_l$ on the fraction of coiled part in a PGA aqueous solution at 25 °C.

$$\vec{M} = \sum_{i=1}^{n} \vec{\mu}_{\perp i} + \sum_{i=1}^{n} \vec{\mu}_{\parallel i} \qquad (20)$$

where $\mu_{\perp i}$ is the dipole moment of ith repeat unit perpendicular to the chain contour and $\mu_{\parallel i}$ is that parallel to the contour. The time-dependent correlation function is given by

$$\langle \vec{M}(0)\vec{M}(t)\rangle = \left\langle \sum_{i=1}^{n} \vec{\mu}_{\perp i}(0) \sum_{i=1}^{n} \vec{\mu}_{\perp i}(t) + \sum_{i=1}^{n} \vec{\mu}_{\parallel i}(0) \sum_{i=1}^{n} \vec{\mu}_{\parallel i}(t) \right\rangle \qquad (21)$$

In general the perpendicular component does not correlate with the parallel one. The perpendicular component brings about a dielectric relaxation independent of the molecular weight of the polymer and the parallel component causes a relaxation depending on it. The first term in eq 21 is rewritten as

$$\left\langle \sum_{i=1}^{n} \vec{\mu}_{\perp i}(0) \sum_{i=1}^{n} \vec{\mu}_{\perp i}(t) \right\rangle = \mu_{\perp}^{2} \left[\sum_{i=1}^{n} \Psi_{\perp i}(t) + \sum_{i=1}^{n} \sum_{l=1}^{n} \Psi_{\perp l}(t) \right] = n\mu_{\perp}^{2} g(t)$$
$$\underset{(i \neq l)}{} \qquad (22)$$

where $\psi_{\perp i} = \langle \vec{\mu}_{\perp i}(0)\vec{\mu}_{\perp i}(t)\rangle / \mu_{\perp}^{2}$ is the autocorrelation function of the ith component, $\psi_{il} = \langle \vec{\mu}_{\perp i}(0)\vec{\mu}_{\perp l}(t)\rangle / \mu_{\perp}^{2}$ is the cross-correlation function between ith and lth components, and $\mu_\perp$ is the magnitude of $\mu_{\perp i}$. The function $g(t)$ at $t = 0$ gives the Kirkwood g factor, which usually takes a value between 0.4 and 0.9 for randomly coiled polymers. Orientation of the ith perpendicular component requires a conformational change of neighboring repeat units (16). In other words, the orientation of the ith component possibly occurs only when a linkage of the neighboring several repeat units forms the coiled part. If the orientation requires the linkage of more than s units, the correlation of the perpendicular components for partially coiled polymer is given by

$$\left\langle \sum_{i=1}^{n} \vec{\mu}_{\perp i}(0) \sum_{l=1}^{n} \vec{\mu}_{\perp l}(t) \right\rangle = n f_c^s \mu_\perp^2 g_s(t; f_c) \tag{23}$$

because the possibility of the linkage of s repeat units in the coiled part is given by f_c^s. If $g_s(0; f_c)$ is assumed to be the same as $g(0)$ in eq 22 and $\phi_\perp(t; f_c)$ is defined as

$$\phi_\perp(t; f_c) = \frac{g_s(t; f_c)}{g(0)} \tag{24}$$

Equation 23 is rewritten as

$$\left\langle \sum_{i=1}^{n} \vec{\mu}_{\perp i}(0) \sum_{i=1}^{n} \vec{\mu}_{\perp l}(t) \right\rangle = (n\mu_\perp^2 g(0)) f_c^s \phi_\perp(t; f_c) \tag{25}$$

where $\phi_\perp(t, f_c) = 1$.

On the other hand, the orientation of $\mu_{\|i}$ reflects the overall rotation of the polymer molecule, because $\sum_{i=1}^{n} \vec{\mu}_{\|i}$ is proportional to the end-to-end distance of the molecule. Therefore, if the polymer is completely random, the dipole correlation is given by

$$\left\langle \sum_{i=1}^{n} \vec{\mu}_{\|i}(0) \sum_{l=1}^{n} \vec{\mu}_{\|l}(t) \right\rangle = n\mu_\|^2 \phi_\|(t) \tag{26}$$

In this case where the excluded volume effect cannot be ignored, eq 26 is rewritten as

$$\left\langle \sum_{i=1}^{n} \vec{\mu}_{\|i}(0) \sum_{l=1}^{n} \vec{\mu}_{\|l}(t) \right\rangle = n\mu_\|^2 h(n) \phi_\|(t; n) \tag{27}$$

where $h(n)$ depends slightly on n if n is not big enough. If the polymer is a rodlike polymer, we get

$$\left\langle \sum_{i=1}^{n} \vec{\mu}_{\|i}(0) \sum_{l=1}^{n} \vec{\mu}_{\|l}(t) \right\rangle = n\mu_{\|}^{2}(n \cos^{2} \theta)\phi_{\|}(t;n) \qquad (28)$$

where θ is the angle between the helical axis and the parallel component of each dipole. In the case of PGA, it changes from the randomly coiled polymer at $f_c = 1$ to the rodlike polymer at $f_c = 0$, and its fraction of the coiled part changes accordingly with the pH value. Therefore, we get the following equation as a total equation for the time correlation:

$$\left\langle \sum_{i=1}^{n} \vec{\mu}_{\|i}(0) \sum_{l=1}^{n} \vec{\mu}_{\|l}(t) \right\rangle = n\mu_{\|}^{2}h(n,f_c)\phi_{\|}(t;n,f_c) \qquad (29)$$

for PGA, where $h(n, f_c)$ coincides with $h(n)$ if PGA is coiled and it coincides with $n \cos^2 \theta$ if PGA is a helix.

Complex permittivity for the intermediate process ε_m^* is related to eq 25 as

$$\varepsilon_m^* = \frac{4\pi Nng(0)\mu_{\perp}^2 f_c^s}{3kT} F_1 \int_0^{\infty} e^{-j\omega t}[-\dot{\phi}_{\perp}(t;f_c)] \, dt \qquad (30)$$

where N is the number of polymer molecules per unit volume, and F_1 is a ratio of the internal field to the applied field. Equation 30 is empirically rewritten as

$$\varepsilon_m^* = \Delta\varepsilon_m \int_0^{\infty} e^{-j\omega t}[-\dot{\phi}_{\perp}(t;f_c)] \, dt \qquad (31)$$

where

$$\phi_{\perp}(t;f_c) = \exp\left[-\left(\frac{t}{\tau_m}\right)^{\beta_k}\right] \qquad (32)$$

and

$$\Delta\varepsilon_m = \frac{4\pi Nng(0)\mu_{\perp}^2 f_c^s}{3kT} F_1 \qquad (33)$$

F_1 is naturally taken as a constant for the process m. Then, we get

$$\Delta\varepsilon_m = \Delta\varepsilon_m^c f_c^s \qquad (34)$$

where $\Delta\varepsilon_m^c$ is the relaxation strength at $f_c = 1$. Experimental results on $\Delta\varepsilon_m$ can be explained satisfactorily by eq 34 as is seen in Figure 10, and we obtain $s = 4.4$ and $\Delta\varepsilon_m^c = 16.3$, respectively. The value obtained for s is

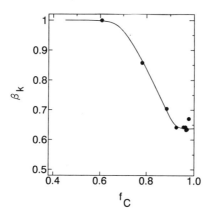

Figure 12. Dependence of the coupling parameter β_k on the fraction of coiled part in a PGA aqueous solution at 25 °C.

in good agreement with that obtained previously for various polymers in organic solvents (7). $\Delta\varepsilon_m$ vanishes at $f_c \sim 0.4$.

The chain motion reflecting orientation of the perpendicular component will be influenced by correlation between motion of the coiled *s* repeat units and those of neighboring chain motions. The correlation is stronger for more coiled chain polymer. The experimental result shown in Figure 12 shows that the coupling parameter increases radically with a decrease of f_c for $0.65 < f_c < 0.95$ and reaches unity for $f_c < 0.65$, where the motion of each *s* repeat unit does not couple with the other.

For the low-frequency process, complex permittivity ε_1^* is given by

$$\varepsilon_1^* = \frac{4\pi Nn\mu_{\parallel}^2 h(n,f_c)}{3kT} F_2 \int_0^{\infty} e^{-j\omega t}[-\dot{\phi}_{\parallel}(t;n,f_c)]\ dt \qquad (35)$$

where F_2 is the ratio of the internal field to the applied field. Equation 35 is empirically given as

$$\varepsilon_1^* = \frac{\Delta\varepsilon_1}{(1 + j\omega\tau_1)^{\alpha_1}} \qquad (36)$$

where α_1 takes unity for $f_c > 0.78$ but takes a slightly smaller value for $f_c < 0.78$. The relaxation time τ_1 extrapolated to $f_c = 1$ is 1×10^{-8} s and is in good agreement with that for the overall rotation of other coiled polymers with the same molecular weight in solution. On the other hand, τ_1 has a value of 5×10^{-6} s, if f_c is extrapolated to 0. This value is in good agreement with that of a helical polymer of the same size in solution. Examples are PBLG (25) and poly (*n*-butyl isocyanate) (26). The relaxation time at $f_c = 0$ is also explained as the overall rotation of the ellipsoidal molecule about the short axis.

If f_c is decreased, the molecule expands and the relaxation time becomes long. Nevertheless, the fact that the relaxation time observed for $f_c < 0.6$ is constant but the relaxation strength increases rapidly evidences the complicated structure of PGA. A ratio of the strength $\Delta\varepsilon_1$ at $f_c = 0$ to that at $f_c = 1$ is 1×10^4. If $F_1 = F_2$ is assumed, the ratio is given by eqs 27 and 28 as $n[\cos^2\theta/h(n)] \sim n$, and it is expected to take a value close to $n = 3.5 \times 10^2$ for the present PGA. The observed value of 1×10^4 is fairly big compared to the predicted one. This large difference will come from migration of counterions along the PGA helix. The ion migration will always bring about a big strength.

Dielectric Behavior of Biopolymers with Helical Structures

Biopolymers such as DNA and tropocollagen take helical structures and the micro-Brownian is prohibited in these polymers. However, aqueous solutions of those polymers exhibit two absorption peaks (11, 27). The high-frequency peak observed near 10 GHz is attributed to the relaxation of bulk water, because the relaxation time and the relaxation strength of dilute solution are in complete agreement with those of pure water. On the other hand another peak at about 100 MHz cannot be due to the micro-Brownian motion or internal motion of the polymer. Furthermore, location of the peak does not change much with the polymer concentration. A dielectric study on tropocollagen showed that the relaxation strength depends largely on the water content and vanishes if water content is extrapolated to 0 (27).

In Figure 13 dispersion and absorption curves for the moist collagen are shown to see the effect of water content on the dielectric behavior. The high-frequency relaxation peak shifts the low frequency as the water content is decreased and coincides with the low-frequency relaxation peak as the content is extrapolated to 0. The low-frequency process can be described by the Cole–Davidson representation

$$\varepsilon_1^* = \varepsilon^* - \varepsilon_m^* - \varepsilon_\infty$$
$$= \frac{\Delta\varepsilon_1}{(1 + j\omega\tau_1)^{\alpha_1}} \tag{37}$$

The relaxation strength $\Delta\varepsilon_1$ for the low-frequency process increases, apparently with increase of the water content. Amount of the bound water increases limitlessly at first sight. However, the relaxation strength depends

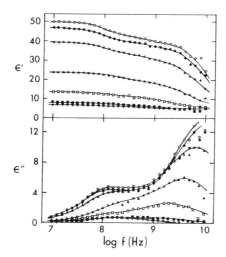

Figure 13. Dielectric dispersion and absorption curves for collagen with various water contents at 25 °C: (○) 1.63, (●) 1.38, (△) 1.03, (▲) 0.55, (□) 0.35, (■) 0.22, and (▽) 0.11 g/g collagen. (Reproduced with permission from reference 27. Copyright 1990 Wiley.)

on the internal field. If the Debye formula of internal field is employed, the relaxation strength is given by

$$\Delta \varepsilon_1 = \frac{4\pi \langle \mu^2 \rangle nN}{9kT} \frac{(\varepsilon_\infty + \Delta \varepsilon_h + 2)^2}{3} \tag{38}$$

where N is number of tropocollagens in unit volume, n is the number of water molecules strongly bound to tropocollagen, $\langle \mu^2 \rangle$ is the square of effective dipole moment of the water molecule, and T is the absolute temperature.

The number n is given by

$$n = \frac{9kT}{4\pi N \langle \mu^2 \rangle} \frac{3\Delta \varepsilon_1}{(\varepsilon_\infty + \Delta \varepsilon_h + 2)^2}$$

$$\propto \frac{\Delta \varepsilon_1}{(\varepsilon_\infty + \Delta \varepsilon_h + 2)^2} \frac{(1 + c)}{d} \tag{39}$$

$$\equiv F$$

where c is the water content (g/g collagen), and d is the density of the moist collagen. Figure 14 shows that F takes a constant value of 0.017 (mL/g) for $c \geq 0.5$. Therefore, the number of bound water molecules per one

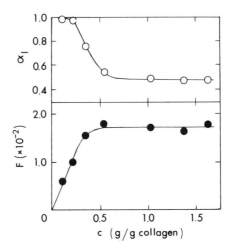

Figure 14. Variation of the Cole–Davidson parameter α_l and F with water content for moist collagen at 25 °C. (Reproduced with permission from reference 27. Copyright 1990 Wiley.)

tropocollagen is saturated at $c \sim 0.5$. The number n cannot be estimated directly from eq 39, because the effective dipole moment μ is not known.

A ratio of $\Delta\varepsilon_h/\Delta\varepsilon_l$ takes a value of 2.0 for the limiting case of zero water content. Two relaxation times τ_h and τ_l have the same value, and the same internal field can be applied to these two dispersions. Furthermore, structure of these two types of water will be indistinguishable for this limiting case. The value of 2.0 represents the ratio of amount of these two kinds of water, and $c/(2.0 + 1)$ is the amount of strongly bound water per 1 g collagen for this case. For a low content region of c $\leq$ 0.35 (g/g), F is linearly proportional to c and F/c takes a value of 4.7×10^{-2} (mL/g) (Figure 14). Therefore, the saturated amount of strongly bound water is obtained as $1.7 \times 10^{-2}/(3.0 \times 4.7 \times 10^{-2}) = 0.12$ g water/g collagen. This value indicates that 21 water molecules are bound to a repeat of the triple helix.

Previously the existence of a stringlike water chain along the fiber axis was suggested (28). In the chain every twelfth molecule is hydrogen bonded to the triple helix. The length of a repeat of the helix is just the same as that of 12 water molecules in the straight line. Rotation of the molecules in the chain occurs possibly about a single bond, accompanied by a hydrogen-bonded breakage. This rotational movement of water molecules is active in dielectric relaxation.

Relaxation time for the movement is given by

$$\tau \cong \nu^{-1} \exp(E/kT) \tag{40}$$

where ν is the $O \cdots O$ stretching frequency of the hydrogen bond, and E is the activation energy required for the breaking of a bond. We set a value of 6×10^{12} Hz to ν and 5 kcal/mol to E, and we obtained a value of 0.8 ns for τ, which is the same order of magnitude as that (1.1 ~ 2.5 ns) observed for the low-frequency process. The experimental results that 21 water molecules are bound to a repeat of the tropocollagen indicate that two stringlike chains are attached to the helix along its fiber axis if the water content is higher than 0.5 g water/g collagen.

The amount of bound water estimated is 0.12 g water/g collagen and seems at first sight to be less than that predicted by other experimental methods. However, the present results indicate the existence of an intermediate water between free and bound water. Relaxation time of the high-frequency process depends strongly on the water content. Further, it reaches the relaxation time of bound water if the content is extrapolated to 0. The intermediate water will be estimated as the bound water by other experiment methods.

In the case of DNA in aqueous solution, the bound water is observed as the relaxation peak at about 100 MHz by the dielectric measurement (*11, 29*). However in this case, the amount of bound water estimated is 0.6–1 g water/1 g DNA and fairly greater than that of collagen (*27*). This result is because DNA has more oxygen atoms and nitrogen atoms to which water molecules are able to attach through hydrogen bonding. The bound water molecules construct a network structure on the DNA surface. DNA generally shows a helix-coil transition. The dielectric measurement demonstrated that the amount of bound water changes at the transition (*29*). This result suggests that the bound water molecules depend on the DNA structure. Calf thymus DNA shows a structural change from B-type helix to A-type helix if ethanol is added to the aqueous solution with an NaCl buffer (*11*). The transition point observed by the CD measurement is 65% (v/v) of ethanol. The dielectric relaxation strength for the bound water process changes at this point. The B-type structure has more bound water than the A-type structure. If the amount of bound water is evaluated by using the effective dipole moment of bound water of the moist collagen, the B-type structure has more than 20 bound water molecules per one nucleotide. If the bound water molecules are removed by adding ethanol and its amount is less than 20 molecules per one residue, DNA prefers the A-type helix.

Poly(dG-dC)poly(dG-dC) shows a transition from B type to Z type and another transition from Z type to A type if ethanol is added to an aqueous solution with NaCl buffer. The critical point of the structural transition from B to Z type is found at 45% (w/w) ethanol by the CD spectroscopy measurement. Dielectric relaxation strength or the amount of water shows two kinds of transition clearly.

Dielectric Behavior of Globule Proteins

Dielectric relaxation of globule protein in aqueous solution is interpreted as a sum of three relaxation processes (*30*). The high-frequency process observed around 18 GHz is due to orientation of bulk water molecules. The intermediate process around 100 MHz as well as that of aqueous DNA solution and the moist collagen is caused by orientation of the bound water molecules on the protein surface, but in this case, it is supplemented by fluctuation of polar side groups on the surface. The relaxation strength is in proportion to the surface of the globule protein except for trypsin and pepsin of hydrolase. The number of bound water molecules estimated is in good agreement with that obtained by other methods such as X-ray analysis. As an example, an albumin molecule has 150 bound water molecules. The low-frequency process in the megahertz region reflects an overall rotation of protein molecule, which involves a counterion process. Its relaxation time is completely proportional to the molecular weight of the protein. The value of relaxation time obtained is quantitatively interpreted as the overall rotation if the molecular dimension is taken into account.

Investigation of freezing processes by dielectric measurement has clarified water structure around globule protein in aqueous solution (*31*). Three kinds of water were found and ascertained. The first is bulk water, which freezes at $-5\,°C$ for 20 wt% albumin solution. The second is unfreezable water, which constructs a shell layer around the protein molecule. The third is bound water, which attaches directly on the protein surface. Amount of the unfreezable water is 0.36 g/g protein at $-5\,°C$, whereas it vanishes at about $-110\,°C$. However, the amount of bound water, 0.04 g water/g protein, does not change with temperature, but the movement is prohibited at $-60\,°C$. This result is interpreted as a result of definite energy difference between two states for the bound water molecules. In the case of DNA aqueous solution, the unfreezable water was found below the freezing point of bulk water (*32*).

References

1. Nozaki, R.; Mashimo, S. *J. Chem. Phys.* **1987**, *87*, 2271.
2. Williams, G.; Watts, D. C. *Trans. Faraday. Soc.* **1970**, *66*, 80.
3. Shioya, Y.; Mashimo, S. *J. Chem. Phys.* **1987**, *87*, 3173.
4. Mashimo, S.; Nozaki, R. *J. Non-Cryst. Solids* **1991**, *131–133*, 1158.
5. Ngai, K. L. *Constants Solid State Phys.* **1979**, *9*, 121; Rajagopal, A. K.; Nagi, K. L.; Teither, S. *J. Phys.* **1984**, *c17*, 6611; Nagi, K. L.; Rajagopal, A. K.; Taitler, S. *J. Chem. Phys.* **1988**, *88*, 6088.
6. Stockmayer, W. H. *Pure Appl. Chem.* **1967**, *15*, 539.
7. Mashimo, S. *Macromolecules* **1976**, *9*, 91.
8. Cole, R. H.; Mashimo, S.; Winsor P., IV *J. Phys. Chem.* **1980**, *84*, 786.

9. Shinyashiki, N.; Asaka, N.; Mashimo, S.; Yagihara, S. *J. Chem. Phys.* **1990**, *93*, 760.
10. Cole, R. H. *J. Phys. Chem.* **1975**, *79*, 1459; Cole, R. H. *J. Phys. Chem.* **1975**, *79*, 1469.
11. Umehara, T.; Kuwabara, S.; Mashimo, S.; Yagihara, S. *Biopolymers* **1990**, *30*, 649.
12. Sjogen, L.; Gotze, W. *J. Non-Cryst. Solids* **1994**, *172–174*, 7.
13. Miura, N.; Shinyashiki, N.; Mashimo, S. *J. Chem. Phys.* **1992**, *97*, 8722.
14. Tanikawa, T.; Mashimo, S.; Chiba, A. *Jpn. J. Appl. Phys.* **1976**, *15*, 219.
15. Adachi, K.; Fujiwara, I.; Ishida, Y. *J. Polym. Sci.: Polym. Phys. Ed.* **1975**, *13*, 2155.
16. Mashimo, S. *J. Chem. Phys.* **1977**, *67*, 2651.
17. Jenkel, E. *Kolloid-Z.* **1942**, *100*, 163.
18. Shinyashiki, N.; Matsumura, H.; Mashimo, S.; Yagihara, S. *J. Chem. Phys.* **1996**, *104*, 6877.
19. Okano, K.; Takada, T.; Kurita, K.; Fukuzaki, M. *Polymer* **1994**, *35*, 2284.
20. Muller, G.; Van del Touw, F.; Zwolle, S.; Mandel, S. *Biophys. Chem.* **1974**, *2*, 242.
21. Mashimo, S.; Ota, T.; Shinyashiki, N.; Tanaka, S.; Yagihara, S. *Macromolecules* **1989**, *22*, 1285.
22. Gestblom, B.; Gestblom, P. *Macromolecules* **1991**, *24*, 5823.
23. Bordi, F.; Cametti, C.; Paradossi, G. *Macromolecules* **1992**, *25*, 4206.
24. Mashimo, S.; Miura, N.; Shinyashiki, N.; Ota, T. *Macromolecules* **1993**, *26*, 6859.
25. Wada, A. *J. Polym. Sci.* **1960**, *45*, 145.
26. Yu, H.; Bur, A. J.; Fetters, L. T. *J. Chem. Phys.* **1966**, *44*, 2565.
27. Shinyashiki, N.; Asaka, N.; Mashimo, S.; Yagihara, S.; Sasaki, N. *Biopolymers* **1990**, *29*, 1185.
28. Berendson, H. J. C. *J. Chem. Phys.* **1973**, *59*, 296.
29. Mashimo, S.; Umehara, T.; Kuwabara, S.; Yagihara, S. *J. Phys. Chem.* **1989**, *93*, 4963.
30. Miura, N.; Asaka, N.; Shinyashiki, N.; Mashimo, S. *Biopolymers* **1994**, *34*, 357.
31. Miura, N.; Hayashi, Y.; Shinyashiki, N.; Mashimo, S. *Biopolymers* **1995**, *36*, 9.
32. Mashimo, S. *J. Non-Cryst. Solids* **1994**, *172–174*, 1117.

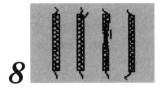

8

Thermally Stimulated Currents of Polymers

G. Teyssedre, S. Mezghani, A. Bernes, and C. Lacabanne*

The main characteristics of thermally stimulated currents (TSC) spectroscopy are presented. Special attention will be paid to its ability to resolve a complex spectrum into elementary components. This procedure allows one to investigate the distribution function of dielectric relaxation times without previous hypothesis. The relaxation modes associated with transitions are characterized by relaxation times following a compensation law reflecting the cooperativity of molecular movements. By using the parameters extracted from TSC experiments, it is also possible to predict the variation of the complex dielectric constant as a function of temperature and frequency.

Molecular motions and relaxation processes of polymers have been widely investigated because they allow correlations between these macroscopic properties and the molecular structure or superstructure.

Among the various available methods of investigation, the thermostimulated depolarization currents, or thermally stimulated currents (TSC), technique is now currently used as a powerful tool for characterizing dielectric relaxation phenomena in polymers. It constitutes an alternative to the conventional techniques in the frequency range. Actually, the low-equivalent frequency (10^{-3} to 10^{-4} Hz) and the high-resolution capabilities of TSC are the most attractive features of this technique.

Relaxation phenomena in polymers are usually described by distributed relaxation processes so that the analysis of the TSC complex mode, as well as dynamic dielectric spectra, is performed by using fitting parameters re-

---

* Corresponding author.

lated to the shape and asymmetry of the distribution of relaxation times characterizing such modes (1–3). To experimentally resolve the complex TSC mode, and therefore the distribution of dielectric relaxation times, several methods have been developed, such as the partial heating technique (4, 5) or fractional polarizations (6–8). Fractional polarization peaks, which are considered in this work, are usually analyzed on the basis of the Bucci–Fieschi–Guidi framework (9, 10), although some alternative methods have been proposed (11).

A general review of the various areas of application of TSC to the characterization of insulating materials has been published recently with some 479 references (12). In this work, we survey some examples of application of this technique to the characterization of the dynamics of relaxation processes in the specific case of polymers, being either amorphous or semicrystalline. The main properties related to the thermal behavior of polymers are covered. Special attention has been paid to the influence of orientation and crystallinity on the dynamics of relaxation processes associated with the glass transition. Also, solid–solid phase transitions in ferroelectric polymers have been considered. In the last section of this chapter, we apply TSC as a useful tool for predicting dynamics of polymeric chains. Not only the frequency dependence of the spectrum as a whole, but also the contribution of isolated modes to the total response can be obtained.

Principle

The TSC method was first developed by Bucci et al. (9, 10). It consists of recording the depolarization current after removal of a static field.

Complex TSC Spectrum

Thermally stimulated currents were carried out with a dielectric spectrometer developed in our laboratory and available from Thermold (Stamford, CT). To obtain a complex TSC spectrum, the sample is polarized by a static field, E_p, at the polarization temperature, T_p, so that the polarization attains equilibrium with the field. Afterward, the sample is cooled down to such a temperature (T_o) that the dielectric relaxation proceeds extremely slowly, so that after removal of the field the sample keeps a frozen-in polarization. The depolarization current, I, due to the return to equilibrium of dipolar units, is then recorded as the temperature is increased from T_o up to the final temperature $T_f > T_p$, at a constant heating rate, resulting in a complex TSC spectrum.

Fine Structure

In polymers, the whole TSC spectrum is in general too broad to be described by only one Debye-like relaxation process (i.e., to only one relaxa-

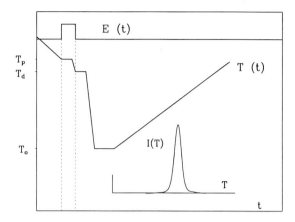

Figure 1. Principle of the fractional polarization procedure.

tion time). However, complex spectrum can be resolved experimentally by using the fractional polarizations (FP) technique (*13, 14*). In this procedure, each elementary spectrum is well approximated by a single relaxation time, allowing Bucci–Fieschi analysis (*9, 10*). In the FP experimental procedure, which is described in Figure 1, the electric field E_p is applied at a temperature T_p for a time Δt_p (≈ 2 min) to allow the dipoles with relaxation time $\tau(T_p)$ lower than Δt_p to orient. The temperature is then lowered by ΔT to T_d ($\Delta T \approx 5$ °C) under field. At T_d, the field is turned off and the sample is then held for Δt_d (≈ 1 min) so that dipoles with relaxation times $\tau(T_d)$ lower than Δt_d relax. The sample is then quenched to $T_o \ll T_d$ ($T_o \approx T_d - 50$ °C), so that only dipoles with relaxation time ($\tau[(T_p - T_d)/2]$ $\approx \Delta t_d$) remain oriented at T_o. The linear heating is then performed during which the rate of decay of the frozen-in polarization (i.e., the depolarization current) is recorded. The heating run is performed up to 50 °C above T_p. The response of this FP experiment is the result of the reorientation of a narrow distribution of relaxation times excited over a ΔT temperature window around T_p. By varying the value of T_p by constant step along the temperature axis, the whole TSC spectrum is resolved into elementary processes.

Analysis

Single Relaxation Time

Basic Equations

When a dielectric material such as a polymer is submitted to a dc field, permanent dipoles orient in the field direction, and the macroscopic polarization is given by

$$\frac{dP}{dt} + \frac{P}{\tau(T)} = \frac{(\varepsilon_s - \varepsilon_\infty)\varepsilon_o}{\tau(T)} E_p \tag{1}$$

where E_p is the polarizing field, $\tau(T)$ is the temperature-dependent relaxation time, ε_o is the dielectric constant in vacuum, and ε_s and ε_∞ are the relative dielectric constants, at low and high frequency, respectively. After an infinite polarization time, the polarization reaches its saturation value, P_o, which is related to the relaxation strength, $\Delta\varepsilon = \varepsilon_s - \varepsilon_\infty$, by

$$P_o = \Delta\varepsilon\varepsilon_o E_p = \frac{N\mu^2}{3kT_p} E_p \tag{2}$$

where N is the dipole concentration (per volume unit), μ is the dipole moment, k is the Boltzmann constant, and T_p is the polarization temperature.

In case of depolarization under zero-field condition, the right-hand term of eq 1 is null, and $P(t)$ is obtained from

$$\frac{dP}{dt} + \frac{P}{\tau(T)} = 0 \tag{3}$$

When the sample is polarized at a T_p and cooled down to $T_o \ll T_p$, such that $\tau(T_o)$ is high enough preventing depolarization in usual time scales, P_o is frozen-in. If the sample is warmed up at a constant rate, β, a thermostimulated depolarization occurs, and the temperature dependence of the polarization is given by

$$P(T) = P_o \exp - \int_{T_0}^{T} \frac{1}{\beta} \frac{dT'}{\tau(T')} \tag{4}$$

where T' is the running temperature variable and $P(T)$ is obtained experimentally by integrating the current density $J(T)$ between T and T_f, the final temperature, at which the current vanishes

$$P(T) = \frac{1}{\beta} \int_{T}^{T_f} J(T') \, dT' \tag{5}$$

The dielectric strength of the relaxation is obtained from

$$\Delta\varepsilon = \frac{1}{\beta\varepsilon_o E_s} \int_{T_0}^{T_f} I(T) \, dT \tag{6}$$

where s is the sample area, and I is the depolarization current

$$I(T) = -s \frac{\mathrm{d}P}{\mathrm{d}T} \tag{7}$$

Relaxation Time

The relaxation time, τ, can be calculated experimentally, without any hypothesis on its temperature dependence. Indeed, from eqs 3, 5, and 7

$$\tau(T) = -\frac{P(T)}{\mathrm{d}P/\mathrm{d}t} = \frac{1}{s}\frac{P(T)}{I(T)} = \frac{1}{\beta I(T)} \int_{T}^{T_f} I(T') \, \mathrm{d}T' \tag{8}$$

The analysis of elementary peaks gives a temperature-dependent relaxation time $\tau(T)$, which often follows an Arrhenius-like dependence

$$\tau(T) = \tau_o \exp \frac{\Delta H}{kT} \tag{9}$$

where ΔH is the activation energy, and τ_o is the preexponential factor. Because the relaxation time is the inverse of the frequency of jump between two activated states, the empirical eq 9 can be approximated from that derived by Eyring in the theory of activated states (15):

$$\tau(T) = \frac{h}{kT} \exp - \frac{\Delta S^*}{k} \exp \frac{\Delta H^*}{kT} \tag{10}$$

where ΔH^* and ΔS^* are the enthalpy and entropy of activation, respectively, and h is the Planck's constant. The variable ΔH^* is related to the barrier height between two activated states, and ΔS^* is related to the ratio of the number of available sites between the activated and inactivated states. The Eyring equation constitutes a theoretical interpretation of the Arrhenius equation because it gives a true thermodynamic sense to Arrhenius parameters. Moreover, the use of the Eyring's formalism allows transformation of the preexponential factor τ_o of the Arrhenius law, which often has unrealistic values, into an entropy factor. Because no difference exists between the Arrhenius and Eyring's formalism in the limit of common accuracy (16), and because experimental results cannot distinguish which model is more appropriate, equivalency between eqs 9 and 10 is obtained from

$$\Delta S^* = k \ln(kT/h) - k \ln(\tau_o) \tag{11a}$$

$$\Delta H^* = \Delta H - kT \tag{11b}$$

In some cases, the analysis of the elementary peaks gives the Arrhenius diagram (log τ versus reciprocal temperature) showing a curvature in the form of a decrease of the apparent activation energy ($\sim$d(log τ)/dT^{-1}) as T increases. Therefore, alternative temperature dependencies of τ have been proposed, such as the Vogel–Tammann–Fulcher–Hesse (VTFH) (17–19) equation:

$$\tau(T) = \tau_{ov} \exp \frac{A}{(T - T_\infty)} \tag{12}$$

where T_∞ has been identified with the true second-order transition (20) corresponding to the vanishing of the excess entropy (21), and A is related either to an activation energy, or to the thermal expansion coefficient of the free volume (22). Equation 12 has been obtained in the framework of the free volume model of the glass transition (22–24), which constitutes one of the most widely used theoretical approaches of molecular mechanisms of the glass transition. Therefore, we recall its basic principles.

In this model, the characteristic magnitudes of molecular transport are driven by the free volume, and they behave according to the empirical VTFH equation. The main features of this model are summarized subsequently (25). A local cell volume v is associated with each molecule. This molecule is restricted in movement within a cell volume v, which is defined by the nearest neighbors. When v reaches some critical value, v_c, the excess volume, $v_f = v - v_c$, can be considered as free volume. Cells having $v > v_c$ are termed liquid-like. No local free energy variation is required for free-volume redistribution. The relaxation time τ is expressed in terms of molecular transport coefficients. Molecular transport is driven by the fraction of liquid-like cells having a free volume greater than a minimum free volume v_f. By using some assumptions, such as the temperature dependence of the averaged fractional free volume, $f = v_f/v = \alpha_f(T - T_\infty)$, eq 12 was obtained, where α_f corresponds to the thermal expansion coefficient of free volume, and T_∞ is the vanishing temperature of v_f.

Discrete Distribution of Relaxation Times

As previously discussed, TSC peaks in polymers generally cannot be analyzed with the hypothesis that they are described by a single relaxation process. The direct analysis of a complex peak would require a hypothesis on the form of the distribution function of relaxation times (DRT), which describes such a mode. Should this distribution be known, the numerical analysis for fitting such data is not easily tractable. The alternative way of resolving such modes consists of applying the FP procedure, which does not require any hypothesis on the DRT but allows it to be obtained experimentally. We first consider the way in which FP peaks are being analyzed.

Fractional Peaks Analysis

FP peaks are often poorly defined in their high-temperature range, so that the baseline used for integrating the current is rather arbitrary. Moreover, experimental peaks are usually more symmetric than would be expected from an Arrhenius-type relaxation time. Therefore, the total area of the peaks, P_o, is approximated from the integration of the left-hand side of experimental peaks, and the polarization at a given temperature is obtained from this total area as

$$P(T) = P_o - \frac{1}{\beta} \int_{T_o}^{T_f} J(T') \, dT' \tag{13}$$

The method we developed for approximating P_o supposes an Arrhenius-like dependence of the relaxation times.

For a pure Debye peak, the maximum current density (at $T = T_m$), $J_m = J(T_m)$, is related to P_m, the polarization at $T = T_m$, by (26)

$$J_m = \frac{\Delta H \beta P_m}{kTm^2} \tag{14}$$

Moreover, J_m is related to the total area of the peak, P_o, by (27)

$$J_m = \frac{\beta P_o}{eT_m} x_m \exp L(x_m); \qquad x_m = \frac{\Delta H}{kT_m} \tag{15}$$

where e is the base of the natural logarithm, and $L(x_m)$ is defined by

$$L(x_m) = 1 - x_m^2 e^{x_m} \int_{x_m}^{\infty} e^{-z} z^{-2} \, dz \tag{16}$$

where z is the running x_m variable. It has been shown that (27)

$$\exp L(x_m) \approx (x_m + 4)/(x_m + 2) = t \tag{17}$$

with a relative error $< 0.4\%$ for $x_m > 20$.

From eqs 14 and 15 one gets the approximate form of P_o as

$$P_o \approx \frac{e}{e - t} (P_o - P_m) = YP_{om} \tag{18}$$

where

$$Y = e/(e - t), \quad \text{and} \quad P_{om} = \frac{1}{\beta} \int_{T_o}^{T_m} J(T) \, dT \tag{19}$$

The value of P_{om} is obtained experimentally. Because x_m is unknown, an averaged value should be used for calculating Y. Considering that ΔH = 1 eV at 273 K, we obtain x_m = 42.5. This value also corresponds to ΔH = 0.73 eV at 200 K, or 1.30 eV at 350 K, which is usual for activation energies of polymer relaxations. With these approximations, the Debye factor is Y = 1.6245. This value is fairly consistent with that found numerically (1.6260) by computing Debye peaks with a large set of (τ_{oi}, ΔH_i) pairs, calculating their integrals, and averaging the obtained value of P_o/P_{om} (28).

The calculation of the relaxation times is performed in a temperature range (T_a, T_b), where T_a is set at a value such that the noise on the TSC current is negligible with regard to $I(T_a)$, and T_b is close to T_m ($T_b < T_m$). In this temperature domain, the relaxation times are in the 50–5000-s range. One can also obtain a rough approximation of the equivalent frequency of the TSC technique from $f_e = 1/2\pi\tau(T_m) \approx 10^{-3}$ Hz. Note that the analysis of the peaks at temperatures higher than T_m would not significantly increase the range of experimental relaxation times because the ratio of P/J (see eq 8) usually does not vary much above T_m in experimental peaks. This analysis gives the distribution of relaxation times by transforming all the elementary peaks into their Arrhenius representation.

Activation Parameters

When using the previously stated analysis, each elementary process is fitted according to an Arrhenius equation (9), where the activation energy and the preexponential factor for the ith component of the distribution of relaxation times are designated by ΔH_i and τ_{oi}, respectively. They are obtained from a least-squares fitting of log τ versus $(1/T)$. In some cases (29) the extrapolated plot of log τ versus $(1/T)$ shows that several Arrhenius lines converge into a single point, which is called a "compensation point" (T_c, τ_c). At the compensation temperature T_c, all relaxation times would take the same value τ_c. This behavior shows that the relaxation times obey a compensation law defined as

$$\tau_i(T) = \tau_c \exp \frac{\Delta H_i}{k}\left(\frac{1}{T} - \frac{1}{T_c}\right) \tag{20}$$

Therefore τ_{oi} is related to the activation energy ΔH_i by

$$\tau_{oi} = \tau_c \exp - \frac{\Delta H_i}{kT_c} \tag{21}$$

Hence, the variation of log τ_{oi} versus ΔH_i constitutes an alternative representation of the compensation rule. The intercept and slope deduced from a linear regression of log τ_{oi} versus ΔH_i allow one to obtain the compensation parameters, T_c and τ_c. As example, TSC experiments performed on a series of phosphocalcic hydroxyapatites (*30, 31*) have shown that the relaxation times corresponding to the dielectric manifestation of the monoclinic-hexagonal transition obey a compensation law. In all cases, T_c corresponds to the solid–solid transition temperature obtained by X-ray diffraction.

An usual interpretation of the compensation law is based on a two-site model proposed by Hoffman et al. (*32*) to describe crystalline relaxations in *n*-paraffins. Molecular movements are supposed to involve an entire short-chain molecule: the length of the molecules corresponds to the thickness of crystallites. Under these assumptions, the relaxation time is expressed by an Eyring equation (eq 10), with $\Delta H^* = \Delta H_{eg}^* + n\Delta H_{cs}^*$ and $\Delta S^* = \Delta S_{eg}^* + n\Delta S_{cs}^*$. The suffixes eg and cs refer to end group contribution and to elementary contribution of constitutive segments, respectively, and n is the number of segments per molecular chain. This model was applied adequately to *n*-paraffins, but also to *n*-ester and *n*-ether: a linear variation of both the activation enthalpy and entropy as a function of n has been obtained experimentally (*32*). By designating as T_c the parameter that joins the intrinsic parameters describing the elementary contributions of the crystalline relaxation, $\Delta H_{cs}^* = \mathrm{Tc}.\Delta S_{cs}^*$, eq 10 may be rewritten as the compensation law of eq 20.

An alternative and somewhat simpler manner of analyzing kinetic data from relaxation processes has been proposed by Starkweather (*33, 34*) and applied by Sauer and Avakian (*35*) to FP analysis. It consists in discriminating between cooperative and noncooperative relaxation processes by comparing the experimental activation energy–temperature dependence with the one predicted with a zero-activation entropy process. Considering eq 10 from the Eyring theory, and the equivalent frequency of TSC technique, f_e, the activation enthalpy for $\Delta S^* = 0$ obeys the relationship

$$\Delta H_o^* = kT \ln\left(\frac{kT}{2\pi h f_e}\right) \tag{22}$$

According to Starkweather's scheme, noncooperative relaxation processes such as secondary relaxations in polymers are characterized by experimental values of ΔH^* that correspond to ΔH_o^* within experimental error. Conversely, for the dielectric relaxation associated with the glass transition of polymers, a significant increase of ΔH^* appears around the glass transition temperature (T_g). Such transitions where ΔH^* values markedly differ from ΔH_o^* values are termed cooperative. The strong increase of the activation energy is also predicted by the compensation equation, which can easily be turned into a transcendental equation between ΔH and T_m as (*36*)

$$\ln\left(\frac{kT_m^2}{\beta\Delta H}\right) = \ln(\tau_c) + \frac{\Delta H}{k}\left(\frac{1}{T} - \frac{1}{T_c}\right) \tag{23}$$

Dielectric Constant

If a relaxation process is described by one single relaxation time, the variations of the real and imaginary parts of the dielectric constant as a function of frequency (ω) and temperature are given by the Debye equations:

$$\varepsilon'(\omega,T) = \varepsilon_\infty + \frac{\Delta\varepsilon}{1 + [\omega\tau(T)]^2} \tag{24a}$$

$$\varepsilon''(\omega,T) = \frac{\Delta\varepsilon\omega\tau(T)}{1 + [\omega\tau(T)]^2} \tag{24b}$$

In case of a discrete DRT, the analysis of the fine structure of the complex model allows one to deduce a set of $\Delta\varepsilon_i$ values. The dielectric strength of the relaxation is obtained from

$$\Delta\varepsilon = \sum_i \Delta\varepsilon_i \tag{25}$$

The real and imaginary parts of the dielectric constant are obtained by summing up the contributions of the elementary processes.

$$\varepsilon'(\omega,T) = \varepsilon_\infty + \sum_i \frac{\Delta\varepsilon_i}{1 + [\omega\tau_i(T)]^2} \tag{26a}$$

$$\varepsilon''(\omega,T) = \sum_i \frac{\omega\tau_i(T)\Delta\varepsilon_i}{1 + [\omega\tau_i(T)]^2} \tag{26b}$$

This model allows the prediction of the variations of the dielectric loss as a function of temperature and frequency. These results can be compared with those obtained by dynamic dielectric analysis.

Continuous Distribution of Relaxation Times

In case of a continuous DRT, the distribution function, $f(\ln\tau)$ must be defined to deduce the complex dielectric constant:

$$\varepsilon'(\omega,T) = \varepsilon_\infty + \Delta\varepsilon\int_{-\infty}^{+\infty} \frac{f(\ln\tau)}{1 + \omega^2\tau^2}\,d(\ln\tau) \tag{27a}$$

$$\varepsilon''(\omega,T) = \Delta\varepsilon \int_{-\infty}^{+\infty} \frac{\omega\tau f(\ln \tau)}{1 + \omega^2\tau^2} \, d\,(\ln \tau) \tag{27b}$$

where $\Delta\varepsilon$ corresponds to the relaxation strength of the whole mode.

By the technique of FP, a discrete distribution of $(\tau_{oi}, \Delta H_i)$ pairs is obtained. For each activation parameters couple, a statistical weight function must be defined. This process can be done through the $\Delta\varepsilon_i T_p$ parameter, which, from eq 2, is proportional to $N_i\mu^2$. Once the distribution function $f(\Delta H)$ and $f(\tau_o)$ are known, the temperature-dependent distribution function for τ_i can be obtained. If τ_{oi} and ΔH_i are correlated by a compensation equation, $f(\tau)$ is obtained from the distribution function of only one activation parameter, either τ_o or ΔH.

Experimental

Arrangement

The block diagram of the apparatus is shown in Figure 2. The electroded sample is sandwiched between stainless steel disks to which the electrical contacts are made to allow polarization and current measurement. The assembly is placed into a controlled temperature chamber. Before the measure, the chamber is pumped out at $1.3\ \omega^{-3}$, and is filled in with a dry He exchange gas. The temperature of the sample can be varied between -180 °C and 300 °C.

Measurements and Temperature Control

The depolarization current is recorded with a Keithley 642 ammeter whose precision is 10^{-16} A. The output signal is treated by a channel including a current–tension converter. The regulation loop is controlled by an Eurotherm 818 regulator. Software for a personal computer allows one to control automatically the thermal cycle, to polarize the sample and to connect it to the electrometer by activating–inactivating the rotary switch, and to record the current as a function of the temperature.

Characterization of Polymeric Materials

Structure of the Amorphous Phase

The classical techniques such as differential scanning calorimetry (DSC) do not show the influence of molecular orientation on transitions (*37*).

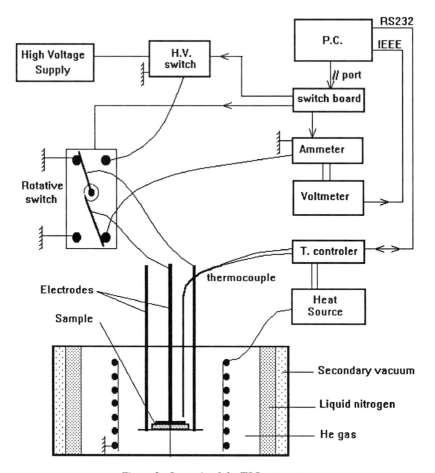

Figure 2. Synopsis of the TSC apparatus.

Therefore, TSC spectroscopy has been applied to the study of local-order distribution (*38*).

Poly(ethylene terephthalate) (PET) was chosen as the model material; the molecular orientation of the amorphous phase was obtained by varying the thermomechanical history. A series of PET films were prepared by Rhône-Poulenc. The biaxially oriented Terphane (10-μm thick) has 39% crystallinity. It was produced in a double-stretching process: the first in the travel direction, the second, of greater strain, in the transverse direction. The film was then heat-set at 180 °C. The biaxially oriented films were compared with uniaxially oriented films (50-μm thick and only 4% crystalline) obtained after the first process in the travel direction. The initial unoriented amorphous PET (150-μm thick film) was also studied as reference material.

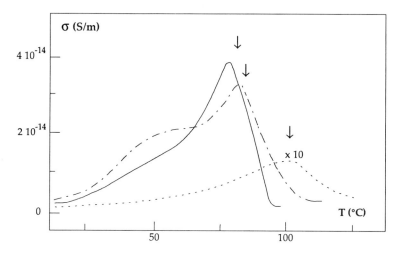

Figure 3. Complex TSC spectra of (———) unoriented PET, (—·—) uniaxially oriented PET, and (· · ·) biaxially oriented PET.

Complex TSC spectra of PET films with various levels of molecular orientation are shown on Figure 3. The current has been normalized to the applied field and to the sample area. The polarization temperature was practically T_g as indicated by the arrows on Figure 3. For sake of clarity, the peak corresponding to biaxially oriented PET has been magnified 10 times. The TSC peak is situated respectively at 82, 85, and 100 °C for the unoriented, uniaxially, and biaxially oriented PET. Because these peaks are situated around T_g, they have been attributed to the dielectric manifestation of the glass transition. Uniaxial orientation induces a shift of the TSC peak toward high temperatures. In case of biaxially oriented film, the shift is much more pronounced, and a strong decrease of the magnitude of the peak is observed. In fact, not only orientation effect, but also the strong crystallinity increase are associated with these changes. Around 50 °C, it is interesting to note the existence of a sub-T_g relaxation mode that we do not discuss herein.

By using fractional polarizations, the complex TSC spectra were resolved into elementary processes. The value of T_p was varied in steps of 5 °C in the temperature range of the glass transition. Each elementary spectrum was analyzed and the temperature dependence of the corresponding relaxation times was deduced from eq 9. Then, the experimental points were plotted on an Arrhenius diagram. For example, Figure 4 show the Arrhenius diagram of unoriented PET. By extrapolation, we see that $\tau(T)$ variations converge at a given temperature: they obey a compensation law (eq 20). Thus, a compensation line is defined on the compensation diagram (Figure 5). We also report the experimental points corresponding to uniaxi-

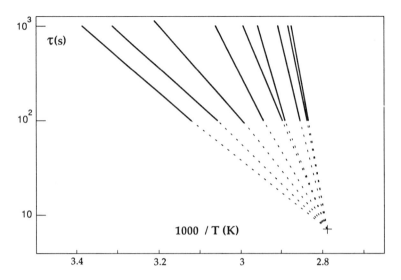

Figure 4. Arrhenius diagram of relaxation times of unoriented PET

ally and biaxially oriented PET. In all cases, the relaxation mode associated with T_g is characterized by a distribution of relaxation times obeying a compensation law.

The slope is practically the same for all PET samples: in other words, molecular orientation does not modify the compensation temperature in a significant way. Contrarily, the width of the distribution function is considerably decreased. Such an evolution in the activation energy distribution

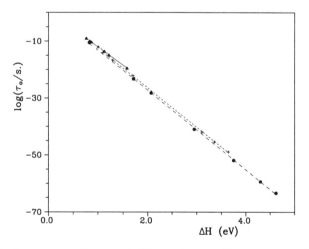

Figure 5. Compensation diagram of (●) unoriented PET, (+) uniaxially oriented PET, and (▲) biaxially oriented PET.

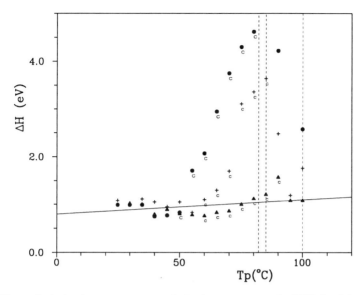

Figure 6. Activation energy versus polarization temperature for PET. Symbols are defined in Figure 5. Labels "c" indicate compensation points. The solid line corresponds to ΔS = 0.

is also revealed in Figure 6 where ΔH is plotted as a function of the polarization temperature. The labels c indicate those processes that follow the compensation law. The activation energy increases up to the maximum temperature of the TSC peak, which is indicated as a vertical dashed line. In the low-temperatures range, activation energies lie close to the zero activation entropy curve deduced from eqs 22 and 11b, and represented as a solid line. In some cases, experimental points are located below the $\Delta S = 0$ curve, which means that the activation entropy is negative. This result would indicate that the activated state is less ordered than the equilibrium state. However, these deviations are mild and are probably partly due to experimental errors in the estimation of activation parameters. Note that the bioriented sample shows a specific behavior in comparison with other materials. Specially, activation energies are strongly lowered. This behavior is probably more related to its semicrystalline nature rather than to orientation effects. The influence of crystallinity on the relaxation kinetics will be discussed in the next section.

Figures 5 and 6 clearly show that the departure from the zero activation entropy prediction is closely associated with the compensation phenomenon and that both processes have the same origin. According to Starkweather, noncooperative relaxations involve short chain sequences acting individually (*39*). Noncooperative versus cooperative behavior would be essentially controlled by the length of mobile sequences involved in a given process. In biaxially oriented PET film, the mobile sequences are of course limited by the presence of crystallites. Such interpretation is consistent with the model of Hoffman et al. (*32*), where the activation enthalpy and entropy vary like the size of the mobile sequences. This consistency emphasizes the cooperative nature of relaxation processes involved in a compensation pattern.

Interaction with Small Molecules

Collagen is characterized by a structural hierarchy, quite unlike anything that has been observed with other proteins. At the molecular level, this biopolymer is a triple helix constituted by regions containing residues with either apolar or polar side chains (*40, 41*). The wealth of amino acids and types of intra- and intermolecular bonds in collagen set this macromolecule apart from even the most complicated synthetic polymers. These considerations account for the fact that molecular movements in collagen are still largely not understood. Nevertheless, the role of collagen in diseases, in tissues aging, and the use of this protein and its derivatives as biomaterials make their physical characterization essential. Measurements of dynamic mechanical properties of collagen have been reported (*42, 43*). However, dynamic mechanical data alone would not permit mechanistic assignments of the modes. Conversely, TSC, which is a water-sensitive spectroscopy, allows scanning dielectric relaxation processes at a lower equivalent frequency and following the water-induced effects on the relaxations. In the following, we consider the interactions between water and collagen, in the temperature range of the liquid nitrogen to 60 °C. TSC experiments have been performed on samples of freeze-dried collagen at three stages of hydration: native collagen (NC), dehydrated collagen (DC), and hydrated collagen (HC). The hydration degree in weight (δ) was 8–10%, 1–2%, and 25% for NC, DC, and HC, respectively.

Figure 7 shows the complex TSC spectra of collagen of different levels of hydration. The TSC spectrum of NC is not reported, because it shows relaxation modes similar to those of DC. The spectrum of DC, represented as a dashed line on Figure 7, is constituted by three peaks located around -150 °C, -50 °C, and 30 °C, which have been labeled γ, β_1, and α, respectively. The solid line spectrum corresponding to HC displays some differences comparatively to TSC spectrum of DC. First, an additive mode was observed around -100 °C and has been labeled β_2. Second, the β_1 peak was shifted to lower temperatures, whereas the α peak was shifted to higher temperatures. Therefore, the presence of water seems to influence the mobility of collagen at different levels. By analogy with previous works (*44–46*), and on the basis of studies undertaken on the influence of water, and on the analysis of relaxation times distribution, the modes assignment has been achieved (*47, 48*).

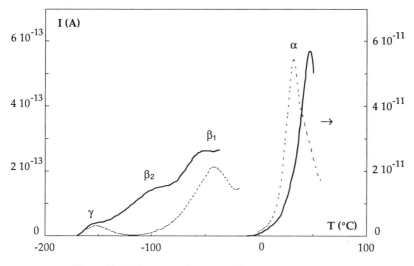

Figure 7. TSC spectra of (———) HC and (– – —) DC.

The α-mode is associated with intermolecular movements along triple helix. On the other hand, molecular motions at the origin of the β_1 and λ modes correspond to movements precursor of the glass transition, of polar and apolar regions, respectively. Finally, the β_2 mode corresponds to movements of water molecules bound to polar side chains in collagen. Overall, TSC experiments distinguish between intrinsic (γ, β_1, α) and extrinsic (β_2) modes. The dielectric manifestation of water depends on the ratio and the location of water molecules. Indeed, the extrinsic mode cannot be directly observed on TSC spectra of DC and NC. It means that water molecules are not free when the hydration level is lower than 10%. By analogy with results from Nomura et al. (*49*), water corresponds to "structural water" and "bound water," for NC and DC, respectively. Finally, a high degree of hydration (δ ~ 25%) induces the presence of free water in collagen (*49*), because the extrinsic mode can be observed on complex spectra. This kind of water increases the activation energy of the α mode: water molecules stiffen the structure. Water molecules probably create hydrogen bonds with polar groups located at the extremity of the molecule. Introduction of water molecules inside collagen fibers is in good agreement with the observations of Pineri et al. (*50*).

Study of Relaxations Associated with Transitions

Fractional polarization analyses undertaken up to now on the dynamics of polymers merely concern the glass transition–relaxation and secondary relaxation processes (*35, 51, 52*). To enlarge the field of application of this technique, we have recently applied it to the study of the ferroelectric-to-paraelectric transition in fluorinated polymers (*53*). In the following, we outline the main characteristics that have been deduced concerning both the glass and solid–solid transitions.

Ferroelectric-to-Paraelectric Transition–Relaxation

For this study, we have considered samples of poly(vinylidene fluoride) PVDF and PVDF-*co*-poly(trifluoroethylene) P(VDF–TrFE) copolymers supplied by Solvay & Co. To investigate the compositional dependence of relaxation processes in these random copolymers, we have considered the following molar contents of TrFE units: 0% (PVDF), 25%, 35%, and 50%. The Curie transition of these copolymers is clearly revealed by different techniques, such as X-ray diffraction, DSC, and switching measurements in high electric fields (*54, 55*), for compositions in the range 50–80% VDF units. Moreover, the ferroelectric-to-paraelectric transition temperature varies strongly, from about 60 °C to 130 °C in the composition range we investigated (*53*). We have applied the technique of fractional polarizations for analyzing the complex TSC mode associated with this transition. In Figure 8, we have drawn the Arrhenius diagram of the distribution of relaxation times obtained for this copolymer. The applied field was $E_p = 5\ MV/m$ on Al-metallized 50-μm thick films. The polarization temperature has been varied from -85 °C to 110 °C by 5 °C steps, and the polarization window was $\Delta T_p = 5$ °C. Two compensation laws are revealed in this diagram. The lower temperature corresponds to the glass transition–relaxation and will be discussed later. The upper temperature compensation pattern is associ-

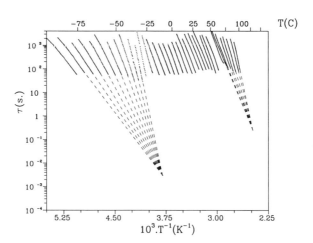

Figure 8. Arrhenius diagram of dielectric relaxation times isolated in P(VDF–TrFE) 75/25 by fractional polarizations.

Table I. Ferroelectric-to-Paraelectric Transition and Compensation
Temperatures as a Function of Chemical Composition
of P(VDF–TrFE) Copolymers

Polymer	mol%	$T_{fp}(°C)$	$T_c(°C)$
PVDF (β)	100/0	—	170
P(VDF–TrFE)	75/25	130	130
P(VDF–TrFE)	65/35	105	109
P(VDF–TrFE)	50/50	65	78

ated with the Curie transition of the copolymer. Here, the important fact is that we obtained a compensation temperature that corresponds to the ferroelectric-to-paraelectric transition temperature as revealed by DSC. A comparison of results obtained by DSC and by FP analysis for various copolymers is given in Table I.

We have previously shown that the Arrhenius law could easily be turned into an Eyring dependence of the relaxation time. Therefore, the compensation law can be discussed in terms of the model of Hoffman, Williams, and Passaglia (*32*). This model has been established for paraffins that are crystalline materials. An extension of this model to the Curie transition of fluorinated copolymers requires some assumptions.

First, the linear increase of ΔH and ΔS on the length of molecular chains was verified for short-chain molecules ($n < 37$), and so for thin crystalline lamellae. For high molecular weight polymers, and thick lamellae (about 250 Å for PVDF (*56*) and up to 0.1 μm for annealed P(VDF-TrFE) copolymers (*57*)), we shall consider chain segments rather than entire macromolecules.

Second, their model was applied to a series of *n*-paraffins with known chemical structure. Contrarily, the compensation laws obtained in this work are characteristic of a given polymer or copolymer. This result implies that an elementary relaxation peak isolated by the technique of fractional polarizations, and the deduced relaxation time, is characteristic of the length of relaxing entities, and more generally, is characteristic of the volume of the crystal involved in an elementary process. Indeed, dipoles are chemically connected to the backbone, and their rotation involves cooperative conformational changes accompanied by at least three bond motions, a characteristic that is required for keeping the crystalline order (*58*). These conformational changes involve not only rotation of one chain with respect to the backbone but also surrounding segments in a cooperative manner.

The increase in activation energies corresponding to the compensation law, for increasing temperature, reflects the increase in the length of mobile units. Furthermore, the strong values of the activation energies may be related to the size of relaxing sequences. Overall, the observed compensation phenomena can be related to cooperative molecular motions involving both intra- and intermolecular interactions. Considering the value of the compensation temperature T_c, T_c corresponds practically to the ferroelectric-to-paraelectric transition temperature, at least for P(VDF-TrFE) 75/25 and 65/35 (*see* Table I). Such a behavior has been observed for another kind of solid–solid transition in crystalline materials, the hydroxyapatites (*30, 31*). In this case, the compensation temperature corresponds to the monoclinic–hexagonal phase transition as revealed by X-ray diffraction. At the transition temperature, all the crystalline dipoles relax with the same characteristic time, τ_c, whatever the height of the barrier is. They are involved in the same macroscopic phenomenon: that is, the ferro-to-paraelectric transition. For P(VDF-TrFE) 50/50, the compensation temperature is slightly above the ferro-to-paraelectric (fp) transition ($T_c = 78\ °C$; $T_{fp} \approx 65\ °C$). This behavior may be partly ascribed to the structural disorder characterizing the ferroelectric phase of this copolymer (*59, 60*). For β-PVDF, the compensation temperature is 170 °C; that is, close to the melting point. For the unpolar α-PVDF, this compensation phenomenon has not been observed (*61*). Therefore, an intrinsic mobility of the β-phase has been identified. By analogy with results obtained in fluorinated copolymers, the Curie transition of β-PVDF has been located at 170 °C; that is, in the vicinity of the melting temperature. This assignment is consistent with previous studies on PVDF and P(VDF-TrFE) copolymers (*59, 62–64*).

Glass Transition–Relaxation

Even though for true thermodynamic phase transitions, the compensation temperature corresponds to the transition temperature with a good approximation, in the case of the glass transition, a significant lag is generally observed between T_c and T_g (*29*). Although the physical origin of the compensation law is still discussed (*65, 66*), it is clear that it arises from the common origin of processes involved in the corresponding complex mode (*51*). With regard to the method of analysis, several criteria have been developed for discriminating between extrathermodynamic (true) and statistical (associated with experimental errors) compensation patterns (*16, 36, 67, 68*). An important feature of (true) compensation phenomena found in the relaxation associated with the glass transition of polymers is that it involves increasing activation energies from low-temperature peaks to high-temperature ones. By itself, this behavior shows that the compensation law is not a peculiar effect associated with the method of analysis. The understanding of such a phenomenon could be improved by considering its dependence on physical or chemical structure of polymers because a large set of materials has now been investigated. Two features related to the compensation effect are considered: the temperature lag between T_c and T_g, and the continuous increase of activation energies involved in the compensation law.

Table II. Parameters for Various Semicrystalline Polymers

Polymer		$\chi_c(\%)^a$	$T_g(°C)$	$T_c-T_g(°C)$	$\tau_c(s)$	Ref.
Poly(propylene)		50	−10	33	2.0×10^{-2}	69
PVDF		50	−42	27	5.8×10^{-3}	61
P(VDF–TrFE)	75/25	55	−36	25	3.0×10^{-3}	
	65/35	52	−33	22	1.6×10^{-2}	70
	50/50	48	−28	30	5.0×10^{-4}	
Polyamides	12	40	40	44	7.8×10^{-3}	
	6.6	40	57	48	2.5×10^{-2}	71
PEEK		12	144	17	2.0	
		31	157	8	2.0	72
PPS		10	94	24	2.55	73
PClTrFE		10	52	74	0.25	52
PET		45	100	15	10	74

^{a} χ_c is crystallinity.

In Tables II and III, we compare the T_c–T_g lag obtained in a wide variety of polymers, by considering the crystalline versus amorphous state as a criterion. As can be shown in these tables, a systematic dependence of T_c–T_g on crystallinity is not observed. The general tendency is an increase of T_c–T_g in crystalline polymers. As example, for fluorinated polymers and copolymers, which have a quite high crystallinity (50–60%), this lag is around 25 °C. The averaged value of T_c–T_g is 30.6 °C for semicrystalline and 17.5 °C for amorphous polymers. If the anomalous high values of T_c–T_g of poly(ethyl methacrylate) (PEMA) (54 °C) and of poly(chlorotrifluoroethylene) (PClTrFE) (74 °C) are excluded, one gets an averaged lag of 14.2 °C for amorphous polymers, which is about half that of semicrystalline materials (26.6 °C).

Table III. Parameters of Various Amorphous Polymers

Polymer	$T_g(°C)$	$T_c-T_g(°C)$	$\tau_c(s)$	Ref.
Poly(methyl methacrylate)	110	18	0.26	75
Polystyrene	100	2	10	38
PET	82	5	6.2	74
Polycarbonate	150	32	0.7	14
PPS	91	2	36	76
Polyurethane	−69	12	0.94	77
DGEBA-DDS^a	133	18	3.2	77
PEMA	54	54	6.8×10^{-3}	35
Poly(vinyl chloride)	74	10	22	78
Polyacrylate	178	25	16	51
Poly(vinyl methyl ether)	248	15	0.6	35, 51
Polysulfone	189	7	4.7	51

^{a} Diglycidyl ether of bisphenol A.

McCrum (79) proposed a method for correcting the perturbation of distributed processes analyzed in terms of the Bucci framework (9, 10). It was suggested that T_c obtained with corrected data would be closer to T_g than initially (79). This method appears quite successful in the analysis of experimental peaks in the temperature range where $P(T) \leq (P_o/2)$. It leads to an increase of the experimental activation energy of the corrected (effective) relaxation time, and to an improvement of the linearization of the Bucci–Fieschi–Guidi plot. As a result, the compensation temperature decreases. This method requires isothermal measurements of the deformation, $\gamma(t)$, and of its derivative with respect to time, $\gamma'(t)$, in thermomechanical experiments. This procedure is experimentally applicable to thermostimulated creep measurements (80, 81), where both γ and γ' can be recorded simultaneously as a function of time and temperature. However, it is not suited to TSC experiments, where only the time derivative of polarization, current density, is available.

Some authors interpreted T_c as the reciprocal of the jump of the thermal expansion coefficient, $\Delta\alpha$, across the T_g range (68, 82, 83). Because $\Delta\alpha \times T_g$ is roughly constant independent of the nature of the polymer (84), this assignment may be fortuitous. The lag between T_c and T_g may have some relation with the experimental setup, but it is obviously not the main explanation to the large T_c–T_g lag observed in some systems (>20 °C). The kinetic character of such transitions may have some relation with this effect. Instead of considering directly this lag, Ibar (85) defined two compensation laws around the glass transition of polymers. The averaged value of the corresponding compensation temperatures (T_c^+ and T_c^-) would allow one to deduce T_g (85). The lower temperature compensation law corresponds to the one we described, and T_c^+ is generally above T_g. The upper one involves decreasing activation energies and is generally observed for processes isolated above the investigated relaxation mode. Although this later compensation effect could be found in some analyses, it is not clear whether it is of real significance, or if it is merely the consequence of a natural decrease of activation energies toward the constant values generally found at high temperatures.

The activation energy that characterizes the relaxation mode associated with T_g is generally strong, whatever the technique used to obtain it is. Recent theories explain this behavior by a coupling of primitive relaxation species with their surroundings (86). Thus, the activation energy is all the more strong when the coupling of relaxation mode—the cooperativity—is strong. For P(VDF-TrFE) copolymers, the main activation energy in the distribution increases as the TrFE unit content increases. This behavior has been ascribed to a stiffening of molecular chains for increasing TrFE unit content (70). The substitution of a hydrogen atom by a fluorine atom induces a sterical hindrance that increases the coupling between relaxation modes. This behavior is consistent with results obtained in poly(ether ether ketone) (PEEK), a rigid chain semicrystalline polymer: Indeed activation energies as high as 6 eV have been obtained in the compensation phenomenon associated with T_g (*72*).

A usual assumption for the compensation law is based on the existence of elementary processes involving chain sequences of increasing length. Chain segments of increasing length relax at increasing temperature until a maximum size is reached at the compensation temperature (*32*). This interpretation is fairly consistent with the variation of activation energy as a function of crystallinity in semicrystalline polymers. Indeed, the width of the distribution of activation energy decreases, mainly because of a lowering of the highest values of ΔH, when the crystallinity increases. This behavior is clearly observed in P(VDF–TrFE) 65/35 copolymer whose crystallinity can be significantly varied. For this copolymer, activation energies versus peak maximum temperature are reported in Figure 9 for crystallinities of about 55% and 80% (*70*). As previously discussed, the activation energy dependence on crystallinity has also been observed in lower crystallinity materials such as PET or PEEK (*87*).

Several models have been built for predicting the typical length scale associated with the glass transition, and they result in a value of some nanometers (*88, 89*). Although a direct verification of these values is presently not available, indirect methods have been applied that consist in squeezing-in the phase of interest into dimensions of the order of the characteristic length. This process has been done recently by considering semicrystalline systems of known thickness of amorphous, interfacial, and crystalline layers (*90, 91*), which can be varied by appropriate thermal treatment. In amorphous polymers, the characteristic length may be dependent on the chain

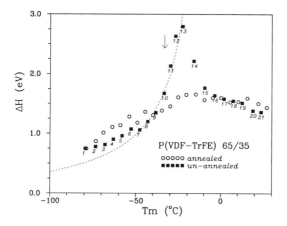

Figure 9. Activation energies versus peak maximum temperature of P(VDF–TrFE) 65/ 35. Arrow indicates T_g. The dotted line has been deduced from the compensation rule (65). Annealing above T_{fp} increased crystallinity.

entanglement spacing (typically 10 nm), whereas in semicrystalline polymers, the thickness of amorphous layer becomes an important parameter
below a critical value (or length). An important concept introduced by
Donth (89) in the fluctuation approach of the glass transition is the mode
length, and the general scaling principle for it: the larger the mode length,
the larger its relaxation time, and the slower its mobility.

The described results on the activation energies of semicrystalline polymers can be well accommodated with the mode length concept: the higher
the activation energy, the longer the relaxation time, and the stronger the
length of chain segments involved in the relaxation. Overall, the broadness
of the distribution in activation energies reflects the length scale of molecular movements. The as-received P(VDF–TrFE) copolymer with moderate
crystallinity shows a compensation pattern similar to that found in amorphous polymers. Therefore, the amorphous phase behaves like in a purely
amorphous polymer, which implies that the thickness of amorphous layers
is higher than the typical length scale of movements associated with T_g.
The strong decrease of ΔH in high-crystallinity materials shows that the
growth of crystallites induces a squeezing-in of the amorphous phase into
dimensions lower than this characteristic length scale.

Prediction of Dielectric Properties

As previously discussed, the temperature and frequency dependence of a
complex mode can be predicted by using FP parameters. This analysis can
be performed by considering either a discrete or a continuous distribution
of relaxation times. For a discrete distribution, the dynamic behavior is
obtained by a direct summation of the isolated elementary contributions
without any hypothesis on the relationships between activation parameters
of several processes. This has been done in PClTrFE (92) and poly(phenylene sulfide) (PPS) (73). In case of a continuous distribution, the analysis
is more complex because each activated process is described by at least
two activation parameters. Therefore, the corresponding distribution, in
activation energies and in preexponential factors in case of Arrhenius-like
processes, must generally be treated independently. A relationship like the
compensation law can overcome this problem because relaxation times are
dependent on only one variable parameters. In that case, an analytical form
of a continuous distribution of only one activation parameter is necessary.
In the following, we apply the compensation law to the study of the contribution of the β-relaxation mode (associated with T_g) of PVDF, a semicrystal-
line polymer. The compensation parameter associated with T_g are $T_c = -15\ °C; \tau_c = 5.8\ 10^{-3} s$.

This compensation effect was entirely obtained from processes isolated
in the complex TSC β-mode. So, the common origin of these processes is

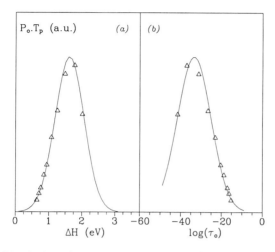

Figure 10. Distribution of activation energies (a) and preexponential factors (b) of PVDF for relaxation times obeying the compensation law. The solid lines correspond to a Gaussian fitting.

confirmed. By the following, we consider that the β-mode of PVDF is fully described by elementary processes obeying a compensation law. This description allows one to deduce a continuous distribution of relaxation times. Use is made of a continuous distribution because the frequency response of dielectric materials is generally analyzed based on this approach (*93–95*).

The statistical function defined for the elementary peaks is $P_{oi}T_{pi} \propto N_i$, where P_{oi} corresponds to the area of the ith peak, and T_{pi} is the polarization temperature. When $kT_p >> \mu E_p$, which is valid for usual TSC experimental procedures (*96*), this function is related to the strength of the relaxation, so that

$$P_{o_i} = \frac{\alpha N_{d_i}\mu^2 E_p}{kT_{p_i}} \tag{28}$$

where E_p is the applied field, N_{d_i} is the dipole concentration (per volume unit), α (= 1/3 for free-rotating dipoles) is a structure factor, and μ is the dipolar moment.

The quantity $P_{oi}T_{pi} \propto N_i$ has been adopted as a statistical function because it is independent on the polarization temperature. In Figure 10, we plotted the variations of $P_{oi}T_{pi}$ as a function of (a) the activation energy, ΔH_i, and (b) the preexponential factor, τ_{oi}. In this figure, data relative to the elementary processes obeying the compensation law are reported. The full lines correspond to a fitting of the distribution according to a Gaussian function. For the activation energy, the distribution function is expressed by

$$\Psi(\Delta H) = \frac{1}{\sigma_H \sqrt{\pi}} \exp - \left[\frac{\Delta H - \Delta H_m}{\sigma_H}\right]^2 \qquad (29)$$

The main value of the activation energy is $\Delta h_m = 1.62$ eV, and the width of the distribution is $\sigma_H = 0.62$ eV. Note that the value of ΔH_m is consistent with that obtained by dynamic analysis: 1.6 to 1.7 eV (5, 97).

Dielectric relaxations in polymers have often been modeled by a Gaussian distribution in activation energy (95, 98, 99). However, the preexponential factor, τ_o, is usually a supposed constant. In the present work, we have shown that ΔH_i and τ_{oi} are coupled by a compensation equation. Because a linear relationship exists between ΔH_i and $\ln(\tau_{oi})$, the $\ln(\tau_{oi})$ is also distributed in a Gaussian manner (100). The width of the distribution for $\ln(\tau_o)$ is

$$\sigma_o = \frac{\sigma_H}{kT_c} \qquad (30)$$

and its main value is defined by

$$\ln(\tau_{om}) = \ln \tau_c - \frac{\Delta H_m}{kT_c} \qquad (31)$$

This Gaussian distribution may be applied to $\ln(\tau)$ because this value is linearly dependent on $\ln(\tau_o)$ and ΔH. In that case, the parameters of the distribution are temperature dependent. Because the slope of $\ln(\tau_o)$ versus ΔH is negative, the width of the distribution for $\ln(\tau)$ is given as

$$\sigma_\tau = \left|\sigma_o - \frac{\sigma_H}{kT}\right| \qquad (32)$$

The mean relaxation time, τ_m, and the distribution function for $\ln(\tau)$ are expressed by

$$\ln(\tau_m) = \ln(\tau_{om}) + \frac{\Delta H_m}{kT} \qquad (33)$$

$$\Psi(\ln \tau) = \frac{1}{\sigma_\tau \sqrt{\pi}} \exp - \left[\frac{\ln(\tau/\tau_m)}{\sigma_\tau}\right]^2 \qquad (34)$$

To illustrate the properties of this model, we plotted in Figure 11 the variations of the imaginary part of the dielectric permittivity as a function of the temperature and frequency, obtained from eqs 27b and 34. The variable ε'' has been normalized to the strength of the relaxation, $\Delta\varepsilon$. In the primary

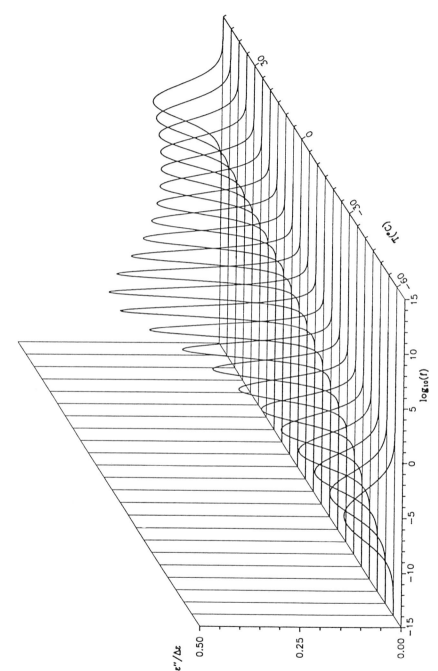

Figure 11. Variation of the normalized loss factor as a function of temperature and frequency for the β-relaxation mode of PVDF.

dispersion of amorphous polymers, $\Delta\varepsilon$ is generally inversely proportional to the absolute temperature: $\Delta\varepsilon = K/T$. However, in PVDF, literature data on K are variable: 300 K and 820 K according to references 101 and 102, respectively. PVDF has a complex relaxational behavior, and the discrepancy obtained for the values of $\Delta\varepsilon$ may arise from contributions of additive relaxation modes such as the α-mode, and possibly from the upper component of T_g. In Figure 11, $\varepsilon_n'' = \varepsilon''/\Delta\varepsilon$ reaches a maximum for a temperature $T = T_c$ and a frequency $f_c = 1/2\pi\tau_c$. Moreover, the width of the distribution is reduced to a single relaxation time.

The strong increase of the loss factor in the low-temperature range is quantitatively consistent with dielectric dynamic analysis (102, 103). However, in the high-temperature range, a continuous increase of ε_{max}'' is generally observed for PVDF (104) (ε_{max}'' is defined as the maximum of ε'' for constant temperature and variable frequency). Several reasons may be invoked for such a behavior.

1. As previously discussed, the strength of the relaxation, $\Delta\varepsilon$, and its variations as a function of the temperature have not been dissociated from the contribution of the distribution function, whereas dielectric relaxation analyses give directly the variations of $\varepsilon''(T, \omega)$.

2. The distribution function has been deduced from measurements at low temperatures and low-equivalent frequency, and the weight function for the isolated relaxation times, namely $P_{oi}T_{pi}$, has been considered as independent of the temperature. However, $P_{oi}T_{pi}$ is possibly temperature dependent, through a temperature dependence of the dipolar moment or the local density of the material as example, resulting in a modification of the shape and magnitude of the loss curves. At this point, the reliability of the extrapolations made from low-frequency measurements to high frequency is questioned, and additive work should be achieved by complementary techniques to define the proper range of validity of these extrapolations.

3. For purely amorphous polymers such as polyvinylacetate, which is considered a standard polymer for testing current theories on dielectric relaxations, ε_{max}'' reaches a maximum for a temperature slightly higher than the glass transition temperature ($T_g \approx 31$ °C, $T_m \approx 37$ °C) (105). Therefore, the continuous increase of ε_{max}'' as a function of the temperature is not a general rule. This behavior of PVDF with regard to its dielectric β-mode may originate from the heterogeneous (i.e., semicrystalline) structure of the polymer. Under this hypothesis, the previously discussed analysis would isolate the response of the purely amorphous phase. Overall, the use

of a continuous distribution of relaxation times allowed us to isolate the contribution of the β-mode to the dielectric relaxation. Conversely, by considering a discrete distribution of processes isolated over a wide temperature range, one can predict the whole dynamic behavior of the polymer, but without further insight into the precise behavior of the various relaxation modes that are expected to contribute, such as semicrystalline polymers where processes related to both crystalline and amorphous phases are expected to be active.

These methods are useful tools, especially for investigating the low-frequency behavior that is one of the limitations of classical dynamic analyses.

Conclusion

TSC spectroscopy is well suited to the characterization of the amorphous polymers. Indeed, the dielectric relaxation mode associated with the glass transition is due to a cascade of molecular movements obeying a compensation law. The corresponding parameters define the "physical structure" of the amorphous phase. The effect of orientation is clearly shown on the "compensation diagram." In semicrystalline polymers, the presence of crystallites is accompanied by an interphase that also has been characterized. The role of small molecules, macromolecules, or fillers inside the polymeric matrix can also be studied. Moreover, the parameters extracted from TSC studies allow for the prediction of the variation of the complex dielectric constant as a function of temperature and frequency.

References

1. Marchal, E. *J. Chem. Phys.* **1992**, *96*, 4676.
2. Havriliak, S.; Negami, S. *J. Polym. Sci.: Polym. Symp.* **1966**, *614*, 99.
3. Schönhals, A.; Kremer, F.; Hofmann, A.; Fischer, E. W.; Schlosser, E. *Phys. Rev. Lett.* **1993**, *70*, 3459.
4. Creswell, R. A.; Perlman, M. M. *J. Appl. Phys.* **1970**, *41*, 2365.
5. Mitzutani, T.; Yamada, T.; Ieda, M. *J. Phys. D: Appl. Phys.* **1981**, *14*, 1139.
6. Chatain, D.; Gautier, P.; Lacabanne, C. *J. Polym. Sci.: Phys. Ed.* **1973**, *11*, 1631.
7. Hino, T. *Jpn. J. Appl. Phys.* **1973**, *12*, 611.
8. Zielinski, M.; Kryszewski, M. *Phys. Stat. Solidi* **1977**, *42a*, 305.
9. Bucci, C.; Fieschi, R. *Phys. Rev. Lett.* **1964**, *12*, 16.
10. Bucci, C.; Fieschi, R.; Guidi, G. *Phys. Rev.* **1966**, *148*, 816.
11. Van Turnhout, J. *Thermally Stimulated Discharge of Polymer Electrets;* Elsevier: Amsterdam, Netherlands, 1975; pp 66–82.
12. Lavergne, C.; Lacabanne, C. *IEEE Elec. Insul. Mag.* **1993**, *9*, 5.
13. Chafai, A.; Chatain, D.; Dugas, J.; Lacabanne, C.; Vayssié, E. *J. Macromol. Sci.: Phys.* **1983**, *22B*, 633.

14. Bernes, A.; Chatain, D.; Lacabanne, C. *Polymer* **1992**, *33*, 4682.
15. Eyring, H. *J. Chem. Phys.* **1944**, *4*, 283.
16. Exner, O. *Coll. Czech. Chem. Commun.* **1972**, *37*, 1425.
17. Vogel, H. *Phys. Z.* **1921**, *22*, 645.
18. Fulcher, G. S. *J. Am. Ceram. Soc.* **1925**, *8*, 339.
19. Tammann, G.; Hesse, G. *Z. Anorg. Allg. Chem.* **1926**, *156*, 245.
20. Adam, G.; Gibbs, J. H. *J. Chem. Phys.* **1965**, *43*, 139.
21. Kauzmann, W. *Chem. Rev.* **1948**, *46*, 219.
22. Cohen, M. H.; Turnbull, D. *J. Chem. Phys.* **1959**, *31*, 1164.
23. Turnbull, D.; Cohen, M. H. *J. Chem. Phys.* **1961**, *34*, 120.
24. Turnbull, D.; Cohen, M. H. *J. Chem. Phys.* **1970**, *52*, 3038.
25. Colmenero, J.; Alegria, A.; Del Val, J. J.; Alberdi, J. M. *Makromol. Chem.: Makromol. Symp.* **1988**, *20/21*, 397.
26. Müller, P.; Teltow, J. *Phys. Stat. Solidi* **1972**, *12a*, 471.
27. Christodoulides, C.; Apekis, L.; Pissis, P. *J. Appl. Phys.* **1988**, *64*, 1367.
28. Ibar, J. P. In *Fundamentals of Thermal Stimulated Current and Relaxation Map Analysis;* SLP Press: New Canaan, CT, 1993; p 67.
29. Lacabanne, C.; Lamure, A.; Teyssèdre, G.; Bernes, A.; Mourgues, M. *J. Non-Cryst. Solids* **1994**, *172–174*, 884.
30. Hitmi, N.; Lacabanne, C.; Young, R. A. *J. Phys. Chem. Sol.* **1984**, *44*, 701.
31. Hitmi, N.; Lacabanne, C.; Young, R. A. *J. Phys. Chem. Sol.* **1986**, *47*, 533.
32. Hoffman, J. D.; Williams, G.; Passaglia, E. *J. Polym. Sci.: Polym. Symp.* **1966**, *14*, 173.
33. Starkweather, H. W. *Macromolecules* **1981**, *14*, 1277.
34. Starkweather, H. W. *Polymer* **1991**, *32*, 2443.
35. Sauer, B.; Avakian, P. *Polymer* **1992**, *33*, 5128.
36. Teyssèdre, G.; Lacabanne, C. *J. Phys. D: Appl. Phys.* **1995**, *28*, 1478.
37. Hagege, R.; Mamy, C.; Thiroine, C. *Makromol. Chem.* **1978**, *179*, 1069.
38. Bernès, A.; Boyer, R. F.; Lacabanne, C.; Ibar, J. P. In *Order in the Amorphous State of Polymers;* Keinath, S. E.; Miller, R. L.; Rieke, J. K., Eds.; Plenum: New York, 1986; p 305.
39. Starkweather, H.W. *Macromolecules* **1988**, *21*, 1798.
40. Hannig, K.; Nordwig, A. In *Treatise on Collagen;* Ramachandran, G. N., Ed.; Academic: London, 1967; Vol. 1, p 73.
41. Parry, D. A. D. *Polymer* **1977**, *18*, 1091.
42. Yannas, I. V. *J. Macromol. Sci. Rev.: Macromol. Chem.* **1972**, *C7*, 49.
43. Chien, J. C. W. *J. Macromol. Sci. Rev.: Macromol. Chem.* **1975**, *12*, 1.
44. Lamure, A.; Hitmi, N.; Lacabanne, C.; Harmand, M. F.; Herbage, D. *IEEE Trans. Elec. Insul.* **1986**, *EI-21*, 443.
45. Fauran, M. J.; Fabre, G.; Oustrin, J.; Lacabanne, C.; Stefenel, M.; Lamure, A. *Biomaterials* **1988**, *9*, 187.
46. Gervais-Lugan, M.; Haran, R.; Lamure, A.; Lacabanne, C. *J. Biomed. Mater. Res.* **1991**, *25*, 1339.
47. Mezghani, S. *Ph.D. Thesis*, University of Toulouse, 1994.
48. Mezghani, S.; Lamure, A.; Lacabanne, C. *J. Polym. Sci.: Phys. Ed.* **1995**, *33*, 2413.
49. Nomura, S.; Hiltner, A.; Lando, J. B.; Baer, E. *Biopolymers* **1977**, *16*, 231.
50. Pineri, M. H.; Escoubes, M.; Roche, G. *Biopolymers* **1978**, *17*, 2799.
51. Colmenero, J.; Alegria, A.; Alberdi, J. M.; Del Val, J. J.; Ucar, G. *Phys. Rev. B* **1987**, *35*, 3995.
52. Shimizu, H.; Nakayama, K. *Jpn. J. Appl. Phys.* **1989**, *28*, L1616.
53. Teyssèdre, G.; Bernès, A.; Lacabanne, C. *J. Polym. Sci.: Phys. Ed.* **1995**, *33*, 879.
54. Furukawa, T. *Phase Transition* **1989**, *18*, 143.

55. Kepler, R. G.; Anderson, R. A. *Adv. Phys.* **1992**, *41*, 1.
56. Nakagawa, K.; Ishida, Y. *J. Polym. Sci.: Phys. Ed.* **1973**, *11*, 2153.
57. Kimura, K.; Ohigashi, H. *Jpn. J. Appl. Phys.* **1986**, *25*, 383.
58. Tanaka, H.; Yukawa, H.; Nishi, T. *J. Chem. Phys.* **1989**, *90*, 6740.
59. Tashiro, K.; Takano, K.; Kobayashi, M.; Chatani, Y.; Tadokoro, H. *Ferroelectrics* **1984**, *57*, 297.
60. Lovinger, A. J.; Davis, G. T.; Furukawa, T.; Broadhurst, M. G. *Macromolecules* **1982**, *15*, 323.
61. Teyssèdre, G.; Bernès, A.; Lacabanne, C. *J. Polym. Sci.: Phys. Ed.* **1993**, *31*, 2027.
62. Royer, M.; Micheron, F. *C. R. Ac. Sci. B, Paris* **1978**, *287*, 145.
63. Furukawa, T.; Johnson, G. E.; Bair, H. E.; Tajitsu, Y.; Chiba, A.; Fukada, E. *Ferroelectrics* **1981**, *32*, 61.
64. Koga, K.; Nakano, N.; Hattori, T.; Ohigashi, H. *J. Appl. Phys.* **1990**, *67*, 965.
65. Del Val, J. J.; Alegria, A.; Colmenero, J.; Barandiaran, J. M. *Polymer* **1986**, *27*, 1771.
66. Crine, J. P. *J. Appl. Phys.* **1989**, *66*, 1308.
67. Krug, R. R.; Hunter, W. G.; Grieger, R. A. *J. Phys. Chem.* **1976**, *80*, 2335.
68. Zielinski, M.; Swiderski, T.; Kryszewski, M. *Polymer* **1978**, *19*, 883.
69. Ronarc'h, D.; Audren, P.; Moura, J. L. *J. Appl. Phys.* **1985**, *58*, 474.
70. Teyssèdre, G;. Bernès, A.; Lacabanne, C. *J. Polym. Sci.: Phys. Ed.* **1995**, *33*, 2419.
71. Sharif, F. Ph.D. Thesis, University of Toulouse, 1984.
72. Mourgues, M.; Bernès, A.; Lacabanne, C. *Thermochim. Acta* **1993**, *226*, 7.
73. Shimizu, H.; Nakayama, K. *J. Appl. Phys.* **1993**, *74*, 1597.
74. Bernès, A.; Chatain, D.; Lacabanne, C. *Thermochim. Acta* **1992**, *204*, 69.
75. Ibar, J. P. *Thermochim. Acta* **1991**, *192*, 91.
76. Mourgues-Martin, M.; Bernès, A.; Lacabanne, C.; Nouvel, O.; Seytre, G. *IEEE Trans. Elec. Insul.* **1992**, *27*, 795.
77. Dessaux, C.; Dugas, J.; Chatain, D.; Lacabanne, C. *Journées d' Etude de Polymères XXIII;* Toulouse, France, 1995.
78. Del Val, J. J.; Alegria, A.; Colmenero, J.; Lacabanne, C. *J. Appl. Phys.* **1986**, *59*, 3829.
79. Mc Crum, N. G. *Polymer* **1982**, *23*, 1261.
80. Monpagens, J. C.; Chatain, D.; Lacabanne, C.; Gautier, P. *Solid State Comm.* **1976**, *18*, 1611.
81. Demont, P.; Chatain, D.; Lacabanne, C.; Glotin, M. *Makromol. Chem.: Macromol Symp.* **1989**, *25*, 167.
82. Mc Crum, N. G. *Polymer* **1984**, *25*, 299.
83. Eby, R. K. *J. Chem. Phys.* **1962**, *37*, 1785.
84. Boyer, R. F. *J. Polym. Sci.: Polym. Symp.* **1975**, *50*, 189.
85. Ibar, J. P. *Thermochim. Acta* **1991**, *192*, 297.
86. Rendell, R. W.; Ngai, K. L.; Mashimo, S. *J. Chem. Phys.* **1987**, *87*, 2359.
87. Mourgues, M.; Bernès, A.; Lacabanne, C. *J. Thermal Anal.* **1993**, *40*, 697.
88. Owen, A. J.; Bonart, R. *Polymer* **1985**, *26*, 1034.
89. Donth, E. *J. Non-Cryst. Solids* **1982**, *53*, 325.
90. Schick, C.; Donth, E. *Phys. Scripta* **1991** *43*, 423.
91. Koy, U.; Dehne, H.; Gnoth, M.; Schick, C. *Thermochim. Acta* **1993**, *229*, 299.
92. Shimizu, H.; Nakayama, K. *Jpn. J. Appl. Phys.* **1990**, *29*, L800.
93. Nowick, A. S.; Berry, B. S. *IBM J. Res. Develop.* **1961**, *5*, 312.
94. Mc Donald, J. R. *J. Appl. Phys.* **1987**, *61*, 700.
95. Muzeau, E.; Perez, J.; Johari, G. P. *Macromolecules* **1991**, *24*, 4713.
96. Chen, R.; Kirsh, Y. *Analysis of Thermally Stimulated Processes;* International Series on the Science of the Solid State 15; Pergamon: Oxford, England, 1981; pp 60–81.

97. Boyer, R. F. *Enc. Polym. Sci. Technol.* **1977,** *2(Suppl.),* 745–839.
98. Nomura, S.; Kojima, F. *Jpn. J. Appl. Phys.* **1973,** *12,* 205.
99. Degli-Esposti, G.; Tommasini, D. *IEEE Trans. Elec. Insul.* **1989,** *25,* 617.
100. Nowick, A. S.; Berry, B. S. *IBM J. Res. Develop.* **1961,** *5,* 297.
101. Anada, Y.; Kakizaki, M.; Hideshima, T. *Jpn. J. Appl. Phys.* **1990** *29,* 322.
102. Abe, Y.; Kakizaki, M.; Hideshima, T. *Jpn. J. Appl. Phys.* **1985** *24,* 1074.
103. Ishida, Y;. Watanabe, M.; Yamafugi, K. *Kolloid Z. Z. Polym.* **1964,** *200,* 48.
104. Rushworth, A. Ph.D. Thesis, University of Leeds, U.K., 1977.
105. Nozaki, R.; Mashimo, S. *J. Chem. Phys.* **1987,** *87,* 2271.

Application of Dielectric
Spectroscopy to Polymer Systems

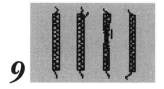

9

Dielectric Relaxation in Polymer Solutions

Keiichiro Adachi

Studies of dielectric relaxation in polymer solutions are reviewed based on the classification of dipole moment of polymers proposed by Stockmayer. For type-A polymers having dipoles in the direction parallel to the chain contour, the molecular weight dependence of the relaxation times in dilute solution is successfully described by the Zimm theory and the dependence in semidilute solutions is described by blob models. From the relaxation strength for type-A polymers, the expansion factor of the end-to-end distance is determined. For type-B polymers having perpendicular dipoles, the relaxation behaviors depend strongly on their chemical structure. The relationship between the relaxation behavior and the chain stiffness is discussed.

In 1967 Stockmayer (*1*) published a pioneering and farsighted review on dielectric relaxation of flexible polymers in solution. Later North (*2*) published an instructive and more detailed review from a view of a chemist. Block (*3*) also reviewed electrical properties of polymers including topics on the dielectric behavior of polymer solutions. After publications of these reviews, the experimental studies of dielectric relaxation for polymer solutions made slower progress than those for solid polymers (*4, 5*). One of the reasons may be the difficulty in dielectric measurements on solutions of polymers having the component of dipole moment perpendicular to the chain contour, because these polymers exhibit dielectric relaxation in the high frequency region ranging from 10^7 to 10^{10} Hz. Until recently commercial apparatuses for dielectric measurements in the gigahertz range were not available. The other reason may be the development of techniques for the study of polymer dynamics such as NMR spectroscopy, dynamic light scattering, or fluorescence depolarization. Therefore, the dielectric spectroscopy was not necessarily used for the study of polymer dynamics. On the other hand, theoretical studies of polymer dynamics were developed in the 1970s. Especially, theories of global motions of polymers in semidilute solutions were described by a scaling theory developed by French scientists (6). Theories of local segmental motion were developed by Monnerie and his co-workers (7), Hall and Helfand (8), and Jones and Stockmayer (9) to describe the results of studies of NMR spectroscopy and fluorescence depolarization. Experimental tests of the scaling theories were made recently by dielectric measurements on solutions of polymers having the component of dipole moment aligned parallel to the chain contour (10).

Although the dielectric method is a classical one, it still gives us direct information about the reorientation of polymer chains and undoubtedly is one of the most important methods for studies of polymer dynamics and conformation. In this chapter the recent progress of dielectric studies of polymer solutions is described and some unsolved problems are discussed.

Dipole Moment of Polymer Chains

Three Types of Polymer Dipoles

Dielectric properties of a polymer depend on the orientation of the dipole vector against the chain contour. Stockmayer (1) classified polymer dipoles into three types: A, B, and C. The dipoles aligned in the direction parallel to the chain contour are classified as type A; those aligned perpendicular to the contour are classified as type B; and those located on mobile side groups are classified as type C. The vector sum P_A of the type-A components of a chain is proportional to the end-to-end vector R:

$$P_A = \mu R \tag{1}$$

where μ is the constant. Thus, the dielectric relaxation of a type-A chain reflects the fluctuation and orientation of the end-to-end vector R. Using dielectric techniques we can obtain information on the global motions of polymer chains not only in dilute solution but also in complex condensed systems such as block copolymers composed of type-A chains and non-type-A chains (4). On the other hand, the relaxation of type-B dipoles reflects the local segmental motion (1, 2). The vector sum P_B of the type-B dipoles of a chain does not have correlation with R provided that the molecular weight is sufficiently high (11):

$$\langle P_B \cdot R \rangle = 0 \tag{2}$$

where the bracket indicates the statistical average. Obviously the dielectric relaxation due to type-C dipoles reflects rotation of the side group.

A polymer possessing only one type of dipole moment is an exceptional case. Often polymers possess different types of dipoles. For example, a polymer possessing the components A and B is called type AB. Usually the dielectric responses for the type A, B, and C dipoles occur in well-separated frequency ranges and therefore the motions due to the different components are observed separately (*1–3, 10*).

Whether a polymer possesses type-A dipoles is judged from the symmetry in the array of the backbone groups such as —CH_2—, —CHR—, and —O—. These groups are connected by either an actual bond such as the C—C bond or a virtual bond such as —$C_6H_4R_2$— and —CH_2—CHR=CH—CH_2— where R indicates a pendant group. Then, these groups and bonds are lined up straight according to the chemical structure. If the chain has the symmetry of mirror image or inversion, then the chain does not have type-A dipoles. For example, all vinyl polymers do not have the intrinsic type-A dipole, whereas polymers composed of the repeat units represented by —X—Y—Z— are intrinsic type-A polymers where X, Y, and Z represent atoms or chemical groups. An exceptional case is where a polymer belonging to type-B polymers according to the above criterion can possess type-A dipole if the polymer has the helical conformation (*2, 12*). Such a polymer is called a *form type-A* polymer (*10*).

Relationship Between Chemical Structure and Dipole Moment

The assignment of the type-A, -B, and -C components is not always straightforward and depends on situations. For instance when a side group rotates much faster than the backbone, the fluctuation of the dipole moment of the type-C component is observed by dielectric measurements. However, when the side group rotates at a rate slower than the local segmental motion of the backbone, the dielectric relaxation due to the rotation of the side group is no more observed and the dipoles attached to the side group can be regarded as the type-B component. Similar cases are encountered for type-AB polymers (*11*). Hereafter, I will leave type-C dipoles out of consideration.

Various type-A and type-AB chains are further classified at least into six types as depicted schematically in Figure 1 (*13, 14*). The thick arrows indicate the dipole vector attached to the bonds or virtual bonds and the dotted lines indicate the nonpolar bonds. Type-A2 and -A2B chains are composed of bonds with type-A dipole and nonpolar bonds. In the final type-A3 and -A3B groups, chains are composed of more than two kinds of parallel dipoles. In addition to these six types, various combinations of these types

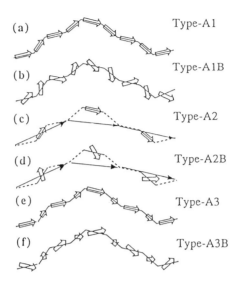

Figure 1. Various type-A chains. Thick arrows indicate dipole vectors and dotted lines are nonpolar bonds.

exist. Examples of type-A1 to -A3B polymers are shown in Table I together with the type-A dipole moment $|\mathbf{P_A}|$ per monomer unit. A typical type-A1 chain is poly(2,6-disubstituted-1,4-phenylene oxide) (15, 16). When a poly(1,4-phenylene oxide) chain has only one substituent, the polymer belongs to type-A1B. In polyacetylene (17) and polyphosphazene derivatives (18), the —C≡C— and —P≡N— bonds can be regarded as virtual bonds having both the type-A and -B components of the dipole moment. Type-A1 and -A1B chains are composed of polar bonds so that eq 1 is satisfied at any moment.

Obviously eq 1 does not hold exactly for type-A2 (19–21) and -A3 polymers (22, 23). However, it is expected that eq 1 still holds for a time average of $\mathbf{P_A}$ and $\mathbf{R}$ averaged for a short period. Then, the statistical average of $\mathbf{P_A^2}$ is proportional to the mean square end-to-end distance $\langle R^2 \rangle$:

$$\langle \mathbf{P_A^2} \rangle = \mu^2 \langle R^2 \rangle \tag{3}$$

For type-A1 and-A1B chains, μ is a constant independent of temperature and solvent quality but μ for A2 or A3 type chains might change with the conditions. This difference is important in the determination of $\langle R^2 \rangle$ from the dielectric data.

The total dipole moment $\mathbf{P}$ of a type-AB chain is the sum of $\mathbf{P_A}$ and $\mathbf{P_B}$. From eqs 1 and 2, μ is given by

$$\mu = \frac{\langle \mathbf{P \cdot R} \rangle}{\langle R^2 \rangle} \tag{4}$$

Table I. Type-A Polymers and Type-A Dipole Moment (P_A)

Type	Polymer	Repeat Unit		P_A/D	Ref.
A1	Poly(phenylene oxide)[a]	$-C_6H_4X_2-O-$	X = Cl	1.56	15
			X = CH$_3$	0.36	16
A1B	Poly(phenylene oxide)[b]	$-C_6H_5X-O-$	X = Cl	0.78	—
	Polyacetylene[c]	$-CH = CX-$	X = C$_6$H$_4$, Cl	2	17
			X = C$_6$H$_4$OCH$_3$	3	17
	Polyphosphazene[d]	$-PX_2 = N-$	X = OC$_6$H$_5$	6.7	18
			X = OCH$_2$CF$_3$		18
A2	Poly(methylene phenylene)[e]	$-(CH_2)_4-C_6H_4X_2-$	X = Cl	1.56	—
A2B	Polydiene	$-CH_2-CX = CH-CH_2-$	X = Cl	1.03 (cis)	
			X = Cl	0.21 (trans)	19
			X = CH$_3$	0.22 (cis)	10
			X = C$_6$H$_5$	0.2	
	Polyester	$-(CH_2)_nO-CO-$	n = 4, 5	0.82	14, 20, 21
	Polypeptide	$-CHX-CO-NH-$			—
	Polylactam	$-(CH_2)_n-CO-NH-$	n = 5	1.0	—
A3B	Polyether	$-CH_2-CHX-O-$	X = CH$_3$	0.18	1, 22, 23
			X = C$_6$H$_5$	0	56

NOTE: D is diffusion coefficient.
[a]Poly(2,6-disubstituted-1,4-phenylene oxide).
[b]Poly(2-substituted-1,4-phenelene oxide).
[c]Poly(substituted acetylene).
[d]Poly(organophosphazene).
[e]Poly[tetramethylene(2,6-dichloro-1,4-phenylene)].

The calculation of eq 4 may be performed based on the statistical method for the rotational isomeric state model developed by Flory (24) when the energy differences among the isomeric states are known. When μ is calculated as a function of the degree of polymerization x, μ changes with x in the range of small x but converges to a certain value with increasing x in a way similar to the characteristic ratio.

For type-B dipoles, no way to classify them exists because we do not know how type-B dipoles on a chain correlate each other. An interesting problem is whether type-A2 polymers exhibit the dielectric relaxation due to local segmental mode. This problem has not been examined experimentally. At present, therefore, we regard all components satisfying eq 2 as type B.

Dielectric Relaxation of Type-A Chain in Solution

Theory of Normal-Mode Relaxation

Complex Dielectric Constant

The complex dielectric constant $\varepsilon^* = \varepsilon' - i\varepsilon''$ (where i is the imaginary number) is generally given by the Fourier–Laplace transform of the decay function ψ of the polarization. For a type-A chain ψ coincides to the autocorrelation function ϕ (t) (where t is time) of the type-A dipole $\mathbf{P}_A$ ($1–3$, 10). Thus, the theoretical description of the dielectric relaxation of type-A dipoles is much clearer than that for type-B dipoles. From eq 1, ϕ (t) becomes equivalent to the autocorrelation function of the end-to-end vector $\langle \mathbf{R}(0){\cdot}\mathbf{R}(t)\rangle/\langle R^2\rangle$. Thus ε^* for a chain with type-A dipoles is given by

$$\frac{\varepsilon^*(\omega) - \varepsilon_\infty}{\Delta \varepsilon} = \int_0^\infty - \frac{d}{dt}\left(\frac{\langle \mathbf{R}(0){\cdot}\mathbf{R}(t)\rangle}{\langle R^2\rangle}\right)\exp(-i\omega t)dt \qquad (5)$$

where ω is the angular frequency, ε_∞ is the high frequency dielectric constant, and $\Delta\varepsilon$ is the dielectric relaxation strength ($= \varepsilon'(0) - \varepsilon_\infty$). For the cases of block copolymers composed of a type-A and a type-B subchain or the bifurcated chains in which the alignment of dipoles is inverted, ε^* is calculated similarly to eq 5 (10).

For dilute solutions, $\langle \mathbf{R}(0){\cdot}\mathbf{R}(t)\rangle$ is expressed by the bead-spring model (25, 26) in which a chain is assumed to be composed of $N + 1$ beads and N springs. According to the free draining model proposed by Rouse (25) the correlation function is given by

$$\langle \mathbf{R}(0){\cdot}\mathbf{R}(t)\rangle = \frac{8\langle R^2\rangle}{\pi^2}\sum_{p:\text{odd}}^\infty \frac{1}{p^2}\exp\left(\frac{-t}{\tau_p}\right) \qquad (6)$$

$$\tau_p = \frac{\zeta N^2 b^2}{3\pi^2 k_B T p^2} \qquad (7)$$

where τ_p is the relaxation time for the pth normal mode, ζ is the friction coefficient per bead, N is the number of the beads, b is the average distance between two adjacent beads, k_B is the Boltzmann constant, and T is temperature.

For the nondraining model, Zimm (26) expressed $\langle \mathbf{R}(0){\cdot}\mathbf{R}(t)\rangle$ by the same form as eq 6, but with different τ_p given by

$$\tau_p = \frac{\pi^{3/2}\eta_s b^3 N^{3/2}}{12^{1/2}k_B T \lambda_p} \qquad (8)$$

when η_s is the solvent viscosity, and λ_p is the pth eigen value ($\lambda_1 = 4.04$, $\lambda_3 = 24.2$, $\lambda_5 = 53.5$). The Zimm model predicts $\tau_1 \propto M^{1.5}$ (where M is the molecular weight).

Dielectric Relaxation of Type-A Chains in Semidilute and Concentrated Solution

In dilute solutions chains are isolated, but with increasing concentration, C, chains begin to overlap. The overlapping concentration C^* is often used as a measure of the critical concentration around which a crossover from the behavior in an isolated state to that in entangled state occurs (6):

$$C^* = \frac{3M}{4\pi N_A S^3} \tag{9}$$

where N_A is Avogadro's number, and S the square root of the mean square radius of gyration. The value of C^* is conveniently given by the inverse of the intrinsic viscosity. Empirically, above about $10C^*$ the chains are entangled with each other (10). In such cases, the bead-spring model is no more applicable and the tube model proposed by De Gennes (27) and developed by Doi and Edwards (28) may be applied. De Gennes predicted that the correlation function $\langle \mathbf{R}(0)\cdot\mathbf{R}(t)\rangle$ in an entangled system is expressed by exactly the same form as eq 6. According to the Doi–Edwards theory, the relaxation time τ_p^e is given by

$$\tau_p^e = \frac{L^2}{\pi^2 D_c} \tag{10}$$

where L and D_c are the length of the tube and the diffusion coefficient of the chain along the tube, respectively. As is well known, this theory predicts $\tau_p^e \propto M^{3.0}$.

Relaxation Strength and End-to-End Distance of Type-A Chain

The dielectric response of a type-A chain is proportional to the end-to-end vector. Therefore, we can determine the mean square end-to-end distance $\langle R^2 \rangle$ from the relaxation strength $\Delta\varepsilon [\alpha \langle P_A^2 \rangle]$ of a type-A chain:

$$\frac{\Delta\varepsilon}{C} = \frac{4\pi N_A \mu^2 \langle R^2 \rangle F}{3 k_B T M} \tag{11}$$

where C is the concentration in wt/vol, and F is the ratio of the internal to external electric fields. Because $\Delta\varepsilon$ is independent of time, eq 11 holds irrespective of the entanglement effects. Thus, if μ is known, $\langle R^2 \rangle$ can be determined from the $\Delta\varepsilon/C$ data (10). This equation provides the method for the study of excluded volume effect of polymer chains in solution as will be seen later.

To determine $\langle R^2 \rangle$, F needs to be expressed as a function of the dielectric constant of the solvent. Co-workers and I (10, 29) indicated that F is close to unity rather than the internal fields used traditionally for low molecular weight compounds or the local segmental mode process of polymers (10). Obviously any of the equations of internal field for low molecular weight molecules cannot be applied to a type-A chain because the assumption of a vacuum cavity of the size of random coil is not realistic. More detailed discussion to support this assumption is given in reference 10.

Experimental Study on Solutions of Type-A Chains

Dilute Solution

As mentioned previously, the dielectric relaxation time of type-A chains depends strongly on M. In the early 1970s only a limited numbers of the dielectric data on type-A chains were reported (1, 2). Among these data, a strong and clear M dependence of the nominal relaxation time τ_n defined by $\tau_n = 1/(2\pi f_m)$ was reported for dilute solutions of stiff rod-like polymers such as poly(γ-benzyl-L-glutamate) (30) or poly(butyl isocyanate) (12), where f_m is the loss maximum frequency. Both polymers have the helical structure, and poly(butyl isocyanate) is a form type-A polymer. Recently Takada et al. (31) reported the M dependence of τ_n for solutions of polyisocyanate over wide M range. They found that when the chain length exceeded the persistence length, the dielectric behavior approaches that for a flexible polymer with increasing M. Jones et al. (20) first reported the dielectric relaxation of a flexible type-A polymer in dilute solution. They used solutions of poly(ε-caprolactone) in benzene and dioxane. At the present time, several dielectric data of solutions of type-A polymers are available as listed in Table I. Among these polymers, cis-polyisoprene (cis-PI) was studied most extensively (10). Therefore, the dielectric behavior of solutions of cis-PI is mainly reviewed hereafter.

Figures 2A and 2B show the double logarithmic plot of τ_n for solutions of cis-PI in a good solvent, benzene, and a theta solvent, dioxane, respectively (32). If the Rouse–Zimm theory (25, 26) or the Doi–Edwards theory (28) describes the normal-mode relaxation, τ_n defined previously for flexible linear polymers is ca 0.97 times the longest relaxation time τ_1. Thus, τ_n corresponds essentially to τ_1. A plot of log τ_n versus log M for solutions in benzene and dioxane conforms well to straight lines with the slope 1.50 and 1.77, respectively. The exponent 1.50 agrees well with the Zimm theory. However, neither the Rouse nor the Zimm theory predicts the $M^{1.77}$ dependence observed for benzene solutions. This result is explained by the excluded volume effect: τ_n in a good solvent has an M dependence stronger than that in a theta solvent because of the expansion of the dimension. As is known for viscoelastic relaxation, the effect of the excluded volume can be incorporated in the intrinsic viscosity [η], and τ_n for the free-draining model is rewritten as (25):

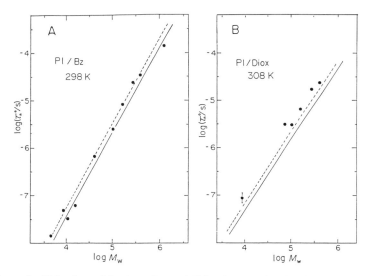

Figure 2. Molecular weight dependence of dielectric normal-mode relaxation time τ_n for dilute solutions of cis-*polyisoprene in a good solvent benzene and a theta solvent dioxane.*

$$\tau_p = \frac{12 M \eta_s [\eta]}{\pi^2 R T p^2} \tag{12}$$

where η_s is the solvent viscosity. The non-free-draining model predicts (26):

$$\tau_p = \frac{3.41 M \eta_s [\eta]}{R T \lambda_p} \tag{13}$$

In Figure 2, the dashed and solid lines represent, respectively, eqs 12 and 13. The theories agree approximately with the experiment. Similar results were observed for solutions of *cis*-PI in the other solvents (*33*) and also for the other flexible polymers such as poly(ε-caprolactone) (*14*) and poly(*p*-henoxy phosphazene) (*18*).

The Rouse–Zimm theory predicts that the τ_n versus M data for various polymers in various solvents fall on a universal curve if τ_n is plotted against $M \eta_s [\eta] / RT$. Figure 3 shows the test of this prediction. As is seen, the data fall on a straight line with the slope one within the experimental error. The best-fit curve (solid line) is expressed as (*14*)

$$\tau_n = 1.4 M \eta_s [\eta] / RT \tag{14}$$

The value of the front factor 1.4 ± 0.3 is slightly larger than the theoretical factor 1.22 given by the Rouse theory and 0.85 given by the Zimm theory.

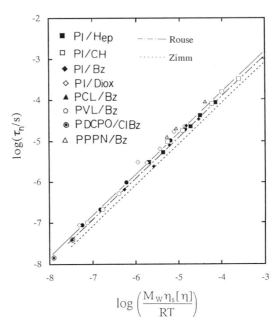

Figure 3. Double logarithmic plot of dielectric normal-mode relaxation times τ_n versus $M_w \eta_s(\eta)/RT$ for dilute solutions of various polymer–solvent systems. Sample codes: PI, cis-polyisoprene; PCL, poly(ε-caprolactone); PVL, poly(valerolactone); PDCPO, poly(2,6-dichloro-1,4-phenylene oxide); PPPN, poly(phenoxy phosphazene); Hep, heptane; CH, cyclohexane; Bz, benzene; Diox, dioxane; and ClBz, chlorobenzene.

Therefore, the M dependence of τ_n is successfully explained by the Rouse–Zimm theory, and the behavior is independent of the types of the alignment of dipoles shown in Figure 1.

Semidilute and Concentrated Solution of Type-A Chains

In the crossover regime between the dilute and semidilute regimes, the concentration (wt/vol) C dependence of the relaxation time τ_p for the pth normal mode is described by the Muthukumar–Freed theory (*34, 35*):

$$\tau_p = \tau_p^0[1 + CAp^{-k} - 2^{0.5}(CAp^{-k})^{1.5} + 2(CAp^{-k})^{2.0} + \ldots] \quad (15)$$

where τ_p^0 is the relaxation time at infinite dilution, A is the constant proportional to $M^{3\nu-1}$, and k is equal to $3\nu - 1$. Because $[\eta]$ is proportional to $M^{3\nu-1}$, the theory predicts that $\ln \tau_n (\cong \ln\tau_1)$ is proportional to $C[\eta] (\cong C/C^*)$ in semidilute regime.

Before the Muthukumar–Freed theory was proposed, Jones et al. (*21*) already examined the $C[\eta]$ dependence of τ_n for solutions of poly(ε-caprolactone) (PCL) and found that log τ_n is linear to $C[\eta]$. Patel and Takahashi (*36*) also studied the C dependence of τ_n for *cis*-PI solutions. Later Urakawa et al. also studied the C dependence of τ_n for *cis*-PI solutions (*33*) and PCL solutions (*14*). More complex ternary systems were recently studied by Baysal and Stockmayer (*37*) for PCL–poly(*p*-chlorostyrene)–dioxane system and by Urakawa et al. (*33*) for *cis*-PI–polybutadiene–heptane system.

With increasing concentration, polymer coils entangle each other and τ_n increases. By using the scaling theory, De Gennes (*38*) discussed the C dependence of the longest relaxation time ε_1^e in semidilute solutions. He assumed that a chain composed of *blobs* moves in a tube whose diameter is equal to the correlation length ξ (*6*). Two extreme cases were assumed: One is a free-draining model in which solvent molecules go through the blob, and the other is the nondraining model. The free-draining model predicts τ_1^e given by

$$\tau_1^e \propto \zeta C^{2(1-v)/(3v-1)} \tag{16}$$

where ζ is the friction per bead, and v is the Flory exponent of $3/5$ in good solvents and $1/2$ in a theta solvent. Thus, this model predicts $C^{1.0}$ and $C^{2.0}$ dependencies of τ_n^e in a good and a theta solvent, respectively. On the other hand, in the nondraining model of De Gennes, the friction coefficient $\zeta_b(=6\pi\eta_0\xi)$ per one blob is given by the Stokes equation where η_0 is the viscosity of the medium and ξ is the correlation length. This model leads to

$$\tau_1^e \propto \zeta_b C^{3(1-v)/(3v-1)} \tag{17}$$

Because ζ_b is proportional to $C^{(1-v)/(3v-1)}$, the relaxation time $\tau_{n\zeta}$ reduced to an iso-friction state has a C dependence given by eq 16. In other words, the two models predict the same C dependence of τ_n in an iso-friction state.

The C dependencies of τ_n in semidilute regime were reported for solutions of *cis*-PI in good (benzene) and theta (dioxane) solvents (*32*). The observed τ_n data were reduced to the values $\tau_{n\zeta}$ in an iso-friction state and compared with eq 16. The results indicated that the double logarithmic plots of $\tau_{n\zeta}$ for the benzene and dioxane solutions are almost straight in the range of $0.03 < C < 0.2$. The slope of the plot for solutions in benzene is 1.3, and that in dioxane is 1.6. Comparing these values with the theoretical slopes 1.0 and 2.0 calculated with eq 16, we see that the observed slope in benzene is larger than the theoretical prediction by 0.3, but in dioxane the slope is smaller than the theory by 0.4. Thus, the De Gennes theory agrees with the experiment qualitatively but does not quantitatively.

Molecular weight dependence of τ_n^e was measured on concentrated solutions of *cis*-PI in toluene with concentrations of 20, 30, and 50 wt% (*39*). The slopes of the double logarithmic plots of τ_n^e versus M for these solutions were as high as 4.0 ± 0.2. This slope is higher than the theoretical power 3.0 predicted by eq 10.

Distribution of Relaxation Times in Solutions of Type-A Polymers

The distribution of the relaxation times for the dielectric relaxation of type-A chains was examined for solutions of PCL (14) and cis-PI (32, 33). In the C range around C^*, the Muthukumar–Freed theory (34, 35) predicts that the relaxation spectrum broadens with increasing C as given by eq 15. On the other hand, the free-draining bead-spring model and the tube model predict exactly the same relaxation spectra.

The most primitive way to test the relaxation spectra is to compare directly the shape of the ε'' curves. Figure 4 shows the C dependence of ε'' curve for benzene solutions of PCL with narrow distribution of molecular weight (14). To see the change of the shape of the ε'' curve, the $\varepsilon''/\varepsilon''_m$ values are plotted against $\log(f/f_m)$ where ε''_m and f_m are the maximum value of ε'' and the loss maximum frequency, respectively. The solid line shows the theoretical ε'' curve calculated from eqs 6 and 8 for the nondraining Zimm theory (26). Experimental ε'' curves for dilute solutions agree approximately with the Zimm theory. With increasing $C[\eta]$, the ε'' curves broaden in the range $C[\eta] > {\sim}1$. Urakawa et al. (14) calculated ε'' curve with the Muthukumar theory (35). The dashed and dash–dot lines indicate the theoretical ε'' curves for $C[\eta]$ are 3.0 and 6.0, respectively. The experimental ε'' curves in the range of $C[\eta] < 3$ agree with the theory but do not when $C[\eta]$ exceeds 6. Similar behavior was observed in cis-PI solutions (32, 33).

End-to-End Distance and Polarization of Type-A Chains

An attempt to determine $\langle R^2 \rangle$ from the dielectric relaxation strength $\Delta\varepsilon$ was not made until recently (10). In this section I review briefly the studies

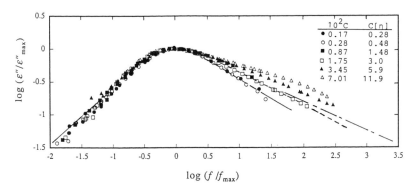

Figure 4. Normalized ε'' curves for solutions of poly(ε-caprolactone) in benzene.

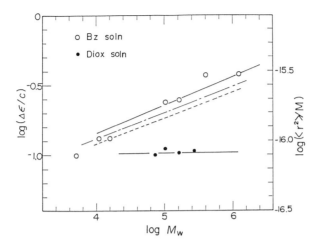

Figure 5. Molecular weight dependence of the relaxation strength $\Delta\varepsilon$ and the expansion factor α_r of cis-*polyisoprene in a good solvent benzene and a theta solvent dioxane.*

of polymer dimension in terms of the dielectric method for dilute solutions of poly(2,6-dichloro-1,4-phenylene oxide) (*29*), poly(organophosphazene) (*18*), and *cis*-PI (*40*). In these studies, the main interest was not in the determination of the absolute value of $\langle R^2 \rangle$ but was in the expansion factor defined by

$$\alpha_r^2 = \frac{\langle R^2 \rangle}{\langle R_0^2 \rangle} \tag{18}$$

where $\langle R_0^2 \rangle$ is the mean square end-to-end distance in the unperturbed state.

Figure 5 shows the double logarithmic plot of $\Delta\varepsilon/C$ versus M for dilute benzene and dioxane solutions of *cis*-PI (*40*). As described in the previous section benzene and dioxane are a good and a theta solvent, respectively. From these data, $\langle R^2 \rangle / M$ was calculated with $\mu = 4.8 \times 10^{-12}$ electrostatic units in the centimeter–gram–second system and with $F = 1$. The value of $\langle R^2 \rangle / M$ in benzene solutions is higher than that in dioxane solution, and it increases with M. The slope of the solid line in the figure is 0.20, which is expected from the Flory theory. The dash and dash–dot lines indicate $\langle R^2 \rangle$ theoretically (*40*).

Recently Urakawa et al. (*41*) studied the dielectric relaxation and light scattering for dilute solutions of *cis*-PI in a good solvent cyclohexane. The expansion coefficient α_s for radius of gyration S is defined by

$$\alpha_s^2 = \frac{\langle S^2 \rangle}{\langle S_0^2 \rangle} \tag{19}$$

as compared with α_r. They found that $\alpha_s^2 = 0.82\alpha_r^2$. The front factor 0.82 is smaller than the factor 0.92 reported by Domb and Hioe (42) who determined the factor with a computer simulation of self-avoiding random flight chain.

A single polymer chain has an expanded conformation in a good solvent. With increasing concentration, the coils overlap and the correlation length of the segment density decreases. Thus, the long range interactions are screened and the dimension of the coil decreases. The study of C dependence of α_r was reported for solutions of *cis*-PI (32, 43).

To determine the absolute value of $\langle R^2 \rangle$, μ and F in eq 11 need to be determined. The value of μ for type-A1 chains is independent of solvent quality, concentration, and temperature, but μ for type-A2 or -A3 polymers depends on them. No systematic studies to test this problem have been reported. The internal field F for type-A chains has not been described theoretically, even though it is known to be close to unity.

Dielectric Study of Type-B Chains

Overview

We have seen in the previous sections that the dielectric behaviors of type-A dipoles do not depend on the chemical structure and fall on universal curves. This fact indicates that the dielectric relaxation of type-A chains is independent of local structures. In contrast, the dielectric relaxation of type-B dipoles depends strongly on the chemical structure, specifically:

1. barrier height for the internal rotation
2. energy difference of the rotational isomeric states
3. shape and bulkiness of the monomeric unit
4. friction coefficient for local segmental motions

Yamamoto and his co-workers (44, 45) termed factor 1 *dynamical stiffness* and factor 2 *static stiffness* to interpret the local dynamics of polymer chains in dilute solutions observed with the fluorescence depolarization. The dynamical stiffness causes the decrease in the rate of the local jump and increases the relaxation time for the segmental motion. On the other hand, the static stiffness affects the correlation length of the type-B dipoles and determines the size of the segmental motion. Factor 3 is closely related to factor 2. Factor 3 plays an important role in undiluted polymers: that is, the glass transition temperature (T_g) of a homologous polymer increases with increasing bulkiness of the rigid side group. With regard to factor 4, in the case of solution of a type-A chain, solvent molecules are small and can be regarded as a continuous medium. But for local motions, this assumption probably does not hold because the size of surrounding solvent molecules is similar to that of the segments. In the following sections, we will focus our attention on these factors.

Theory of Local Motion of Flexible Polymers in Solution

Many molecular theories of segmental motions in dilute solution are based on the models of crankshaft motion and three bond jump (7). The crankshaft model assumes that the local jump occurs by rotation of a subchain around a coaxial axis that is formed by the two bonds at the ends locating on the same straight line. Theoretical consideration on crankshaft motions was developed by Helfand (46) and Iwata (47). Helfand used the Kramers theory (48) in the high viscosity limit to describe the correlation time τ_c for the local jumps:

$$\tau_c = G\eta_0 \exp\left(\frac{E_a}{k_B T}\right) \tag{20}$$

where G is the constant, η_0 is the viscosity of the medium, and E_a is the barrier height for the internal rotation.

Monnerie and his co-workers (7) proposed the model of the three-bond motion. In this motion, gauche$^+$ conformation transforms into gauche$^-$ or vise versa. In this transition a linkage of three bonds a—b—c changes into c—b—a, where a, b, and c represent the bond vectors. In this way the replacement of the orientations of two bonds a and c occurs without creating new orientations. Thus, the time-correlation function of the orientation of the bonds is essentially given as a solution of the diffusion equation. For local motions two kinds of correlation functions are used depending on the experiment:

$$P_1(t) = \langle\cos\theta(t)\rangle \tag{21}$$

$$P_2(t) = \frac{3\langle\cos^2\theta(t)\rangle - 1}{2} \tag{22}$$

where $\theta(t)$ is the angle between the directions of the bond at $t = 0$ and $t = t$. The variable P_1 represents dielectric relaxation, and P_2 represents the NMR spectroscopy (49) or fluorescence depolarization (45, 46, 50, 51). Monnerie and his co-workers (7) calculated P_2 assuming that the segmental motion occurs as the results of the three-bond and the four-bond motions. In long time region, P_2 is given by a form:

$$P_2(t) = \left(\frac{A}{t^{1/2}}\right) \exp(-Bt) \qquad (23)$$

where A and B are constant. Jones and Stockmayer (9) and Hall and Helfand (8) developed the theory of Monnerie et al. (7, 52). As pointed out by Jones and Stockmayer (9) eq 23 cannot be used to describe the dielectric relaxation because the total polarization of an infinitely long chain does not change by the three-bond motion.

A more promising theory for the description of the dielectric relaxation of type-B dipoles is the theory based on the rotational isomeric state (RIS) model (24). Jernigan (53) and Bahar and Erman (54) developed the RIS model for the time-dependent processes. In these theories it is assumed that a segment is composed of n bonds and both ends of the segment are fixed. Its conformation $\mathbf{P}_n$ is represented by a row vector whose components are the angles of internal rotation $\phi_1, \phi_2, \phi_3, \ldots \phi_n$. Then, the time dependence of $\mathbf{P}_n$ is represented by

$$\mathbf{P}_n(t') = \mathbf{A} \cdot \mathbf{P}_n(t) \qquad (24)$$

where $\mathbf{A}$ is the $3n \times 3n$ matrix whose elements are the transition probabilities among the rotational isomeric states. Bahar and Erman expressed reorientation of a bond vector $\mathbf{m}$ at the middle point of a segment assuming the probabilities of the occurrence of trans, gauche, and gauche$^-$ conformations are time dependent.

Average Relaxation Time for Type-B Dipoles

The nominal relaxation time τ_s for the segmental mode is also defined by $1/(2\pi f_m)$, where f_m is the loss maximum frequency for the relaxation of the type-B polymer. In Table II the data of τ_s and the activation energy ΔH_a in dilute solution are listed. As is seen in this table, τ_s ranges from 10^{-8} to 10^{-11}s. The solvent viscosities η_s and their activation energies ΔH_η are also listed.

If the Kramers equation (eq 20) holds, τ_s should be proportional to η_s. Mashimo and Chiba (58) studied the dielectric relaxations in dilute solutions of poly(vinyl chloride) (PVC), poly(p-chlorostyrene) (PPCS), poly(p-bromostyrene) (PPBS), and poly(vinyl acetate) (PVAc). They indicated that eq 20 holds well and determined the barrier height, E_a ($= \Delta H_a - \Delta H_\eta$), to be 8.8, 11.7, 5.0, and 6.7 kJ/mol for PVC, PPCS, PPBS, and PVAc, respectively. Baysal et al. (60) reported that ΔH_a values for PPCS and poly(p-fluorostyrene) (PPFS) in toluene are 20.1 and 18.8 kJ/mol, respectively. From these data the values of E_a for PPCS and PPFS result in 11.3 and 10.0 kJ/mol, respectively. In Table II, data of ΔH_a reported by

Table II. Data for Type-B Dipoles

Polymer	Solvent	T (°C)	log τ_s (s)	ΔH_a (kJ/mol)	η_s (cP)	ΔH_η (kJ/mol)	E_a (kJ/mol)	C_∞	N_s	Ref.
PEO	Bz	25	−10.88	10.0	0.61	10.5	−0.5	3.6	4	55
PPO	Bz	25	−10.2	11.3	0.61	10.5	0.8	5.1	14	23
PSO	Bz	20	−9.45	—	0.61	10.5	—	6.9	6	56
PVBr	Chxon	25	−8.92	21.3	2.02	15.1	6.2	6.6	—	57
PVC	Diox	30	−8.59	21.3	1.10	12.6	8.7	6.7	—	58
PPCS	Bz	25	−8.42	—	0.61	10.5	—	9.3	54	11
	Bz	29.6	−8.3	22.6	0.58	10.5	12.1			59
	Diox	30	−8.0	—	1.10	12.6	—		72	58
	Diox	23.1	−7.9	20.9	1.24	12.6	8.3			59
	Tol	25	−8.30	20.1	1.20	8.8	11.3		74	60
PPFS	Bz	25	−8.38	—	0.61	10.5	—			11
	Tol	13.5	−8.30	18.8	0.64	8.8	10.0			60
PPBS	Tol	24	−8.19	13.4	0.56	8.8	4.6	—	82	58
	Diox	24	−7.94	18.0	1.21	12.6	5.4		—	58
PMFS	Bz	25	−8.52	—	0.61	10.5	—		42	11
PVAc	Tol	23	−8.85	15.9	0.57	8.8	7.1	10	46	58
	Diox	30	−8.78	19.7	1.10	12.6	7.1		34	58
PMMA	Tol	25	−8.39	27.2	0.56	8.8	18.4	7.0	102	2
	Tol	23	−8.36	—	0.57	8.8				61
	Bz	20	−8.4		0.66	10.5			70	62
PMA	Bz	6.6	−9.65		0.81	10.5		7.7		61
	Bz	20	−9.6	23.0	0.66	10.5	12.5		26	2, 62
PMVK	Diox	20	−8.58	—	1.31	12.6			—	63
PCL	Diox	25	−10.08	30.1	1.20	12.6	17.5	5.9	5	20, 64

NOTE: PEO, poly(ethylene oxide), PPO, poly(propylene oxide), PSO, poly(styrene oxide), PVBr, polyvinyl bromide, PVC, polyvinyl chloride, PPCS, poly(p-chlorostyrene), PPFS, poly(p-fluorostyrene), PPBS, poly(p-bromostyrene), PMFS, poly(m-fluorostyrene), PVAc, polyvinyl acetate, PMMA, poly(methyl methacrylate), PMA, polymethylacrylate, PMVK, poly(methyl vinylketone), PCL, poly(ε-caprolactone), Bz, benzene, Chxon, cyclohexanone, Diox, 1,4-dioxane, and Tol, toluene.

other authors are also listed. Values of E_a for most polymers are about 8 kJ/mol and less than 20 kJ/mol. The fact that E_a is fairly low was already pointed out by Stockmayer (1) and North (2) in their reviews. The activation energy for the reorientation of segments determined by the other methods such as NMR spectroscopy and fluorescence depolarization is of the same order (44, 45, 49–51).

Later Mashimo (59) studied the η_s dependence of τ_s for PPCS and found that τ_s is proportional to η_s. Yamamoto and his co-workers also reported that τ_c for the fluorescence depolarization is proportional to η_s. On the other hand Ediger and his co-workers (51) reported that τ_c for the fluorescence depolarization is proportional to $\eta_n^{0.74}$.

If crankshaft motions are assumed, E_a should be twice the barrier height for the internal rotation. For carbon–carbon single bonds, the barrier height of the internal rotation is about 10.5 kJ/mol (65). This result leads $E_a \approx 21$ kJ/mol. As already pointed out by North (2), E_a for some systems such as solutions of poly(ethylene oxide) becomes 0 or negative. This difficulty may be partly avoided by assuming the $\eta_s^{0.74}$ dependence.

Because the activation energy for inversions is lower than that for internal rotations (65), the low values of E_a can be explained by assuming that the three-bond motion is the dominant mechanism for local motions. However, the applicability of the three-bond model to the dielectric data is doubtful as pointed out previously. In Table II, little correlation is shown between E_a and τ_s. Therefore, the dynamic stiffness does not have an important role at least in the dielectric segmental modes.

To assess the effect of static stiffness, the average size of segment has been estimated as follows. As pointed out by Stockmayer (1) and North (2), a competition exists in the relaxation rate between the segmental motions and the overall rotation of the random coil. If the segmental motions are slower than the overall rotation, the dielectric relaxation of type-B dipoles occurs through overall rotation of the whole molecule and τ_s depends on molecular weight (11). The M dependence of τ_s was investigated by Stockmayer and Matsuo (11) for dilute solutions of PPFS, poly(m-fluorostyrene), and PPCS in benzene. They indicated that τ_s decreased with decreasing M when M is less than a critical molecular weight M_c. Thus, M_c is a measure of the size of the motional unit for segmental motions and the larger the motional unit, the higher the M_c. We use the normal-mode relaxation time τ_n given by eq 14 as a measure of the overall rotation of a chain. Then, M_c is determined as the molecular weight at which $\tau_s = \tau_n$:

$$M_c = \left(\frac{\tau_s RT}{1.4 \eta_s K} \right)^{1/(1+a)} \tag{25}$$

where K and a are the front factor and the exponent of the Mark–Houwink–Sakurada equation. In Table II the number of the backbone atoms

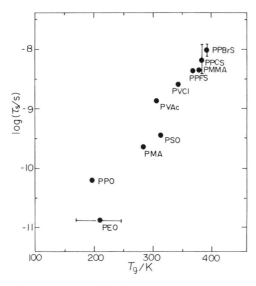

Figure 6. Correlation between the dielectric relaxation time τ_s for the local segmental mode in dilute solutions and the glass transition temperature T_g in the undiluted state.

N_s corresponding to M_c is listed instead of M_c. Therefore, N_s thus calculated depends strongly on chemical structure. The value of N_s is of the order of 100 for PPCS, PPBS, and poly(methyl methacrylate) (PMMA) but less than 10 for poly(ethylene oxide) and PCL. Also, τ_s increases with increasing N_s. Thus, the static stiffness has an important role in the local dynamics of polymer chains in solution.

This method of the estimation of N_s may be rationalized by a computer simulation carried out by Verdier and Stockmayer (*66*). They indicated that the longest relaxation time τ_1 of a chain composed of N beads and N − 1 bonds is equal to $v(N - 1)^2 \tau_0$ where v is the constant ranging from 0.85 to 1.1; and τ_0 is the correlation time for the reorientation of the local bond. This result indicates that M_c approximately corresponds to an elementary process for motions of a chain.

The characteristic ratio C_∞ is often used to assess the flexibility of a chain and is another measure of the static stiffness. In Table II, the values of C_∞ are also listed. A tendency shows that τ_s increases with increasing C_∞ but the correlation is weak.

In Table II, a strange tendency occurs: τ_s of a polymer in dilute solution has a correlation with T_g in the bulk state. Figure 6 shows the log τ_s versus T_g plot in which we see that log τ_s is linear against T_g. The T_g of a bulk polymer is the temperature at which τ_s becomes the order of the experimental time scale for measurements of T_g. Thus we see that the τ_s of a polymer in the bulk and in dilute solution is closely correlated. This result indicates that both the static stiffness and the bulkiness of the side group govern τ_s in dilute solution.

Distribution of Relaxation Time

The simplest measure of the distribution of relaxation times is $\delta = 2\varepsilon''_{max}/\Delta\varepsilon$. North (2) reported in his review that δ for solutions of PMMA is very close to unity but that for PVC is 0.7. Mashimo and Chiba (58) reported that the shape of the ε'' curves for PVC, PPCS, PPBS, and PVAc in various solvents is superposable and $\delta = 0.62 \pm 0.02$. This value is approximately the same as δ for bulk PVAc (~0.65) reported by Ishida et al. (67). They also indicated that the ε'' curve is broader on the high frequency side of the loss peak as observed commonly in bulk amorphous polymers. Thus, we see that the distribution of relaxation times is almost independent of concentration for PVAc. Baysal et al. (60) reported that δ for solutions of PPCS and PPFS is 0.78 ± 0.04. Phillips and Singh (68) reported the concentration dependence of the dielectric behavior of toluene solutions of PMMA. They reported that δ depends on the molecular weight: $\delta = 0.7$ at $M = 47,000$, but $\delta = 1.0$ in the range $M > 115,000$. Cole et al. (61) reported that the Cole–Cole plots for PMMA and PVAc solutions (5 wt%) are of the Davidson–Cole type and δ of PMMA solutions is 0.79. They also reported that a plot for poly(methylacrylate) can be regarded as a circular arc within an error ($\delta = 1$). These data indicate that the distribution of relaxation times for type-B chains depends strongly on the chemical structure and changes from polymer to polymer. However as is seen previously, some data conflict with each other and more accurate measurements are needed to discuss the relationship between the distribution of relaxation times and the chemical structure.

Concentration Dependence of the Dielectric Relaxation of Type-B Chains

As the concentration C of a polymer solution is increased, the interaction between the polymer chains affects the segmental motion. Mashimo (69) attempted to explain the C dependence of τ_s based on the Kramers theory with the friction coefficient expressed by the free-volume theory. The C dependence of τ_s was successfully explained. Bahar et al. (70) applied the Bahar–Erman theory to undiluted cis-PI and explained the dielectric behavior by taking into account the intermolecular interactions. This approach may explain the C dependence of τ_s observed by Mashimo (69). Co-workers and I (71) discussed the relationship between τ_n and τ_s in concentrated solutions of cis-PI. The friction ζ for the normal-mode relaxation is proportional to $C^2\tau_s$. From this result we indicated that the size of the segment increased with decreasing C. For concentrated solutions the model similar to the bulk polymers may be used. Matsuoka and Quan (*72*) proposed the conformer model. I (*73*) proposed the gear model to express the cooperative reorientation of the segments.

References

1. Stockmayer, W. H. *Pure Appl. Chem.* **1967**, *15*, 539–554.
2. North, A. M. *Chem. Soc. Rev.* **1972**, *1*, 49–73.
3. Block, H. *Adv. Polym. Sci.* **1979**, *33*, 93–167.
4. McCrum, N. G.; Read, B. E.; Williams, G. *Anelastic and Dielectric Effects in Polymeric Solids;* Dover: New York, 1991.
5. Williams, G. *Adv. Polym. Sci.* **1979**, *33*, 59–92.
6. De Gennes, P. G. *Scaling Concepts in Polymer Physics;* Cornell University: Ithaca, NY, 1979.
7. Valeur, B.; Jarry, J.-P.; Geny, F.; Monnerie, L. *J. Polym. Sci. Polym. Phys. Ed.* **1975**, *13*, 667–674.
8. Hall, K.; Helfand, E. *J. Chem. Phys.* **1982**, *77*, 3275–3282.
9. Jones, A. A.; Stockmayer, W. H. *J. Polym. Sci. Polym. Phys. Ed.* **1977**, *15*, 847–861.
10. Adachi, K.; Kotaka, T. *Prog. Polym. Soc.* **1993**, *18*, 585–622.
11. Stockmayer, W. W.; Matsuo, K. *Macromolecules* **1972**, *5*, 766–770.
12. Yu, H.; Bur, A. J.; Fetters, L. J. *J. Chem. Phys.* **1966**, *44*, 2568–2576.
13. Urakawa, O.; Adachi, K.; Kotaka, T. *Rep. Prog. Polym. Phys. Jpn.* **1993**, *36*, 113–116.
14. Urakawa, O.; Adachi, K.; Kotaka, T.; Takemoto, Y.; Yasuda, H. *Macromolecules* **1994**, *27*, 7410–7414.
15. Adachi, K.; Kotaka, T. *Macromolecules* **1983**, *16*, 1936–1941.
16. Adachi, K.; Ohta, K.; Kotaka, T. *Polym. J.* **1986**, *18*, 371–374.
17. North, A. M.; Phillips, P. J. *Trans. Faraday Soc.* **1968**, *64*, 3235–3241.
18. Uzaki, S.; Adachi, K.; Kotaka, T. *Macromolecules* **1988**, *21*, 153–156.
19. Adachi, K.; Kotaka, T. *Macromolecules* **1985**, *18*, 294–297.
20. Jones, A. A.; Brehm, G. A.; Stockmayer, W. H. *J. Polym. Sci. Polym. Symp.* **1974**, *46*, 149–159.
21. Jones, A. A.; Stockmayer, W. H.; Molinari, R. J. *J. Polym. Sci. Polym. Symp.* **1976**, *54*, 227–235.
22. Baur, M. E.; Stockmayer, W. H. *J. Phys. Chem.* **1965**, *43*, 4319–4325.
23. Mashimo, S.; Yagihara, S. *Macromolecules* **1984**, *17*, 630–634.
24. Flory, P. J. *Statistical Mechanics of Chain Molecules;* Interscience: New York, 1969.
25. Rouse, P. E., Jr. *J. Chem. Phys.* **1953**, *21*, 1272–1289.
26. Zimm, B. H. *J. Chem. Phys.* **1956**, *24*, 269–278.
27. De Gennes, P. G. *J. Chem. Phys.* **1971**, *55*, 572–579.
28. Doi, M.; Edwards, S. F. *The Theory of Polymer Dynamics;* Clarendon: Oxford, England, 1986.
29. Adachi, K.; Okazaki, H.; Kotaka, T. *Macromolecules* **1985**, *18*, 1486–1491.
30. Wada, A. *J. Chem. Phys.* **1959**, *30*, 329–330.
31. Takada, S.; Itou, T.; Chikiri, H.; Einaga, Y.; Teramoto, A. *Macromolecules* **1989**, *22*, 973–979.
32. Adachi, K.; Kotaka, T. *Macromolecules* **1988**, *21*, 157–164.
33. Urakawa, O.; Adachi, K.; Kotaka, T. *Macromolecules* **1993**, *26*, 2042–2049.
34. Muthukumar, M.; Freed, K. F. *Macromolecules* **1978**, *11*, 843–852.

35. Muthukumar, M. *Macromolecules* **1984**, *17*, 971–973.
36. Patel, S. S.; Takahashi, K. M. *Macromolecules* **1992**, *25*, 4382–4391.
37. Baysal, B.; Stockmayer, W. H. *J. Mol. Liq.* **1993**, *56*, 175–181.
38. De Gennes, P. G. *Macromolecules* **1976**, *9*, 594–598.
39. Adachi, K.; Imanishi, Y.; Kotaka, T. *J. Chem. Soc., Faraday Trans 1* **1989**, *85*, 1065–1074.
40. Adachi, K.; Kotaka, T. *Macromolecules* **1987**, *20*, 2018–2023.
41. Urakawa, O.; Adachi, K.; Kotaka, T., submitted for publication in *Macromolecules*.
42. Domb, C.; Hioe, F. *J. Chem. Phys.* **1969**, *51*, 1915–1928.
43. Adachi, K.; Imanishi, Y.; Shinkado, T.; Kotaka, T. *Macromolecules* **1989**, *22*, 2391–2395.
44. Sasaki, T.; Arisawa, H.; Yamamoto, M. *Polymer J.* **1991**, *23*, 103–115.
45. Ono, K.; Ueda, K.; Yamamoto, M. *Polym. J.* **1994**, *26*, 1345–1351.
46. Helfand, E. *J. Chem. Phys.* **1971**, *54*, 4651–4661.
47. Iwata, K. *J. Chem. Phys.* **1973**, *58*, 4184–4202.
48. Kramers, H. A. *Physica* **1940**, *7*, 284–304.
49. Glowinkowski, S.; Gisser, D. J.; Ediger, M. D. *Macromolecules* **1990**, *23*, 3520–3530.
50. Ediger, M. D. *Annu. Rev. Phys. Chem.* **1991**, *42*, 225–250.
51. Adolf, D. B.; Ediger, M. D.; Kitano, T.; Ito, K. *Macromolecules* **1992**, *25*, 867–872.
52. Geny, F.; Monnerie, L. *J. Polym. Sci., Polym. Phys. Ed.* **1977**, *15*, 1–9.
53. Jernigan, R. L. In *Dielectric Properties of Polymers;* Karasz, F. E., Ed.; Plenum: New York, 1972; pp 99–128.
54. Bahar, I.; Erman, B. *Macromolecules* **1987**, *20*, 1368–1376.
55. Davies, M.; Williams, G.; Loveluck, G. D. *Z. Electrochem.* **1960**, *64*, 575–580.
56. Matsuo, K.; Stockmayer, W. H.; Mashimo, S. *Macromolecules* **1982**, *15*, 606–509.
57. Kryszewski, M.; Marchal, J. *J. Polym. Sci.* **1958**, *29*, 103–116.
58. Mashimo, S.; Chiba, A. *Polym. J.* **1973**, *5*, 41–48.
59. Mashimo, S. *Macromolecules* **1976**, *9*, 91–97.
60. Baysal, B.; Lowry, B. A.; Yu, H.; Stockmayer, W. H. In *Dielectric Properties of Polymers;* Karasz, F. E., Ed.; Plenum: New York, 1972; pp 329–341.
61. Cole, R. H.; Mashimo, S.; Winsor, P. *J. Phys. Chem.* **1980**, *84*, 786–793.
62. Nakamura, H.; Mashimo, S.; Wada, A. *Jpn. J. Appl. Phys.* **1982**, *21*, 467–474.
63. Mashimo, S.; Winsor, P.; Cole, R. H.; Matsuo, K.; Stockmayer, W. H. *Macromolecules* **1983**, *16*, 965–967.
64. Nakamura, H.; Mashimo, S.; Wada, A. *Jpn. J. Appl. Phys.* **1982**, *21*, 1022–1024.
65. Lister, D. G.; Macdonald, J. N.; Owen, N. L. *Internal Rotation and Inversion;* Academic: Orlando, FL, 1998; Chapter 5, pp 103–144.
66. Verdier, P. H.; Stockmayer, W. H. *J. Chem. Phys.* **1962**, *36*, 227–235.
67. Ishida, Y.; Matsuo, M.; Yamafuji, K. *Kolloid Z.* **1962**, *180*, 108–114.
68. Phillips, P. J.; Singh, G. *J. Polym. Sci., Polym. Phys. Ed.* **1975**, *13*, 1377–1386.
69. Mashimo, S. *J. Chem. Phys.* **1977**, *67*, 2651–2657.
70. Bahar, I.; Erman, B.; Kremer, F.; Fischer, E. W. *Macromolecules* **1992**, *25*, 816–825.
71. Adachi, K.; Imanishi, Y.; Kotaka, T. *J. Chem. Soc., Faraday Trans.1* **1989**, *85*, 1083–1089.
72. Matsuoka, S.; Quan, X. *Macromolecules* **1991**, *24*, 2770.
73. Adachi, K. *Macromolecules* **1990**, *23*, 1816–1821.

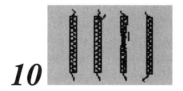

10

Dielectric Studies of Polymer Blends

James P. Runt

This chapter presents a survey of the application of dielectric spectroscopy as a probe for characterizing the phase behavior and dynamics of polymer blends. Several examples of typical analysis methods applied to dielectric spectra of both single and multiphase amorphous polymer mixtures are presented. A discussion of Maxwell–Wagner–Sillars interfacial polarization and correlation parameters is included. The dielectric relaxation behavior of semicrystalline blends is also reviewed, with emphasis on the relationship between microstructures and dielectric relaxations for crystalline, melt-miscible blends.

Polymer blends—physical mixtures of two or more polymers—have been under intense scrutiny in both industry and academia during the past two decades. This activity has resulted in advances in our fundamental understanding of this group of materials as well as a large number of important industrial products. In fact, as of 1992, approximately 30% of all polymers were sold in mixed form (*1*). Interest has burgeoned in polymer blends as a result of the ability to readily tailor properties to specific applications, their great versatility, and the promise of enhanced performance. A significant portion of this activity has been focused on enhancement of mechanical properties, particularly toughness.

As noted previously, great advances have occurred in the science of polymer blends in these two decades but only a few pertinent issues will be discussed here and it is suggested that the interested reader consult one or more of the books on the subject (for example, *1–5*). The vast majority of polymer blends (~90%) are multiphase—either completely immiscible or multiphase with some limited mixing between the component polymers.

This behavior is primarily the result of the low combinatorial entropy of mixing of the dissimilar polymer chains. The remaining fractions are single-phase, melt-miscible mixtures. The terms "compatibility" and "incompatibility" will be avoided here as these have been used in a number of different contexts in the literature.

Despite the general flurry of activity on polymer blends, interest in their dielectric properties, either in their own right or as a probe of phase behavior or polymer dynamics, has only relatively recently received considerable attention. The study of the loss process associated with main-chain segmental motion (this will be called the α-relaxation for amorphous systems), which is closely associated with the glass transition, provides a potentially convenient means for assessing miscibility in amorphous polymer blends. This is seen in the occurrence of one or multiple α loss peaks, the case of one peak corresponding to a miscible mixture. (Care must be taken when using this criteria, however. *See* for example the later discussion on melt-miscible blends that crystallize on cooling.)

The degree to which two polymers having quite different α-transition temperatures perturb their respective local environments to produce a single α-loss peak, intermediate in temperature between those of the component polymers, also illustrates how studies of molecular motion in polymers can in principle be used to probe polymer molecular environment, specifically intra- and intermolecular interactions. In polymer systems with permanent dipole moments, dielectric relaxation measurements have proven to be particularly valuable due to the wide range of frequencies readily available (*see* Chapter 2 by Kremer). Our focus in the following discussion will be exclusively on dielectric relaxation of binary polymer–polymer mixtures. The dielectric properties of polymer solutions (i.e., polymers dissolved in low molecular weight solvents) are covered in Chapter 9 in this volume.

Amorphous Blends

We will begin our discussion by examining several examples where dielectric loss measurements have provided insight into the phase behavior of blends of two amorphous polymers. Figure 1 shows the isochronal dielectric loss curves (at 400 Hz) for a blend of a chlorinated poly(vinyl chloride) containing 67 wt% chlorine (CPVC) and a random copolymer of ethylene and vinyl acetate (containing 70 wt% of vinyl acetate) (EVA 70) (6). A single α-transition is observed for all compositions, whose location falls between those of the neat polymers in a manner consistent with the compositions of the blends. This behavior suggests that CPVC and EVA 70 form a single homogeneous phase over the entire compositional range to temperatures at least up to 120 °C (this was the temperature at which the blends

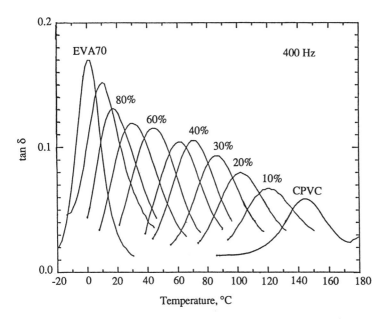

Figure 1. Isochronal tanδ versus temperature behavior for EVA 70–CPVC mixtures and component polymers (400 Hz). Blend compositions are in increments of 10% (by weight). (Reproduced with permission from reference 6. Copyright 1989.)

were heat treated). Similar behavior is also observed for blends of EVA 70 and unmodified PVC (7–9).

However, blends of PVC with an ethylene–vinyl acetate copolymer containing 45 wt% vinyl acetate (EVA 45) were found to be miscible only at the compositional extremes; at intermediate compositions these blends are two-phase with some intermixing between the polymers. As an example, Figure 2 presents the dielectric loss for the 50/50 EVA 45–PVC blends as a function of temperature history. By recognizing that the α-relaxation of neat EVA 45 occurs at significantly lower temperatures than that of PVC, it is evident from the behavior in Figure 2 that the lower temperature relaxing phase becomes increasingly rich in EVA 45 with increasing exposure temperature, whereas the composition of the higher temperature phase stays relatively constant (near that of PVC).

In the situation where one observes two distinct α-transitions in a binary mixture, the amount and composition of each of the phases can be readily estimated (i.e., from data like that in Figure 2, one can get an estimate of the phase diagram) (7, 10). To do so, we define $\omega_{\alpha1}$ and $\omega_{\alpha2}$ as the weight fractions of the lower and upper temperature relaxing phases, respectively. The weight fractions of the two components of the blend are given by $\bar{\omega}_{ij}$ and ω_{ij}. The barred notation represents the weight fraction of the *i*th component (either polymer 1 or 2) in the *j*th phase (either the lower or

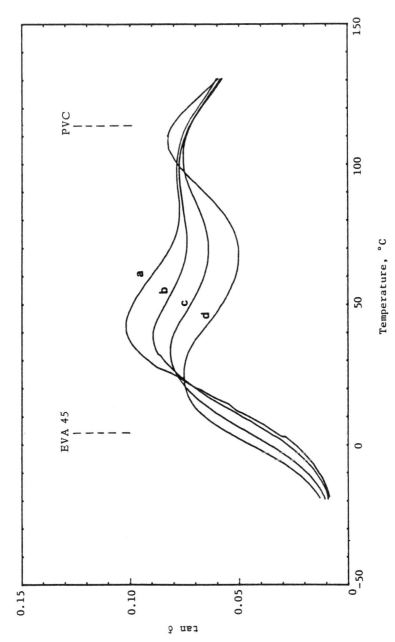

Figure 2. *Loss curves at 100 kHz for 50/50 EVA 45/PVC blends having different thermal histories: (a) no heating, (b) heated to 55 °C for 24 h, (c) heating to 70 °C for 24 h, and (d) heating to 105 °C for 12 h. The dashed vertical lines denote the peak transition temperatures of neat EVA 45 and PVC at this frequency, respectively. (Reproduced with permission from reference 7. Copyright 1986.)*

upper temperature relaxing phase) referred to 1 g of each phase. The various ω_{ij} unbarred values are the weight fractions referred to 1 g of total sample.

A simple series of algebraic equations can be constructed based on the relationships between the weight fractions defined previously as well as the (known) weight fractions of the polymers in the mixture (7). (For a crystalline blend, these weight fractions refer to the amorphous phases only.) The assumption that is required is that the various $\bar{\omega}_{ij}$ values can be calculated from a specific mixing rule (e.g., the Fox equation (11)) from knowledge of the experimental transition temperatures. Such calculations on the data contained in Figure 2 essentially substantiate the qualitative observations noted previously; that is, the higher temperature phase is nearly pure PVC and it increases in concentration by migration of PVC from the lower temperature phase. The relative concentration of EVA 45 in the lower temperature phase increases but the overall amount of EVA 45 stays effectively constant for the different extents of phase separation. From these simple calculations it becomes clear that, for these blends, the change in phase composition with heating is by net migration of PVC out of the lower T_α phase into the higher T_α phase.

Analysis of the shape or breadth of the loss curve for a particular neat polymer or blend can only be conducted properly in the frequency plane by using the loss factor, ε'' (or normalized loss factor). It is desirable in such a situation to acquire the loss factor data isothermally (over as broad a frequency range as possible) at one or more temperatures (near the glass transition temperature (T_g) in the case of the α-relaxation). In a number of studies (like the case of EVA 70–CPVC blends noted previously (6)), however, temperature is continuously ramped and some frequency range is swept during the temperature excursion. Approximate frequency-plane curves can be obtained by cross-plotting the temperature-plane data. However, the available frequency range is sometimes insufficient to construct full loss curves. When this is the case, it may be possible to shift the partial curves at nearby temperatures to construct a master curve for a given reference temperature. Great caution must be exercised, however, because time–temperature superposition is not generally applicable to polymer blends (for example, *12, 13*).

For the case of the EVA 70–CPVC blends noted previously, the breadth of the loss curves of the blends lies between those of the neat polymers (*6*). Similar behavior has been observed in a few cases (*14*) for other miscible blends, but more frequently the blend α-relaxation is broader than those of the component polymers (*8, 15–20*). As an example of the more typical behavior, loss factor curves (normalized and referenced to the same $T - T_g$) are presented for miscible EVA 70–PVC blends in Figure 3 (*8*). It appears that the higher EVA 70 content blends lose the high frequency skewness characteristic of the unblended polymers (and amorphous polymers in general). In fact, the better defined loss curves in Figure 3 (i.e.,

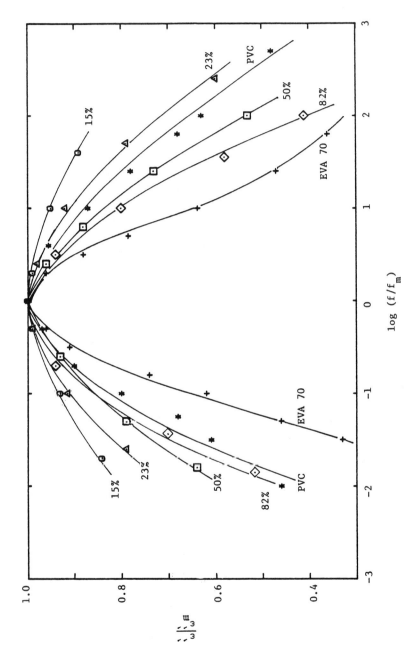

Figure 3. Normalized loss factor curves for EVA 70–PVC blends and neat polymers. (Reproduced with permission from reference 8. Copyright 1986.)

for the 82% and 50% EVA 70 blends) suggest that the asymmetry is shifted to the low frequency side of the dispersion.

The reversal of the asymmetry of the dielectric or mechanical α-relaxation characteristic of neat amorphous polymers has been observed for a number of other miscible blends (*12, 17, 18*). As seen in Figure 3, the loss curve for the 82% EVA 70 blend is broader on the low frequency side, but intermediate in breadth on the high frequency side. As the PVC content in the blends increases, the curves become progressively broader and this broadening correlates with an increased tendency for phase separation. This latter characteristic is inferred from the phase separation observed for the 15% EVA 70 blend on heating (*7*) and the lower critical solution temperature behavior reported for EVA 63–PVC blends (*21*). The broadness of the loss curves is frequently taken as a measure of the range of local molecular environments that the segments experience and, as such, the 15% EVA 70 blends exhibit a more heterogeneous local environment.

As detailed in Chapter 5 by Riande and Diaz-Calleja, interactions or correlations between neighboring dipoles are reflected in the Kirkwood–Fröhlich correlation parameter (*g* factor) (*22, 23*). Work and colleagues (*24, 25*) in particular have explored correlation parameters in copolymers and their research has provided direction for similar studies of polymer mixtures. The familiar form of the expression for the correlation parameter is:

$$\mu_0^2 g = \frac{9kT(\varepsilon_R - \varepsilon_u)(2\varepsilon_R + \varepsilon_u)}{4\pi N\varepsilon_R(\varepsilon_u + 2)^2} \tag{1}$$

where ε_R and ε_u are the relaxed and unrelaxed dielectric constants, respectively; μ_0 is the gas-phase dipole moment, N is the number of dipoles per unit volume, and k is the Boltzmann constant. The parameter g as determined is interpreted as a measure of the extent to which restricted internal rotation and interaction of neighboring dipoles influence dipole alignment. Correlation parameters have been determined for several miscible blends by using this approach (*14–16*). For example, Alexandrovich et al. (*16*) found the calculated g values for the miscible polystyrene–poly(2-chlorostyrene) blend to be independent of composition and concluded that the local conformational states of poly(2-chlorostyrene) chains are unperturbed by mixing with polystyrene.

To assess the effect that blending has on the polarization in a blend relative to that in the unblended state, g factors for blends have been defined differently (*8*):

$$g(b) = \frac{\dfrac{kT(2\varepsilon_R + \varepsilon_u)(\varepsilon_R - \varepsilon_u)}{4\pi N\varepsilon_R}}{g(1)\,n(1)\,\mu_0^2(1)\,\dfrac{[\varepsilon_u(1) + 2]^2}{9} + g(2)\,n(2)\,\mu_0^2(2)\,\dfrac{[\varepsilon_u(2) + 2]^2}{9}} \tag{2}$$

Here, n is the mole fraction and the parenthetical notation is used to designate the blend components. Note that the numerator is the effective squared dipole moment in the blend and the denominator is the squared dipole moment assuming a linear combination of the properties of the component polymers. It follows then that $g(b)$ is a measure of the polarization in the blend relative to the unblended environment. This approach has been used successfully in references 6 and 8. Relaxation strength ratios have also been used as a measure of dipole correlations in blends (6, 8).

As noted previously, relaxation spectra associated with T_g, as derived from a variety of experimental methods, are generally broader for miscible polymer–polymer (as well as solvent–polymer) mixtures. In addition, their shapes are often strongly temperature dependent, a characteristic that negates the construction of master curves for miscible polymer blends (for example, 12). The fundamental understanding of this behavior has received considerable attention in recent years. Two principal models have been proposed and the salient points will be summarized. The interested reader is encouraged to explore the accompanying references.

As noted, the broadening observed for the α-relaxation of miscible polymer blends has often been qualitatively associated with concentration fluctuations present in such mixtures, and these fluctuations result in chain segments experiencing differing local environments (7, 15, 16, 26). A theoretical fluctuation model that permits calculation of concentration fluctuations in miscible polymer blends has been proposed by Zetsche and Fischer (20, 27). This model describes the dynamics of the α-process by including a relaxation time distribution that is designed to characterize the effect of mixing on the relaxation function (which is assumed to be dominated by one of the component polymers). The model also permits determination of the size of cooperatively relaxing units in the mixture. Katana et al. (28) recently used this model to analyze the influence of concentration fluctuations on the dynamics of miscible polystyrene–poly(cyclohexylacrylate-stat-butyl methacrylate) blends. In general, the fluctuation model appears to reproduce the characteristics of the dielectric α-process in the blend, as well as blends of polystyrene and poly(vinyl methyl ether) (20).

The second model is based on the coupling model of relaxation in which the shape of the relaxation function and temperature dependence of the relaxation time are governed by intermolecular cooperativity (29, 30). The intermolecular cooperativity is characterized in this model by a coupling parameter that depends on chemical structure (31, 32). By using this approach, local compositional fluctuations are argued to give rise not only to a distribution of relaxation times but also to dynamic heterogeneity, which is caused by intrinsic differences in the mobility of the component polymers (13, 33). Analysis of the dielectric relaxation data of a number of miscible blend systems (17, 18, 34) illustrates that the predictions of this model are consistent with the usual experimental observations: that is, the increase in breadth and the asymmetric low frequency broadening of the α-relaxation on blending and the failure of time–temperature superposition.

Maxwell–Wagner–Sillars (MWS) interfacial polarization is a phenomenon that occurs in multiphase systems (e.g., multiphase blends or polymer composites) with nonidentical dielectric properties, as a result of accumulation of charge at the interfaces of the phases (*35–38*). Such polarization gives rise to relaxation behavior that can be very difficult to distinguish from dipole relaxation. The position, strength, and shape of an MWS relaxation depend essentially on the volume fraction, geometry, and conductivity of the dispersed phase and the permittivity of the phases. North et al. (*39*) summarized various theoretical expressions for different dispersed-phase geometries and volume fractions, but the morphology of multiphase polymer mixtures is often complex or not well defined and makes modeling difficult. However, the experimental dielectric properties of very well-defined, model polyethylene oxide–polycarbonate blends were found to agree with those predicted from MWS theory (*40*).

A more realistic case was explored recently by Steeman et al. (*41*): injection-molded blends of ABS [a blend of poly(styrene-*co*-acrylonitrile) and poly(styrene-*co*-butadiene) elastomer] and polycarbonate. In addition to relaxations due to the α-transitions of the components, the authors also observed a strong dielectric dispersion at intermediate temperatures (which was absent in mechanical measurements). This dispersion was attributed to MWS interfacial polarization, and its characteristics are described well by dielectric models for multiphase materials (*41*). The shape factor used in the modeling agrees well with the morphology observed by using scanning electron microscopy. Thus, as proposed by Hayward et al. (*40*), these results illustrate the possibility of using dielectric measurements to provide information on the morphology of phase-separated polymer systems.

Finally, there have been a number of other studies of the phase behavior of amorphous polymer blends and some of these can be found in references 42–52.

Crystalline Blends

Characterization of amorphous-phase transitions to evaluate miscibility versus multiphase character in mixtures containing crystalline polymers is often very difficult to accomplish by traditional differential scanning calorimetry due to the breadth of the transition and the magnitude of the heat capacity change. Consequently, dynamic mechanical analysis has often been employed for such characterization. More recently, however, dielectric spectroscopy has become increasing popular for routine as well as more detailed characterization of crystalline multicomponent polymer systems.

As an example, consider the dielectric relaxation behavior of the family of blends based on poly(butylene terephthalate) (PBT), a semicrystalline polyester, and a series of poly(ester-ether) (PEE) multiblock copolymers (53, 54). The multiblock copolymers consist of crystallizable hard segments composed of tetramethylene terephthalate units and polyether soft segments (M_w ~1000). Blends of PBT with copolymers having hard-segment concentrations ranging from 40 to 90 wt% have been examined. Unblended PBT and the copolymers exhibit two dielectric transitions in the temperature range of interest (see Figure 4a). The higher temperature relaxation is associated with cooperative segmental motion (i.e., T_g like) in the noncrystalline phase. For such crystalline materials this transition will be referred to as the β-relaxation. The lower temperature process (γ) has been assigned to local motions of the ester group in noncrystalline regions for PBT, whereas for the copolymers it is believed to be a combination of the local motions of the tetramethylene oxide units of the soft segment and the γ-relaxation of uncrystallized hard segments (55).

Blends of PBT and copolymers containing from 40 to 51 wt% hard segments exhibited two β-transitions that correspond to those of the neat polymers, and amorphous-phase immiscibility was concluded for the range of compositions studied. When blends were prepared with PBT and PEE copolymers containing ≥80 wt% hard segments, one β-process located at a temperature intermediate between those of the component polymers was observed indicating amorphous phase miscibility (see Figure 4b for an example). Additionally, the location of the β-transition varies with blend composition as one would generally expect for a miscible system. (See the next section for a discussion of why *multiple* transitions would be expected for melt-miscible, crystalline blends in certain circumstances.) In blends of PBT with copolymers containing 58–75 wt% hard segments, two β-relaxations are observed: the high temperature process shifts to lower temperatures with increasing copolymer concentration. For example, in the spectrum of a 50/50 PBT–PEE-65 blend shown in Figure 5a, T_β of the PBT-rich phase is depressed by ~10 °C. By assuming that a mixing rule like the Fox–Flory rule applies, the PBT-rich amorphous phase contains about 8% PEE-65.

Because of the overlap of the β-relaxation of PEE-65 and the γ-process of PBT at this measurement frequency, it is difficult to evaluate if there is mixing of PBT in the copolymer-rich amorphous phase. However, because the β- and γ-processes have much different frequency dependencies (i.e., they have different effective activation energies) improved separation is expected at lower frequencies, which are readily accessible with many conventional measuring systems. The dielectric relaxation behavior of PBT–PEE-65, PBT–PEE-70, and PBT–PEE-75 blends were determined at frequencies down to 0.1 Hz, and as demonstrated in Figure 5b for a 50/50 mixture of PBT and PEE-65, there is no change in the lower temperature β-transition on mixing in any of these blends. Therefore, in these partially

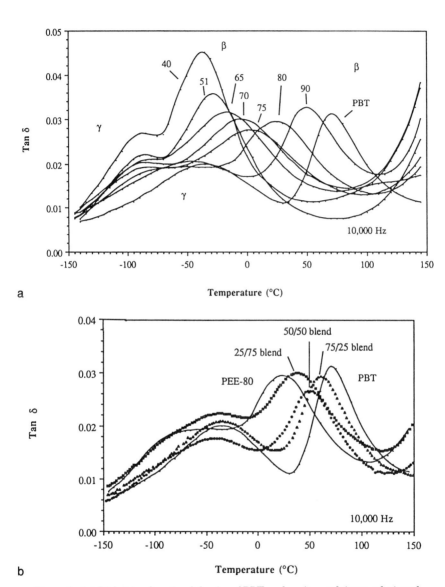

a

b

Figure 4. (a) Dielectric relaxation behavior of PBT and various poly(ester–ether) multiblock copolymers at 10 kHz. The numbers refer to the wt% hard segments in the copolymers. (b) Dielectric tanδ versus T for PBT, PEE-80, and selected blends of PBT and PEE-80 at 10 kHz. (Reproduced from reference 53. Copyright 1993 American Chemical Society.)

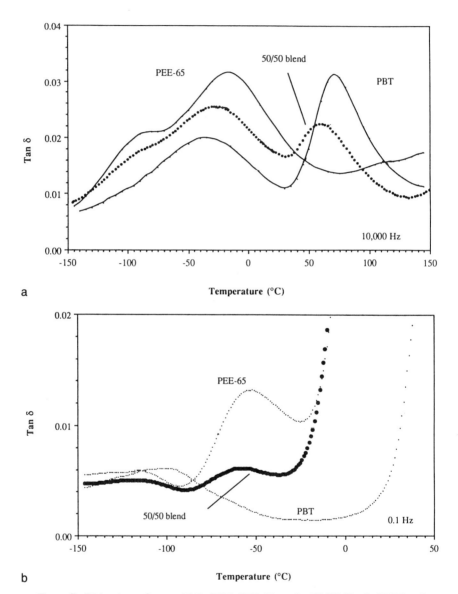

a

b

Figure 5. Dielectric tanδ versus T for PBT, PEE-65, and a 50/50 blend of PBT and PEE-65 at 10 kHz (a) and at 0.1 Hz (b) (data from reference 56).

miscible blends, two amorphous phases exist: an apparently pure copolymer phase and a PBT-rich phase.

Figure 5b also illustrates one of the possible shortcomings of the dielectric method in the quest for characterization of blends with multiple amorphous phases (whether they are crystalline or not). Note that at these low frequencies, dc conduction and possibly electrode polarization (which results from accumulation of charge at the sample electrode interface) become so important in the blend at relatively low temperatures that the higher temperature β-process is obscured. Although it is common for such losses to be modeled by using a power law or multiple power law and subtracted from the experimental loss, the conduction losses can be so large as to render this ineffective. The contribution of conduction and electrode polarization to the dielectric loss is inversely dependent on frequency and, as illustrated, is of no particular difficulty in this particular case because higher frequency measurements clearly unmask the higher temperature β process. This result is not always the case, however. In such situations turning to dynamic mechanical analysis may be appropriate, although there are limitations here as well. For the case of dc conduction alone (where there is no capacitive contribution), one can in principle characterize such transitions by focusing on the behavior of the dielectric permittivity.

In the discussion on multiphase amorphous polymer mixtures in the preceding section, the possibility of the contribution of MWS interfacial polarization to the observed dielectric loss was noted, as well as the difficulty associated with modeling the MWS contribution to the dielectric spectra due to insufficient knowledge, or complexity, of the phase morphology. Such behavior, of course, could also complicate the dielectric loss of thermodynamically immiscible blends that contain one or more crystalline polymers (57, 58). In addition, for melt-miscible blends that crystallize on cooling to ambient temperature (*see* the following section)—or in neat semicrystalline polymers for that matter (59, 60)—the multiphase character below the melting point (i.e., crystals plus mixed amorphous phase) can lead to contributions to the dielectric loss from MWS polarization.

Melt-Miscible Crystalline Blends

An intriguing subset of melt-miscible polymer blends is that containing one or more polymers that are capable of crystallizing. The influence of the second polymer (it is assumed to remain amorphous) on crystallization and crystalline microstructure is expected to be significant. In general, the T_g of the mixture can be raised or lowered, thereby expanding or contracting the temperature window over which the mixture may crystallize. A reduction in the equilibrium melting point would also be expected, whose magnitude will depend, among other factors, on the polymer–polymer interaction parameter (χ) and blend composition (*61*, *62*). In addition, one generally expects a reduction in crystallization rate with increasing diluent concentration (i.e., proportional to the volume fraction of the crystallizable polymer in the mixture).

In the solid state, semicrystalline polymers generally consist of chain-folded lamellar crystals that are separated by regions of disorder. These crystals are about 5–20-nm thick and the polymer chain axis is more or less normal to the lamellar faces. Generally speaking, chain segments residing in amorphous regions have restricted mobility as a consequence of connections to the crystallites (*see* Chapter 4 by Boyd). One of the important considerations in miscible blends containing a crystalline polymer is the location of the amorphous diluent in the microstructure. As illustrated in Figure 6 (*63*), the diluent could reside in interlamellar, interfibrillar, interspherulitic regions, or some combination of these. What controls this placement is an open question. From the point of view of interlamellar incorporation, the diameter of gyration of the diluent polymer is often comparable to or greater than the separation between crystalline lamellae (particularly at high crystalline polymer concentrations) (*64*). Confinement of the diluent would lead to an entropic driving force for migration from the interlamellar zones. In contrast, it has been argued that the presence of strong intermolecular interactions between the crystalline and amorphous polymers will promote interlamellar incorporation.

The location of the amorphous polymer in the microstructure appears to be dominated by the mobility of the diluent at the crystallization temperature (T_c) and its relationship to the crystal growth rate (*65*). At least for weakly interacting mixtures, when the T_g of the amorphous diluent (T_g^a is high compared to T_c, the amorphous polymer generally resides in interlamellar zones. When T_g^a is low compared to T_c, at least some of the amorphous polymer resides outside of the interlamellar regions. In any event, depending on the microstructure (even in the absence of a possible relaxa-

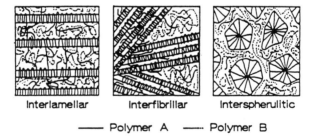

Interlamellar   Interfibrillar   Interspherulitic

———— Polymer A   ------- Polymer B

Figure 6. Possible solid-state microstructures of a melt-miscible binary polymer blend containing a crystalline component. Polymers A and B represent the crystallizable and amorphous polymers, respectively. (Reproduced from reference 63. Copyright 1991 American Chemical Society.)

tion associated with the crystal-amorphous interphases), multiple locations for the amorphous diluent lead naturally to multiple T_g-like transitions although the polymers themselves are miscible at this temperature (*66–70*).

Crystal-Amorphous Interphases

Even in neat semicrystalline polymers, the amorphous regions are not expected to be uniform. Flory et al. (*71*) argued that the transition from the highly ordered crystalline state to the isotropic amorphous state cannot occur abruptly and, hence, predict the existence of a crystal-amorphous interphase that arises from the need to dissipate order at the crystal surface. This interphase is envisioned as a region in which the chain segments retain some traits of crystalline order and lack rigid lattice constraints. Using a lattice theory, they predicted (*71*) an interphase thickness of ~1.5 nm for crystalline homopolymers. A more recent extension of this model (*72*) (in which the chains are no longer assumed to be fully flexible but can have various fold energies) results in an increase in the predicted interphase thickness (to ca. 1.5–3.0 nm).

These predictions are in good agreement with the results of a variety of experimental studies on neat semicrystalline polymers (*73–76*), which have confirmed the existence and generality of crystal-amorphous interphases and have estimated their size to be on the order of 0.5–3.5 nm depending on crystallization conditions, molecular weight, etc. Recent dielectric and thermal analysis findings described in terms of a rigid amorphous phase also appear to be related to such crystal-amorphous interphases (for example, *77–79*). Considering that typical thicknesses of the crystalline and amorphous regions in semicrystalline polymers are about 5–20 nm, we see that the interphase constitutes an appreciable portion of the system volume and will consequently make a significant contribution to the macroscopic properties.

The lattice model noted previously has been extended to the case of a binary miscible blend containing a crystalline and amorphous component, the amorphous component being confined to the interlamellar regions (*80*). For blends that are miscible yet exhibit only weak intermolecular interactions (i.e., $\chi \approx 0$) the theory predicts an order–disorder interphase that spans about 1.5–3.0 nm (i.e., the same as for the neat crystalline polymer) and that is comprised only of chain segments of the crystallizable component. The predicted exclusion of the amorphous polymer arises from packing considerations because crystalline polymer order must be sufficiently low to permit mixing.

This view suggests that the amorphous interlamellar region consists of two more-or-less distinct environments (i.e., an interphase composed of segments of the crystalline polymer only and a mixed phase consisting of the amorphous polymer and relatively unconstrained segments (disordered) of the crystallizable polymer) and it is possible that these regions would relax differently. When the interaction parameter becomes more favorable—when χ is relatively large and negative—the model predicts penetration of the amorphous polymer into the region of partial order. Following similar reasoning, the molecular weight of the amorphous polymer would also be expected to play an important role: as molecular weight decreases there is an increased entropic driving force for penetration. The extreme case has been verified numerous times: solvents are well-known to swell the fold surfaces of polymer crystals (for example, *81, 82*).

Before we proceed to a discussion of experimental studies relating to interphases in crystalline blends, it is instructive to review the general relaxation behavior of neat, semicrystalline polymers (*see* again Chapter 4). When viewed as an isochronal plot versus temperature at a fixed frequency, several dielectric (or mechanical) relaxations are typically observed for crystalline materials. These transitions are designated α, β, γ, etc. with decreasing temperature. As noted earlier, the β-relaxation is associated with cooperative, T_g-like motion. This transition is amorphous in origin but indications are that it does not include a contribution from all of the noncrystalline material in a particular system. For example, the mechanical relaxation behavior of *trans*-1,4-polybutadiene single crystal mats annealed at different temperatures was studied by Takayanagi et al. (*83*). The strength of the β-process decreased with increasing annealing temperature and eventually disappeared. In light of this behavior, and additional small angle X-ray (SAXS) and broadline NMR evidence, loose loops (folds) and cilia were concluded to be transformed to tighter folds as annealing temperature increases, and these tighter folds do not exhibit a cooperative β-process. From this result emerges the view that the mechanical β-relaxation (and presumably the dielectric β-process as well) in neat, bulk-crystallized polymers arises from relatively loose loops and other reasonably mobile amorphous material (i.e., the relaxation is associated with the motion of segments in the interphase (e.g., looser chain folds) as well as relatively unconstrained segments further removed from the crystal surface).

Most experimental investigations of interphases in miscible mixtures have focused on blends containing poly(vinylidene fluoride) (PVF$_2$) as the crystalline component, particularly in combination with poly(methyl methacrylate) (PMMA) (*84–87*). As seen in Figure 7, a relaxation persists in the dielectric spectra of PVF$_2$–PMMA blends at the same location as the dielectric β-transition for neat PVF$_2$, and the relaxation strength decreases as PMMA concentration in the blend increases. Moreover, the transition disappears at high PMMA concentrations: that is, when the blends are amorphous. Although this behavior was originally interpreted as resulting from phase separation in the melt, SAXS experiments showed that the correlation length of the concentration fluctuations is on the order of ~1 nm in

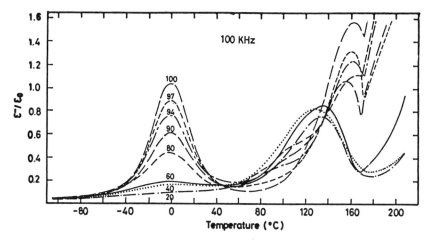

Figure 7. Dielectric loss factor versus T at 100 kHz for PVF₂–PMMA blends. The weight percent of PVF₂ in the mixtures is noted in the figure. (Reproduced with permission from reference 86. Copyright 1991.)

the melt (*88*). Also, on relatively rapid crystallization, PMMA resides between the PVF_2 lamellae (*85, 89*). These observations led to the suggestion that the low temperature relaxation was associated with segmental motions of PVF_2 chains in the interphase region.

The high temperature process seen in Figure 7 is related to that of the mixed, interlamellar PMMA–PVF_2 phase but interpretation is complicated due to overlap with local mode processes of PVF_2 and PMMA. Hahn et al. (*85*) estimated the interphase fraction from measured relaxation strengths, and extrapolation to neat PVF_2 led to the conclusion that all of the noncrystalline material in PVF_2 is interphase-like and the interphase thickness is ~2–2.5 nm for PVF_2 and the blends. Similar dielectric loss characteristics have also been reported for blends of PVF_2 with poly(ethyl methacrylate) (*90*) and poly(vinyl pyrolidone) (*86*). The apparent observation of a pure PVF_2 interphase in blends that are characterized by favorable values of χ is not well understood but it has been conjectured that they behave in a manner similar to blends with small negative values of χ due to the relatively large fraction of head-to-head and tail-to-tail defects in PVF_2(*86*). The presence of any upper or lower critical solution temperature behavior, as well as the environments of the diluent, must be clearly established before attempting interpretation of the relaxation behavior in terms of the interphase structure.

Blends of polyethylene oxide (PEO) and PMMA are well known to be miscible across the compositional spectrum, to interact weakly, and the PMMA resides in the interlamellar regions at typical crystallization temperatures (*64, 91*). Russell et al. (*64*) studied these blends by using small-angle

X-ray and neutron scattering and found a ~2.0 nm interphase that was comprised of only PEO segments (*64*). The dielectric relaxation behavior of isothermally crystallized and quickly cooled PEO–PMMA blends is reminiscent of that observed for crystalline PVF_2–PMMA, in that a relaxation in crystalline blends was found near that of T_β of neat PEO. The relaxation strength of this process scaled with PEO concentration (*92*). Therefore, this transition was postulated to be associated with relaxation of PEO segments in the interphase region. Care must be exercised in interpreting the dielectric ε'' or $\tan\delta$ behavior in these and similar blends due to overlap of this process and that of a secondary relaxation of the diluent polymer. Thus, it is often desirable to focus on the behavior of ε'.

The remainder of the noncrystalline PEO and the PMMA is mixed in the interlamellar zone and is expected to exhibit a transition at significantly higher temperatures. (Because of large losses arising from conduction effects and electrode polarization, this mixed phase transition is not observed by using a dielectric probe but is clearly observed in the mechanical loss–temperature profile.) These results are consistent with the view that, provided the intermolecular interactions are relatively weak, only those relatively unconstrained PEO segments at some distance from the crystal surface are capable of mixing with the amorphous diluent. Finally, it should be noted that the observation of a transition near that of the low T_g component has been suggested to arise from concentration fluctuations in the mixed amorphous phase: that is, that the relaxation is not specific to the crystalline-amorphous interphases (*93*).

References

1. Utracki, L. A. *Polymer Alloys and Blends, Thermodynamics and Rheology;* Oxford University: Hanser, New York, 1990.
2. Olabisi, O.; Robeson, L. M.; Shaw, M. T. *Polymer–Polymer Miscibility;* Academic: Orlando, FL, 1979.
3. *Polymer Blends;* Paul, D. R.; Newman, S., Eds.; Academic: Orlando, FL, 1978, Vols. 1 and 2.
4. *Polymer Blends and Mixtures;* Walsh, D. J.; Higgins, J. S.; Maconnachie, A., Eds.; Martinus Nijhoff: Dordrecht, Germany, 1985
5. Coleman, M. M.; Graf, J. F.; Painter, P. C. *Specific Interactions and the Miscibility of Polymer Blends;* Technomic: Lancaster, PA, 1991.
6. Angeli, S. R.; Runt, J. *Contemporary Topics in Polymer Science;* B. M. Culbertson, Ed.; 1989; Vol. 6, p 289.
7. Rellick, G. S.; Runt, J. *J. Polym. Sci. Polym. Phys. Ed.* **1986**, *24*, 279.
8. Rellick, G. S.; Runt, J. *J. Polym. Sci. Polym. Phys. Ed.* **1986**, *24*, 313.
9. Rellick, G. S.; Runt, J. *J. Polym. Sci. Polym. Phys. Ed.* **1988**, *26*, 1425.
10. Runt, J.; Zhang, X.; Miley, D. M.; Gallagher, K. P.; Zhang, A. *Macromolecules* **1992**, *25*, 3902.
11. Fox, T. G. *Bull. Am. Chem. Soc.* **1956**, *1*, 123.
12. Zetsche, A.; Kremer, F.; Jung, W.; Schulze, H. *Polymer* **1990**, *31*, 1883.

13. Roland, C. M.; Ngai, K. L. *Macromolecules* **1991,** *24,* 2261.
14. Malik, T. M.; Prud'homme, R. E. *Polym. Eng. Sci.* **1984,** *24,* 144.
15. Wetton, R. E.; MacKnight, W. J.; Fried, J. R.; Karasz, F. E. *Macromolecules* **1978,** *11,* 158.
16. Alexandrovich, P. S.; Karasz, F. E.; MacKnight, W. J. *J. Macromol. Sci.–Phys.* **1980,** *B17,* 501.
17. Roland, C. M.; Ngai, K. L. *Macromolecules* **1992,** *25,* 363.
18. Ngai, K. L.; Roland, C. M.; O'Reilly, J. M.; Sedita, J. S. *Macromolecules* **1992,** *25,* 3906.
19. Katana, G.; Kremer, F.; Fischer, E. W.; Plaetscke, R. *Macromolecules* **1993,** *26,* 3075.
20. Zetsche, A.; Fischer, E. W. *Acta Polym.* **1994,** *45,* 168.
21. Nolley, E.; Paul, D. R.; Barlow, J. W. *J. Appl. Polym. Sci.* **1979,** *23,* 623.
22. McCrum, N. G.; Read , B. E.; Williams, G. *Anelastic and Dielectric Effects in Polymeric Solids;* Wiley: London, 1967.
23. Frohlich, H. *Theory of Dielectrics;* Oxford University: London, 1949.
24. Work, R. N.; Trehu, Y. *J. Appl. Phys.* **1956,** *27,* 1003.
25. Smith, F. H.; Corrado, L. C.; Work, R. N. In *Dielecrtric Properties of Polymers;* Karasz, F., Ed.; Plenum: New York, 1972.
26. Shears, M. F.; Williams, G. *J. Chem. Soc., Faraday Trans.* **1973,** *69,* 608.
27. Fischer, E. W.; Zetsche, A. *ACS Polym. Prepr.* **1992,** *33,* 78.
28. Katana, G.; Fischer, E. W.; Hack, Th.; Abetz, V.; Kremer, F. *Macromolecules* **1995,** *28,* 2714.
29. Ngai, K. L.; Rendell, R. W.; Rajagopal, A. K.; Teitler, S. *Ann. N. Y. Acad. Sci.* **1986,** *484,* 150.
30. Ngai, K. L.; Rendell, R. W. *J. Non-Cryst. Solids* **1991,** *942,* 131–133.
31. Roland, C. M. *Macromolecules* **1992,** *25,* 7031.
32. Ngai, K. L.; Roland, C. M. *Macromolecules* **1993,** *26,* 6824.
33. Roland, C. M.; Ngai, K. L. *J. Rheology* **1992,** *36,* 1691.
34. Alegria, A.; Colmenero, J.; Ngai, K. L.; Roland, C. M. *Macromolecules* **1994,** *27,* 4486.
35. Maxwell, J. C. *Electricity and Magnetism I,* 3rd ed.; Oxford University: London, 1892.
36. Wagner, K. W. *Arch. Elektrotech. Berlin* **1914,** *2,* 371.
37. Sillars, R. W. *Proc. R. Soc. London, Ser. A* **1939,** *169,* 66.
38. Hedvig, P. *Dielectric Spectroscopy of Polymers;* Wiley: New York, 1977.
39. North, A. M.; Pethrick, R. A.; Wilson, A. D. *Polymer* **1978,** *19,* 913.
40. Hayward, D.; Pethrick, R. A.; Siriwittayakorn, T. *Polymer* **1992,** *25,* 1480.
41. Steeman, P. A. M.; Maurer, F. H. L.; van Turnout, J. *Polym. Eng. Sci.* **1994,** *34,* 697.
42. MacKnight, W. J.; Stoelting, J.; Karasz, F. E. In *Multicomponent Polymer Systems;* Gould, R. F., Ed.; Advances in Chemistry 99; American Chemical Society: Washington, DC, 1971; pp 29–41.
43. Bank, M.; Leffingwell, J.; Thies, C. *Macromolecules* **1971,** *4,* 43.
44. Feldman, D.; Rusu, M. *Eur. Polym. J.* **1974,** *10,* 41.
45. Naito, K.; Johnson, G. E.; Allara, D. L.; Kwei, T. K. *Macromolecules* **1978,** *11,* 1260.
46. Pathmanathan, K.; Johari, G. P.; Faivre, F. P.; Monnerie, L. *J. Polym. Sci. Polym. Phys. Ed.* **1986,** *24,* 1587.
47. Pathmanathan, K.; Cavaille, J. Y; Johari, G. P. *Polymer* **1988,** *29,* 331.
48. Quan, X.; Johnson, G. E.; Anderson E. W.; Lee, H. S. *Macromolecules* **1991,** *24,* 6500.

49. Liang, K.; Banhegyi, G.; Karasz, F. E.; MacKnight, W. J. *J. Polym. Sci. Polym. Phys. Ed.* **1991**, *29*, 649.
50. O'Reilly, J. M.; Sedita, J. S. *J. Non-Cryst. Solids* **1991**, *131–133*, 1140.
51. Alegria, A.; Telleria, I.; Colmenero, J. *J. Non-Cryst. Solids* **1994**, *172–174*, 961.
52. Ngai, K. L.; Roland, C. M. *Macromolecules* **1995**, *28*, 4033.
53. Gallagher, K. P.; Zhang, X.; Runt, J.; Huynh-ba, G.; Lin, J. S. *Macromolecules* **1993**, *26*, 588.
54. Runt, J.; Du, L.; Martynowicz, L. M.; Hancock, M. E.; Brezny, D. M.; Mayo, M. *Macromolecules* **1989**, *22*, 3908.
55. Lilaonetkul, A.; Cooper, S. *Macromolecules* **1979**, *12*, 1146.
56. Gallagher, K. P. Ph.D. Thesis, The Pennsylvania State University, 1992.
57. La Mantia, F. P.; Valenza, A.; Acreino, D. *Colloid Polym. Sci.* **1985**, *263*, 726.
58. Pratt, G. J.; Smith, M. J. A. *Polymer* **1989**, *30*, 1113.
59. Hayward, D.; Pethrick, R. A.; Siriwittayakorn, T. *Polymer* **1992**, *25*, 1480.
60. North, A. M.; Pethrick, R. A.; Wilson, A. D. *Polymer* **1978**, *19*, 923.
61. Nishi, T.; Wang, T. T. *Macromolecules* **1975**, *8*, 909.
62. Runt, J.; Gallagher, K. P. *Polym. Commun.* **1991**, *32*, 180.
63. Crevecoeur, G.; Groeninckx, G. *Macromolecules* **1991**, *24*, 1190.
64. Russell, T. P.; Ito, H.; Wignall, G. D. *Macromolecules* **1988**, *21*, 1703.
65. (a) Barron, C. A. Ph.D. Thesis, The Pennsylvania State University, 1994. (b) Talibuddin, S.; Wu, L.; Runt, J.; Lin, J. S. *Macromolecules* **1996**, *29*, 7527.
66. Hudson, S. D.; D. D. Davis, D. D.; Lovinger, A. J. *Macromolecules* **1992**, *25*, 1759.
67. Huo, P. P.; Cebe, P.; Capel, M. *Macromolecules* **1993**, *26*, 4275.
68. Sauer, B. B.; Hsiao, B. S. *J. Polym. Sci. Polym. Phys. Ed.* **1993**, *31*, 901.
69. Jones, A. M.; Russell, T. P.; Yoon, D. Y. *Proc. ACS Div. Polym. Mater. Sci. Eng.* **1994**, *70*, 394.
70. Bristow, J. F.; Kalika, D. S. *Proc. Soc. Plastic Eng.* **1995**, 1725.
71. Flory, P. J.; Yoon, D. Y.; Dill, D. A. *Macromolecules* **1984**, *17*, 862.
72. Kumar, S. K.; Yoon, D. Y. *Macromolecules* **1989**, *22*, 3468.
73. Bergmann, K. *J. Polym. Sci. Polym. Phys. Ed.* **1978**, *16*, 1611.
74. Alamo, R. G.; McLaughlin, K. W.; Mandelkern, L. *Polym. Bull.* **1988**, *22*, 29.
75. Muira, H.; Hirschenger, J.; English, A. D. *Macromolecules* **1990**, *23*, 2169.
76. Mandelkern, L.; Alamo, R. G.; Kennedy, M. A. *Macromolecules* **1990**, *23*, 4721.
77. Cheng, S. Z. D.; Wu, Z.; Wunderlich, B. *Macromolecules* **1987**, *20*, 2802.
78. Huo, P.; Cebe, P. *Macromolecules* **1992**, *25*, 902.
79. Kalika, D. S.; Krishnaswamy, R. K. *Macromolecules* **1993**, *26*, 4252.
80. Kumar, S. K.; Yoon, D. Y. *Macromolecules* **1991**, *22*, 5414.
81. Ergoz, E.; Mandelkern, L. *J. Polym. Sci. Polym. Lett. Ed.* **1972**, *10*, 631.
82. Harrison, I. R.; Weaver, T. J.; Runt, J. *Polym. Commun.* **1985**, *26*, 244.
83. Tasumi, T.; Fukushima, T.; Imada, K.; Takayanagi, T. *J. Macromol. Sci., Phys.* **1967**, *81*, 459.
84. Hahn, B.; Wendorff, J.; Yoon, D. Y. *Macromolecules* **1985**, *18*, 718.
85. Hahn, B. R.; Herman-Schonherr, O.; Wendorff, J. W. *Polymer* **1987**, *28*, 201.
86. Yoon, D. Y.; Ando, Y.; Rozstaczer, S.; Kumar, S. K.; Alfonso, G. C. *Macromol. Chem. Macromol. Symp.* **1991**, *50*, 183.
87. Ando, Y.; Yoon, D. Y. *Polym. J.* **1992**, *24*, 1329.
88. Wendorff, J. H. *J. Polym. Sci. Polym. Lett. Ed.* **1980**, *18*, 445.
89. This is not necessarily the case however. Based on mobility arguments, if PVF$_2$–PMMA blends are relatively slowly crystallized, at least some PMMA would be expected to reside outside of the interlamellar regions. This has been demonstrated recently by Saito, H.; Stuhn, B. *Macromolecules* **1994**, *27*, 216.
90. Ando, Y.; Hanada, T.; Saitoh, K. *J. Polym. Sci. Polym. Phys. Ed.* **1994**, *32*, 179.
91. Alfonso, G. C.; Russell, T. P. *Macromolecules* **1986**, *19*, 1143.
92. Runt, J.; Barron, C. A.; Zhang, X.; Kumar, S. K. *Macromolecules* **1991**, *24*, 3466.
93. Kumar, S. K.; Colby, R. H.; Anastasiadis, S. H.; Fytas, G. *J. Chem. Phys.* **1996**,

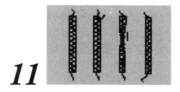

11

Dielectric Monitoring of Polymerization and Cure

David E. Kranbuehl

Frequency dependent dielectric measurements have become an effective instrumental means for monitoring a variety of polymer resin processing properties. Measurements can be made in situ during cure and continuously throughout the cure process as the resin changes from a viscous monomer to a gel and finally to a cross-linked glass. This chapter discusses monitoring the cure of epoxies, and then the application of the technique to composites, adhesives, and intelligent automated processing.

Frequency dependent dielectric measurements, often called frequency dependent electromagnetic sensing (FDEMS), made over many decades of frequency (hertz to megahertz) provide a sensitive, convenient automated means for characterizing the processing properties of thermosets and thermoplastics (*1–21*). By using a planar wafer-thin sensor, measurements can be made in situ in almost any environment. Through the frequency dependence of the impedance, this sensing technique is able to monitor chemical and physical changes throughout the entire cure process. Dielectric sensing techniques have the advantage that measurement can be made both in the laboratory and in situ in the fabrication tool during manufacture. As such, one application of this work is measurement of the resin processing properties in situ during fabrication in the manufacturing mold and as input for on-line intelligent closed-loop process control (*22–30*).

The FDEMS technique (*1, 2*) and related lower frequency work (*3*) can be effective for monitoring a variety of resin-cure processing properties such as reaction onset, viscosity, point of maximum flow, degree of cure

(α), buildup in glass transition temperature (T_g), and reaction completion, as well as detecting the variability in processing properties due to resin age and exposure to moisture (31–41). The dielectric technique monitors similar processing properties in thermoplastics such as T_g, melt temperature (T_m), recrystalization, and solvent–moisture outgassing (2, 42). The technique has the particular advantage over other chemical characterization measurements of being able to monitor these processing properties continuously and in situ as the resin changes from a polymer of varying viscosity to a cross-linked insoluble solid. Another advantage is that measurements can be made simultaneously on multiple samples or at multiple positions in a complex part. Of particular importance is the ability of dielectric sensing to monitor the changing properties of polymeric materials in composites, films, and coatings (43, 44) as adhesives in the bondline and to detect phase separation in toughened systems (20, 21, 45–47).

At the heart of dielectric sensing is the ability to monitor the changes in the transitional mobility of ions and changes in the rotational mobility of dipoles in the presence of a force created by an electric field. These variations in molecular position due to an electric field force are a very sensitive means of monitoring changes in macroscopic mechanical properties such as viscosity, modulus, T_g, and degree of cure. Mechanical properties reflect the response in displacement on a macroscopic level due to a mechanical force acting on the whole sample. The reason why dielectric sensing is quite sensitive is rooted in the fact that changes on the macroscopic level originate from changes in force displacement relationships on a molecular level. Indeed, these molecular changes in force–displacement relationships are monitored by dielectric sensing as the resin cures and are the origin of the resin's macroscopic changes in flow, degree of cure, and other mechanical properties.

One may ask when and why does one need in situ on-line dielectric sensing. A number of important reasons and examples exist:

1. Dielectric sensing allows one to monitor or see the actual state of the material in the tool at all times during the cure process. Temperature and pressure do not provide direct information about the state of the resin. Thus, dielectric sensing is one of the few means by which the operator can actually monitor the state of the material during processing and tell what the material is doing throughout the entire fabrication process.

2. By actually monitoring the state of the material, it is possible to control the fabrication process by data rather than a procedure such as a set time–temperature sequence. This fact means one can have a self-correcting automated intelligent cure process that can adapt to variations in material age, fabric permeability, tool heat-transfer characteristics, etc.

3. In situ sensing is needed to verify the veracity and logic of a model's predictions or an operator's reasoning. Making a composite part that passes mechanical tests does not verify that the modeling equations or operator thinking are correct and can be trusted to make predictions.

4. Modeling and individual thinking that lead to a procedure-driven cure cycle are beset with operating difficulties. Most notably, as has been described by George Springer (*30*), modeling requires extensive material data characterization of resin properties, as well as fabric and tooling properties that are time consuming to measure and, most importantly, that will vary from day to day, batch, and layup to layup.

Further, results are limited generally to a particular or a simplified geometry. Fabric preform properties will vary from preform to preform, with layup, with bagging, with position within the preform, etc. Heat-transfer characteristics will similarly vary with the tool, the autoclave, the position within the autoclave, etc. Thus, given the time and material cost, it is critical to monitor or see what is actually happening and to have the potential to detect, verify, and even correct the processing properties as the cure proceeds. Both monitoring and modeling are essential.

In summary, dielectric sensing provides valuable insight in observing the state of the resin during the process, verifying and reducing the time in developing a cure process, as well as providing an automated self-correcting intelligent control system. Further, dielectric monitoring has, at the same time, the potential to provide on-line quality verification of the fabrication process, thereby increasing product reliability and reducing postfabrication test costs.

In this chapter I will discuss using the frequency dependence in the hertz to megahertz range to separate and determine parameters governing ionic and dipolar mobility. I will discuss the quantitative relationship of the ionic and dipolar mobility to monitoring processing parameters such as viscosity and degree of cure during the reaction. I will describe several applications of in situ sensing. Finally, I conclude with a discussion of FDEMS used for closed-loop intelligent process control of the cure of a high temperature polyimide (*6–10*).

Instrumentation

Frequency dependent complex dielectric measurements are made by using an impedance analyzer controlled by a microcomputer (*15*). In the work discussed here, measurements at frequencies from 5 Hz to 5×10^6 Hz are taken continuously throughout the entire cure process at regular intervals and converted to the complex permittivity, $\varepsilon^* = \varepsilon' - i\varepsilon''$. The measurements are made with a geometry independent DekDyne microsensor, which has been patented and is now commercially available, and a DekDyne automated dielectric measurement system. This system is used with either a Hewlett Packard or a Schlumberger impedance bridge. The system permits multiplexed measurement of nine sensors. The sensor itself is planar, 1 × 1/2 in. in area and 5-mm thick. This single sensor–bridge microcomputer assembly is able to make continuous uninterrupted measurements of both ε' and ε'' over 10 decades in magnitude at all frequencies. The sensor is inert and has been used at temperatures exceeding 400 °C and pressures over 1000 psi.

Theory

Frequency dependent measurements of the dielectric impedance of a material as characterized by its equivalent capacitance, C, and conductance, G, are used to calculate the complex permittivity, $\varepsilon^* = \varepsilon' - i\varepsilon''$, where i is the imaginary number, $\omega = 2\pi f$, f is the measurement frequency, and C_0 is the equivalent air replacement capacitance of the sensor.

$$\varepsilon'(\omega) = \frac{C(\omega)\text{material}}{C_0}$$

$$\varepsilon''(\omega) = \frac{G(\omega)\text{material}}{\omega C_0}$$

(1)

This calculation is possible when using the sensor whose geometry is invariant over all measurement conditions. Both the real and the imaginary parts of ε^* can have dipolar (d) and ionic (i)-charge polarization components.

$$\varepsilon' = \varepsilon'_d + \varepsilon'_i$$

$$\varepsilon'' = \varepsilon''_d + \varepsilon''_i$$

(2)

Plots of the product of frequency (ω) multiplied by the imaginary component of the complex permittivity $\varepsilon''(\omega)$ make it relatively easy to visually determine the permittivity when the low frequency magnitude of ε'' is dominated by the mobility of ions in the resin and when at higher frequencies the rotational mobility of bound charge dominates ε''. Generally, the magnitude of the low frequency overlapping values of $\omega\varepsilon''(\omega)$ can be used to measure the change with time of the ionic mobility through the parameter σ where

$$\sigma(\text{ohm}^{-1}\text{cm}^{-1}) = \varepsilon_0\omega\varepsilon''_i(\omega)$$

$$\varepsilon_0 = 8.854 \times 10^{-14} C^2 J^{-1}\text{cm}^{-1}$$

(3)

The changing value of the ionic mobility is a molecular probe that can be used to quantitatively monitor the viscosity of the resin during cure. The dipolar component of the loss at higher frequencies can then be determined by subtracting the ionic component.

$$\varepsilon''(\omega)\,(\text{dipolar}) = \varepsilon''(\omega) - \frac{\sigma}{\omega\varepsilon_0} \qquad (4)$$

The peaks in ε''(dipolar) (which are usually close to the peaks in ε'') can be used to determine the time or point in the cure process when the *mean* dipolar relaxation time has attained a specific value $\tau = 1/\omega$, where $\omega = 2\pi f$ is the frequency of measurement. The dipolar mobility as measured by the mean relaxation time τ can be used as a molecular probe of the buildup in T_g. The time of occurrence of a given dipolar relaxation time as measured by a peak in a particular high frequency value of $\varepsilon''(\omega)$ can be quantitatively related to the attainment of a specific value of the resin T_g. Finally, the tail of the dipolar relaxation peak as monitored by the changing value of $(d\varepsilon''/dt)/\varepsilon''$ can be used for in situ monitoring during processing the buildup in degree of cure and related end-use properties such as modulus, hardness, etc., during the final stages of cure or post cure.

Isothermal Cure

The variation in the magnitude of ε'' with frequency and with time for the diglycidylether bisphenol-A amine-cured epoxy held at 121 °C is shown in Figure 1. The magnitude of ε'' changes four orders of magnitude during the course of the polymerization reaction. A plot of $\omega\varepsilon''$ is a particularly informative representation of the polarization process because as discussed from eqs 1–4, overlap of $\omega\varepsilon''(\omega)$ frequency for differing frequencies indicates when and at what time translational diffusion of charge is the dominant physical process affecting the loss. Similarly, the peaks in $\omega\varepsilon''(\omega)$ for individual frequencies indicate when dipolar rotational diffusion processes are contributing to ε''.

The frequency dependence of loss ε'' is used to find σ by determining from a computer analysis, or a log–log plot of ε'' versus frequency, the frequency region where $\omega\varepsilon''$ is a constant. In this frequency region the value of σ is determined from eq 3.

Figure 2 is a plot of $\log(\sigma)$ versus $\log$(viscosity [η]) constructed from dielectric data and measurements on a dynamic rheometer. The figure shows that at a viscosity less than 1 Pa·s, (10 P), σ is proportional to $1/\eta$ because the slope of $\log(\sigma)$ versus $\log(\eta)$ is approximately -1. The gel point of the polymerization reaction occurs at 90 min based on the cross-over of G' and G'' measured at 40 rad/s. This result is very close to the time

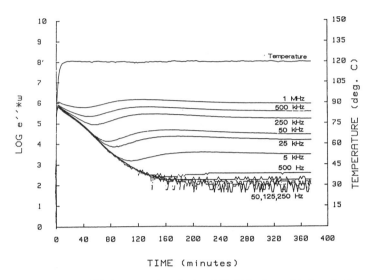

Figure 1. Log(ωε″[ω]) versus time during 121 °C isothermal polymerization.

at which η achieves 100 Pa·s, which is also often associated with gel. The region of gel marks the onset of a much more rapid change in viscosity than with σ. This change is undoubtedly because as gel occurs the viscoelastic properties of the resin involve the cooperative motion of many chains while the translational diffusion of the ions continues to involve motions over much smaller molecular dimensions.

Figure 3 is a plot of $\log(\sigma)$ versus degree of cure (α) determined from differential scanning calorimetry (DSC) measurements. The approximate

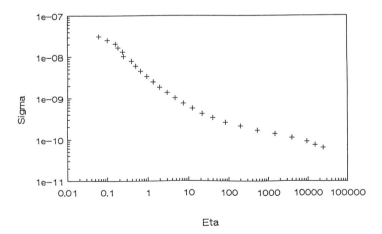

Figure 2. Log(σ) versus log viscosity (η) during 121 °C polymerization.

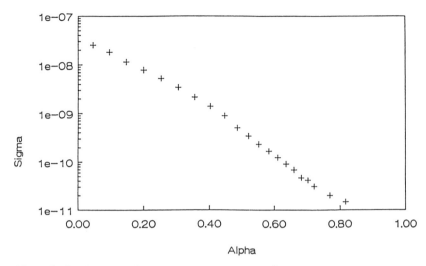

Figure 3. Log(σ) versus degree of cure (α) during 121 °C isothermal polymerization.

exponential dependence of σ on α is not surprising because α is often exponentially related to viscosity through $\eta = \eta_0 \exp(E/RT + K\alpha)$, when E is the activation energy, R is the gas law constant, and K is a fitting parameter. Again, a break occurs in the dependence of σ on α as the value of σ drops from 10^{-9} to $10^{-10} \, \omega^{-1} \, \text{cm}^{-1}$ in the region of gel.

The time of occurrence of a *characteristic* dipolar relaxation time can be determined by noting the time at which $\varepsilon''_{\text{dipolar}}$ achieves a maximum for each of the frequencies measured where $\tau = 1/2\pi f$ at the time at which $\varepsilon''_{\text{dipolar}}$ achieved a maximum for the frequency, f. Values of τ can be measured over a range of frequencies and temperatures.

Figure 4 reports values of $\log(\tau)$ versus $1/T$ at incremental changes in the degree of advancement α. Like $\log(\sigma)$, the values of $\log(1/\tau)$ are related to the viscosity and the extent of cure as is observed on Figures 2 and 3. Because the values of τ occur in the postgel, preglass region of cure, 90–300 min, these values of τ are appropriately analyzed as an α-relaxation associated with the T_g to $T_g + 100°$ region.

A Vogel–Fulcher–Tammann–Hesse equation can be used to characterize the temperature dependence of the relaxation times for six different degrees of cure: 0.70, 0.75, 0.80, 0.825, 0.90, and 0.95:

$$\log \tau = A + \frac{B}{T - T_\infty} \tag{5}$$

The best fit occurs at T_∞ values 50 °C below the respective values of T_g for

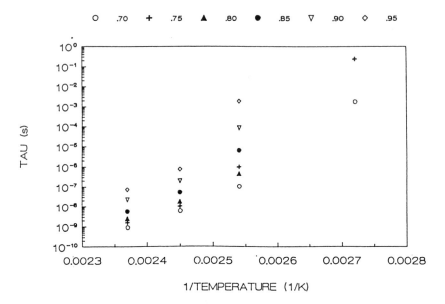

Figure 4. Log(τ) versus 1/T (K) at comparable degrees of advancement: ○, *0.70;* +, *0.75;* ▲, *0.80;* ●, *0.85;* ▽, *0.90;* ◇, *0.95.*

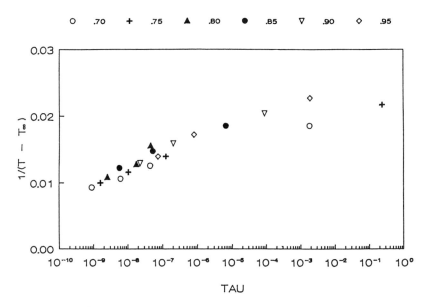

Figure 5. Log(τ) versus 1/(T $-$ T$_\infty$) where T$_\infty$ = T$_g(\alpha)$ $-$ 50 K at comparable degrees of advancement: ○, *0.70;* +, *0.75;* ▲, *0.80;* ○, *0.85;* ▽, *0.90;* ◇, *0.95.*

each of these values of α. The value of T_g was determined by advancing the reaction to a particular value of α based on the time–temperature DSC kinetic data. The softening temperature, T_g, was determined from the onset of the drop in G'. Figure 5 is a plot of $1/(T - T_\infty)$ versus τ, where $T_\infty(\alpha)$ = $T_g(\alpha) - 50\ °C$. The observation that the value of τ changes with T_∞ suggests that the parameters A and B show little variation in this preglass region. Further, the results show that the time of occurrence of successive increases in the relaxation time monitors the value $(T_{rx} - T_g)$, where T_{rx} is the reaction temperature. Thus, through the Fulcher equation, the ε'' relaxation peaks can be used to measure the buildup in T_g and degree of cure as the polymerization proceeds.

Monitoring Cure in Multiple Time–Temperature Processing Cycles

As a representative example of cure monitoring using a common commercially used aerospace resin and a complex cure cycle, Figure 6 displays the output of $\omega\varepsilon''(\omega)$ for a two-stage, 121 °C and 177 °C, ramp-hold sequence used to cure a commercial and widely used MY720 aromatic epoxy system. This resin consists of a tetraglycidyl-4,4'-diaminodiphenyl methane (TGDDM) and diaminodiphenyl sulfone (DDS). This system with catalyst is sold by the Hercules Corporation as 3501-6.

Figure 6 shows two peaks of the overlapping $\omega\varepsilon''(\omega)$ lines. These peaks

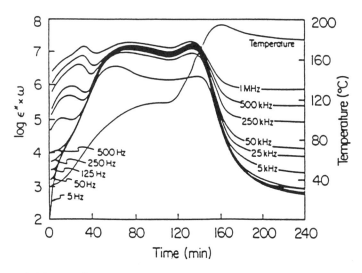

Figure 6. Plot of $\omega\varepsilon''(\omega)$ versus time of the sensor output at the 64th ply of the thick epoxy–graphite laminate during the cure in the autoclave.

indicate the times and magnitude of maximum flow as monitored by the ionic mobility. The first peak, the highest degree of flow, occurs at the beginning of the first hold. The second point of high flow, high ionic mobility, occurs midway up the ramp between the 121 °C hold and the 177 °C hold. As the temperature rises the fluidity increases until such time as the rate of reaction and thereby the degree of cure, which is also increasing during the temperature ramp, overwhelm the temperature effect on fluidity. At this point the fluidity begins to drop and a peak in $\omega\varepsilon''(\omega)$ occurs indicating the second occurrence of maximum flow.

The gradual drop in the magnitude of $\varepsilon''(\omega)$ during the final hold and its rate of change $d\varepsilon''/dt$ monitor the buildup in modulus and its rate of buildup. When ε'' attains a constant value, the system has reached its final lowest value of dipolar ionic mobility. Thus, when no further changes in mobility can be detected, $d\varepsilon''/dt = 0$, the system is fully reacted at that hold temperature. Monitored in this way, the changing values of ε'' are a very sensitive means of detecting the final small changes in degree of cure and the buildup in end-use properties such as T_g and modulus. Figure 6 suggests even after a 2-h hold, for this fresh 3501-6 resin the mobility is still decreasing and final cure or end use properties have not been attained.

Figure 7 shows the correlation between the viscosity and the ionic mobility based on isothermal runs for this system as monitored by the value of $\varepsilon'' = 5$ kHz. A representative calibration curve relating the FDEMS output to degree of cure is shown in Figure 8. Unlike viscosity, separate calibration curves for different temperatures must be generated from the isothermal runs because they are temperature dependent.

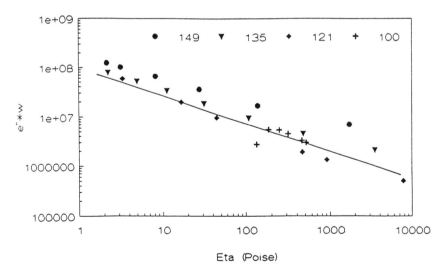

Figure 7. Log $\omega\varepsilon''(\omega)$ versus log(η) for the TGDDM epoxy based on 4 isothermal runs (°C).

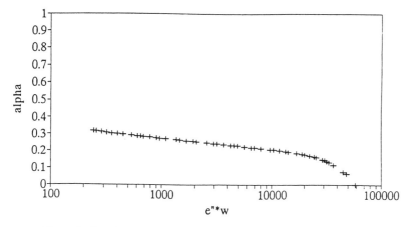

Figure 8. Correlation curve relating degree of cure to $\omega\varepsilon''(\omega)$ at 135 °C.

The buildup in final-curve properties such as degree of cure during the last hold is monitored with a high degree of sensitivity by using the value of $(d\varepsilon''/dt)/\varepsilon''$. Figure 9 is a correlation plot of the normalized rate of change in ε'' (5 kHz) compared with model predicted values (based on numerous DSC runs) of the buildup in the degree of cure.

Figures 10 and 11 show two important applications of FDEMS as applied to processing of an epoxy system. Figure 10 shows the effect on processing properties as monitored by the value of $\omega\varepsilon''(\omega)$ for this system when the first hold temperature drifts 10 °C higher, 131 °C. Figure 11 shows the output for the original 121 °C, 177 °C ramp hold sequence but for a batch

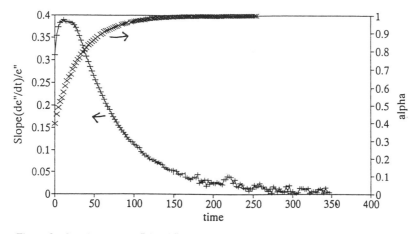

Figure 9. Correlation of $(d\varepsilon''/dt)/\varepsilon''$ with time and α showing sensitivity of normalized rate of change in ε'' to changes in final α near end reactions.

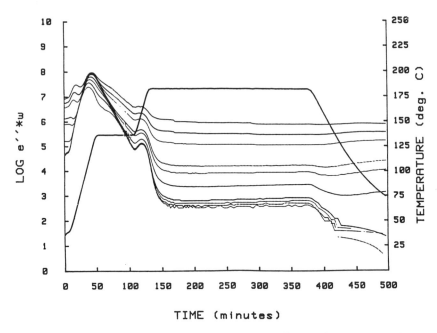

Figure 10. TGDDM epoxy cured in a press with a 135 °C, 177 °C cure cycle. Values of $\omega\varepsilon''(\omega)$ are displayed for frequencies of 1 MHz; 500, 250, 125, 50, and 5 kHz; and 500, 250, 125, and 50 Hz (top to bottom).

of 3501-6 epoxy resin after it has been left to age at room temperature for 30 days.

Even without the calibration relations the effect on cure-processing properties can be clearly seen from the sensor output. In Figure 10, the value of ε'' peaks a little higher but drops much more rapidly. The second peak in ε'' during the ramp is much lower. Thus, the effect of the 131 °C hold is to cause a slightly higher fluidity initially. However, the high fluidity region lasts for a much shorter time. Equally important at the second point of high fluidity, which is usually critical for composite-prepreg consolidation, the fluidity is significantly lower. The overall effect of aging is seen as decreasing the level of fluidity throughout the cure procedure. Full cure is achieved much sooner in the final hold for the aged resin because the value of $d\varepsilon''/dt$ approaches 0 much sooner.

Monitoring Cure in a Thick Laminate

A major advantage of the FDEMS is the capability of monitoring and quantitatively measuring processing properties in situ during cure in complex shaped parts during processing in autoclaves, presses, and pultruders. For

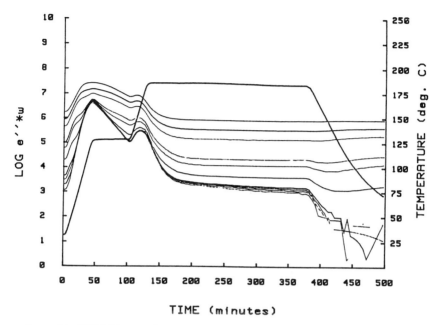

Figure 11. TGDDM epoxy that was aged at room temperature for 30 days after which it was cured in a press by using a 121 °C intermediate hold. Values of ωε″(ω) are displayed for frequencies of 1 MHz; 500, 250, 125, 50, and 5 kHz; and 500, 250, 125, and 50 Hz (top to bottom).

example, sensors can be placed in a thick 192-ply Hercules 3501-6 epoxy–graphite laminate at multiple positions (*see* Figure 12) such as the tool surface, at 32 plies, at 64 plies, and in the center at the 96th ply. This 192-ply composite layup with the embedded sensors was cured in an autoclave. The output at each sensor was measured automatically at intervals throughout the cure cycle by multiplexing the four sensors through a computer-controlled switch to the impedance bridge. The multiplexed sensor–bridge–computer system can make more than 100 permittivity measurements in less than 2 min. Measurements are recorded continuously throughout the cure cycle without interruption over the 10^6–10^{-2} range in magnitude of ε' and ε''.

Figure 13 is a plot of the viscosity determined by the sensor at each of the four positions in the thick laminate during cure in the autoclave. The FDEMS data show that the middle ply achieves its first viscosity minimum 20 min after the plies on the surface of the tool plate. The middle plies continue to lag the surface ply until the second ramp to 177 °C. At this point the exothermic 3501-6 epoxy reaction starts heating the laminate from the inside. As a result, the heat generated at the center ply causes the viscosity to catch up with the viscosity of the surface plies. In this way, FDEMS

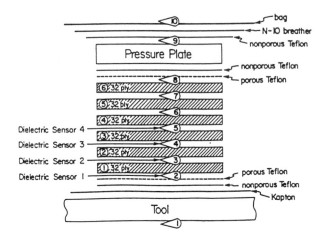

Figure 12. Layup of a thick TGDDM epoxy (Hercules 3501-6) graphite laminate autoclave run showing the positions of the dielectric sensors and thermocouples.

output measurements of the viscosity at the center and surface can be used to evaluate the cure cycle. In this example the center and surface plies achieve their viscosity minimum in the high temperature ramp at roughly the same time. At all earlier times the center lags behind the reaction at the surface.

Thus, the sensor not only reveals what is occurring during cure throughout the part but it also reveals the exact time–temperature–magnitude dependence of processing properties such as viscosity for that tool, geometry, and particular run. The sensor output can be used to test the validity

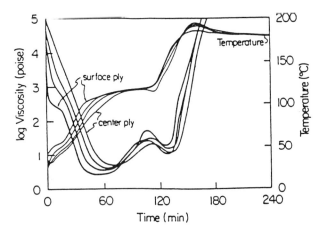

Figure 13. Viscosity of the thick laminate as determined from the frequency dependence of the FDEMS sensors at the surface, 32nd, 64th, and center piles.

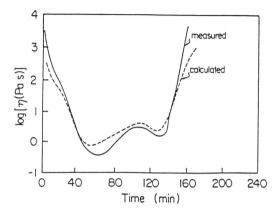

Figure 14. Comparison of viscosity at the 64th ply as predicted by the FDEMS sensors and the Loos–Springer model.

of processing models such as the Loos–Springer model (*11*). Sensor-measured values of η can be compared with the Loos–Springer model predictions. Figure 14 is a comparison of the model predictions and the measured values at the 64th ply. Agreement in the viscosity time dependence and magnitude with the predictions of models is essential if the model is to be verified and used with confidence.

The sensor output can also provide useful input for controlling the cure cycle. For this epoxy the most widely recommended and successful cure cycle forces the second viscosity minima to occur at the same time throughout the laminate thickness (Figure 13). Thus, one can hypothesize that an efficient and effective cure cycle causes the viscosity minima at the surface and center to occur at the same time, and as rapidly as possible. Accordingly, one might propose raising rapidly the air temperature in the FDEMS-controlled run until the exotherm at the center causes the viscosity at the center ply to start to catch up to the surface viscosity. At this point the air temperature could be rapidly lowered, to hold the surface viscosity in this high flow condition while the center viscosity catches up. At such point as the center viscosity goes through its viscosity minimum and advances beyond the surface plies, the air temperature would be set to the final 177 °C hold. The 177 °C hold would continue until the sensor output indicated, through $d\varepsilon''/dt$ approaching 0, that the reaction was complete.

Figure 15 shows the results from such a sensor-controlled run. FDEMS measured viscosities from two sensors at the center ply and one sensor at the surface ply. Air autoclave temperatures and the temperatures at the surface and center ply are also shown. The starting time for the FDEMS-controlled autoclave run and the manufacturer's cure cycle run is defined as the time at which the tool surface temperature starts to increase.

The FDEMS-controlled run significantly reduced the time lag and vis-

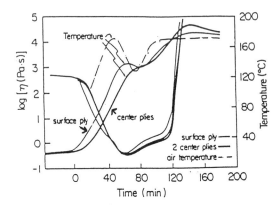

Figure 15. Viscosity at each sensor position of a 192-ply graphite–epoxy composite during an FDEMS sensor-controlled autoclave cure.

cosity difference between the center ply and the tool surface ply. The amount of flow as measured by the magnitude of the viscosity minimum was greater in the FDEMS-controlled run. The approach of $d\varepsilon''/dt$ to 0 was used to determine cure completion. The total cure time of 200 min in this FDEMS-controlled run is 40 min less than the conventional cure cycle.

Monitoring Cure of Coatings, Films, and Adhesives

Because the FDEMS sensors are thin, planar, and active on site, they are ideal for monitoring the cure and buildup in end-use properties of coatings, films, or an adhesive in the bondline. For the case of coatings the sensor is simply placed on the surface to be coated and the coating applied in the user's preferred manner. The planar thin sensors can also be placed in the adhesive between the surfaces of the two materials to be bonded. Thereby cure is monitored in the bond line. Because most coatings, films, or adhesives involve resins that are very similar to traditional thermosets, the sensor output, use of calibration graphs based on laboratory correlation studies, and ability to make measurements in situ in the field or on the manufacturing plant floor are identical to the thermoset work described previously. Often in these applications the terminology or end-use properties are different. Thus, correlation among ionic mobility as measured by σ or $\omega\varepsilon''(\omega)$, time of occurrence of dipolar relaxation peaks, and rate of change in $(d\varepsilon''/dt)/\varepsilon''$ is associated with terms such as dry to touch; dry to hard; peel strength; Barcol hardness; resistance to blistering, fading, peeling, etc.; and lap shear strength in the case of an adhesive.

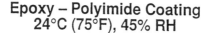

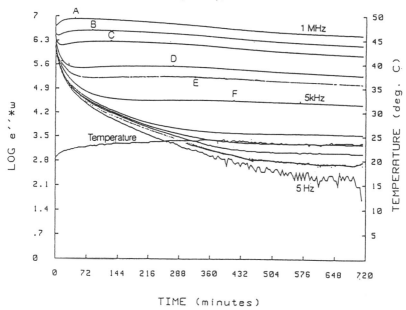

Figure 16. Epoxy–polyamide marine paint undergoing cure over 6 h at 23 °C and 45 % relative humidity.

Figure 16 shows the cure with time of an epoxy–polyamide marine paint. At the time of application the coating is in the wet stage. In this stage the viscosity is very low and solvent evaporation occurs rapidly. Solvent loss from the coating was measured by thermogravimetric analysis (TGA) and correlated with the FDEMS output. The wet stage of solvent loss was seen as rapid weight loss in the first 144 min. The evaporation slows down markedly after 6 h when the change in weight is barely noticeable. The combination of solvent loss and cross-linking of the resin rapidly increases the viscosity of the film, thus lowering the value $\omega\varepsilon''(\omega)$ (Figure 16). At 100 min into the run the coating reaches its set-to-touch point as defined by American Society for Testing and Materials guidelines. Using the 50-Hz line to monitor the initial cure, one attains and can calibrate a value for $\log(2\pi f\varepsilon'') = 2.8$ at dry-to-touch point.

The long term cure process can be monitored by following $\omega\varepsilon''(\omega)$ on a non-log scale. A continual decrease in $\omega\varepsilon''(\omega)$ was observed through the third day. On the fourth day $\omega\varepsilon''(\omega)$ was constant and indicated that the fullest cure possible under 24 °C and 45% relative humidity conditions was achieved.

The high frequency, dipolar component values of $\omega\varepsilon''(\omega)$ in Figure 16 monitor the buildup in T_g as well as the corresponding final-use properties of the curing coating. The time of occurrence in a maximum value of ε'' at each frequency in the 5-kHz to 1-MHz $\omega\varepsilon''(\omega)$ lines, points A–F, indicates the time when the characteristic relaxation time for dipolar relaxation $\tau = 1/\omega$ has occurred. Thus, point A, 15 min, marks the time when $\tau = (2\pi \times 10^6)^{-1}$ s; and the point F, 360 min, marks the time when the rotational relaxation time has slowed to $\tau = (2\pi \times 5 \times 10^3)^{-1}$ s. The relationship between the value of τ and the value of T_g or any other use property can be quantitatively determined by correlating measurements of τ with time measurements of T_g or the use property of interest as discussed earlier for epoxy resins.

Figures 17 and 18 show two applications. Figure 17 shows the ability of FDEMS to monitor the difference in rate of cure under two different conditions of humidity. Figure 18 shows the FDEMS sensor coated with a first coat and allowed to cure for 24 h. On the second day a second coat was applied. The sharp rise in $\omega\varepsilon''(\omega)$ from 10^2 to 10^5, or over three decades, clearly indicates the ability of the second coat to resoften the first coat in this paint system.

Turning to latex paints, Figure 19 shows the FDEMS sensor monitoring cure of a black metallic latex paint used in the appliance industry. The

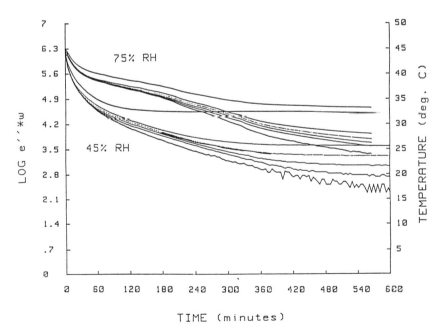

Figure 17. Dielectric sensor monitoring variation in cure rate of 75% and 45% relative humidity at 24 °C.

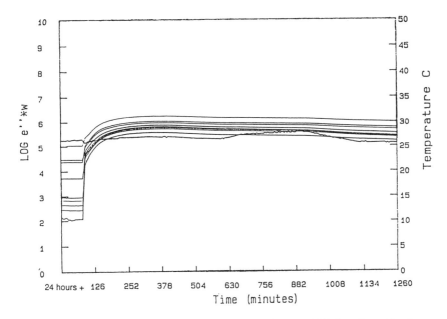

Figure 18. Dielectric sensor monitoring softening of first coat 24 h later by application of a second coat. Frequencies are the same as Figure 1.

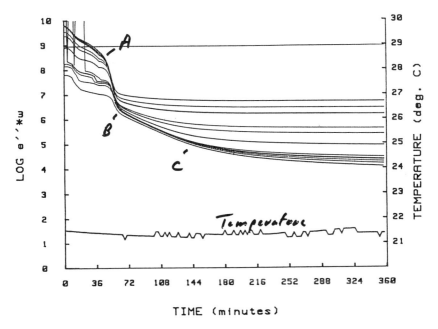

Figure 19. Dielectric sensor monitoring cure of a metallic latex paint. Frequencies are the same as Figure 1.

changing value of the overlapping $\omega\varepsilon''(\omega)$ lines, which monitor changes in ionic mobility, clearly shows three regions of cure. First, up to 50 min solvent is evaporating. Secondly, at 50 min, a rapid drop occurs in mobility (three decades) monitors latex coalescence. Third, the gradual buildup in final end-use durability is seen over time in the gradual drop in $\omega\varepsilon''(\omega)$ during the remaining 5 h.

Major advantages of FDEMS monitoring of coating cure are that

- it can be done on site in the field or in the plant dryer
- multiple samples (e.g., with different pigment formulations) can be simultaneously and continuously monitored under any conditions
- cure can be monitored in complex environments of temperature humidity and air flow

The technique is also very helpful in monitoring coating cure under other types of radiation drying such as UV.

Intelligent Automated-Sensor Cure Control

An intelligent automated FDEMS cure-control system developed at the College of William and Mary was used to process fresh and aged polyimide prepreg materials (6–10). By responding on a molecular level to prepreg variations resulting from resin or batch variation, aging or moisture absorption, and tool heat-transfer phenomena, the sensor-model expert system is capable of successfully processing materials that may not be processable by using a strict time–temperature recipe approach. At the heart of the automated sensor cure-control system is the coupling of the sensor output with an artificial intelligence software control system. A knowledge base is created that contains the material process goals. These goals are usually based on the output of processing models and previous processing experience. This knowledge base consists of a collection of rules, heuristics, and actions. The actions include reading the sensor input and other parameters such as temperature and pressure. Then, based on the goals, the rules and heuristics for achieving these goals are used to analyze the input and make decisions regarding the remainder of the cure process. Prior experimentation allows the expert system to optimize in real time the ongoing cure process.

The expert sensor system is used to control the cure of PMR-15–graphite prepreg provided by ICI Fiberite. Conventionally formulated, PMR-15 is a stoichiometric mixture of the nadic ester of 5-norbornene-2,3-dicarboxylate (NE), the dimethyl ester of 3,3′,4,4′-benzophenone tetracarboxylate (BTDE), and 4,4′-methylendianiline (MDA) in the monomer ratio

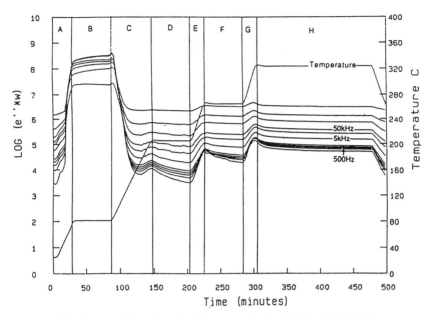

Figure 20. Values of T and $\omega\varepsilon''(\omega)$ for a standard cure of fresh PMR-15 prepreg. See text for definition of regions A–H. Frequencies are the same as Figure 1.

$2(NE):(n + 1)MDA:n(BTDE)$ where $n = 2.087$. The intelligent control software used in these runs is called QPAL, Qualitative Process Automation Language, developed cooperatively by Wright-Patterson Air Force Base, Ohio and Lawrence Associates, Dayton, Ohio.

Figure 20 shows the FDEMS output sensor as it monitors a typical fresh PMR-15 prepreg (ICI Fiberite Lot No. 10502 S) cure by using a convention time–temperature cure cycle. Region A shows FDEMS sensor wet-out on a slow 2 °C/min ramp to 80 °C and the achievement of a viscosity minimum for the system. Region B is a 1-h hold at 80 °C to allow for solvent elution and maximum part wet-out. Region C shows the onset of imidization as characterized by a drop in resin fluidity. Region D shows the continuation of the imidization reaction during a 1-h hold at 200 °C. Normally one would want to hold until the imidization is complete as evidenced by the decrease in the slope of the FDEMS signal, that is, $(d\varepsilon''/dt)/\varepsilon''$ approaching 0. Here we see that the use of sensor feedback would have shown the imidization reaction to be incomplete and therefore the hold would have been extended by the intelligent, automated cure-control system. Regions E and F show the viscosity minimum achieved during a ramp to 265 °C and the subsequent decrease in FDEMS signal as residual solvent and other volatiles are eluted during the hold. Finally, regions G and H show the ramp to and hold at 320 °C, the final cross-linking hold. Note that after approximately

75 min in the 320 °C hold the change in slope of ε'', $(d\varepsilon''/dt)/\varepsilon''$ is small and close to 0, and full cure based on sensor criteria is reached. Had the intelligent, automated cure-control system controlled this run by using the criterion of $(d\varepsilon''/dt)/\varepsilon'' \leq 3.0 \times 10^{-5}$, it could have terminated the run after a total cure time of 375 min, thus saving 105 min of the 480 min shown here.

Figure 21 shows the dielectric and temperature output for the intelligent sensor-model automated cure of fresh PMR-15 prepreg (ICI Fiberite Lot No. 11769 S). Region A shows the panel as it warms from room temperature to approximately 80 °C. FDEMS sensors wet out is indicated by the sharp increase in the $\log(\varepsilon''\omega)$ signal at 81 °C. Note that the max in $\log(\varepsilon''\omega)$ for the prepreg sample is almost one decade lower than that seen for prepreg Lot No. 10502 S in Figure 20. Region B, imidization onset, is a hold near 80 °C to allow for solvent elution before beginning a slow ramp to the imidization hold temperature. Region B requires that the magnitude of $\log(\varepsilon''\omega)$ at 500 Hz, 5 kHz, and 25 kHz be less than 8.0 for four consecutive readings, that the change in FDEMS signal at 25 kHz $[(d\varepsilon''/dt)/\varepsilon'']$ be decreasing (i.e., negative) for four consecutive cycles, and that the magnitude of $\log(\varepsilon''\omega)$ at 25 kHz decreases by 10% from the maximum value determined at the beginning of the hold. On the basis of these criteria,

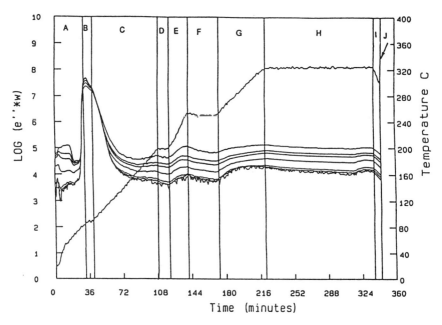

Figure 21. Values of T *and* $\omega\varepsilon''(\omega)$ *for a sensor-model intelligent automated cure of fresh PMR-15 prepreg. See text for definition of regions A–H. Frequencies are the same as Figure 1.*

region B is an 11-min hold at 81 °C. Region C is the slow 2–3 °C/min ramp to the 200 °C hold temperature. Region D shows the imidization hold at 200 °C lasting for only 12 min before meeting the criterion that the value of $(d\varepsilon''/dt)/\varepsilon''$ at 25 kHz be less than 3×10^{-4} for four consecutive data acquisition cycles. This short dwell time indicates the virtual completion of the imidization reaction before entering imidization hold. Obviously, then, the required 1-h hold seen in region D of Figure 21 would have been wasted processing time for this composite cure. Regions E and F are the ramp to the minimum viscosity consolidation temperature and the preset 30-min hold at that temperature, respectively. The knowledge base determined the minimum viscosity temperature for this run to be 260 °C based on the criterion that the ionic conductivity of the resin is decreasing (i.e., $\log(\varepsilon''\omega)$ signal is folding over at a temperature that is quite close to the optimal model generated 265 °C compaction temperature seen in Figure 21). Finally, region G shows the slow ramp to the 320 °C final cross-linking cure temperature, and region H shows the achievement of full cure, based on the criterion that $(d\varepsilon''/dt)/\varepsilon''$ at 25 kHz be less than 3×10^{-5} for four consecutive data acquisition cycles, after 115 min. Regions I and J show system cool down and shut down, respectively. The total cure time for this fresh PMR-15 prepreg processing run was 338 min. Thus we see that resin–prepreg batch variability significantly affects the way a given standard material will process.

Figure 22 shows the FDEMS data for a PMR-15–graphite prepreg sample aged 7 months under freezer conditions (ICI Lot No. 10502 S). The labeled zones correspond to those previously described. Comparisons with the dielectric and temperature data for the optimized cures of fresh PMR-15 prepreg seen in Figures 2–4 show several significant differences as a result of the aging process. In region A, the viscosity minimum defined by the maximum in $\log(\varepsilon''\omega)$ at 25 kHz occurred at 8.5, one decade higher than was seen for the expertly cured fresh PMR-15–graphite prepreg. In fact, the low frequency $\log(\varepsilon''\omega)$ lines at 125, 250, and 500 Hz lie below the overlapping ionic frequency band seen centered around region B. This phenomenon results from charge polarization effects that are significant only in highly fluid systems and is indicative of water absorption during the aging process. Region B, imidization onset, was a 30-min hold at the intelligently determined minimum viscosity temperature of 83 °C.

Region D, the imidization hold, lasted only 4 min, the minimum time necessary to meet the criterion that the dielectric signal $(d\varepsilon''/dt)/\varepsilon'' \leq 3.0 \times 10^{-4}$ at 25 kHz for four consecutive data acquisition cycles and thus signifying completion of the imidization process. This result is indicative of the advancement of the reaction processes leading to complete imidization as a function of the aging process. The third major indicator of material variability is the minimum viscosity consolidation temperature determined by the sensor-model intelligent cure control system in region E. For this 7-

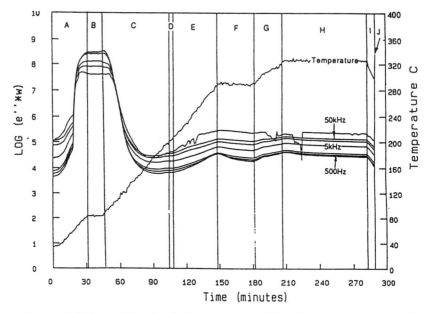

Figure 22. Values of T and ωε″(ω) for a sensor-model intelligent automated cure of PMR-15 prepreg aged 7 months under freezer conditions. See text for definition of regions A–H. Frequencies are the same as Figure 1.

month freezer-aged PMR-15 prepreg, this cross-linking minimum viscosity temperature, as determined by the normalized decrease or folding over in the FDEMS signal occurred at 290 °C. This 25 °C increase from the model determined 265 °C optimized softening temperature again seems to indicate that the prepreg has aged considerably. Region H, the final cure hold at 320 °C, requires only 75 min to meet the cure-complete criterion that $(d\varepsilon''/dt)/\varepsilon'' \leq 3.0 \times 10^{-5}$ at 25 kHz for four consecutive data acquisition cycles. This hold time was much shorter than those seen for the prepreg materials in Figures 1–4, the final indicator of the age of the starting prepreg. The expert dielectric system does a good job in responding in real time to material variability as a function of batch variations, aging, transfer phenomena, and differences in matrix resin systems.

Acknowledgment

David E. Kranbuehl appreciates partial support from the National Science Foundation, Science and Technology Center at Virginia Polytechnic Institute and State University under Contract No. DMR91–2004, a NASA Langley grant NAGI–23, and support from the Northrop Corporation.

References

1. Kranbuehl, D. *Developments in Reinforced Plastics;* Elsevier Applied Science: New York, 1986; Vol. 5, pp 181–204.
2. Kranbuehl, D. In *Encyclopedia of Composites;* Lee, S. M., Ed.; VCH: New York, 1989, pp 531–543.
3. Senturia, S.; Sheppard, S. *Appl. Polym. Sci.* **1986,** *80,* 1–48.
4. *Polymer Materials Science and Engineering;* May, C., Ed.; ACS Symposium Series 227; American Chemical Society: Washington, DC, 1983.
5. Hedvig, P. *Dielectric Spectroscopy of Polymers;* Wiley: New York, 1977.
6. Mijovic, J.; Bellucci, F.; Nicolois, L. *Electrochem. Soc.* **1995,** *142(4),* 1176–1182.
7. Mijovic, J.; Winnie Tee, C. F. *Macromolecules* **1994,** *27,* 7287–7293.
8. Bellucci, F.; Valentino, M.; Monetta, T.; Nico demo, T. L.; Kenny, J.; Nicolais, J.; Mijivic, J. *J. Polym. Sci., Part B.: Polym. Phys.* **1995,** *33,* 433–443.
9. Mangion, M. B. M.; Johari, G. P. *Macromolecules* **1990,** *23,* 3687–3695.
10. Mangion, M.; Johari, G. *J. Polym. Sci., Polym. Lett. Ed.* **1991,** *29,* 1117–1125.
11. Parthun, M. B.; Johari, G. *Macromolecules* **1992,** *25,* 3254–3263.
12. Boiteux, G.; Dublineau, P.; Feve, P.; Mathieu, C.; Seytre, G.; Ulanski, J. *Polym. Bull.* **1993,** *30,* 441–447.
13. Mathieu, C.; Boiteux, G.; Seytre, G.; Villain, R.; Dublineau, P. *J. Non-Cryst. Solids* **1994,** *172–174,* 1012–1016.
14. Xu, X.; Galiatsatos, V. *SPE Tech. Pap. ANTEC '93* **1993,** *39,* 2875.
15. Xu, X.; Galiatsatos, V. *Makromol. Symp.* **1993,** *76,* 137.
16. Deng, Y.; Martin, G. *Macromolecules* **1994,** *27,* 5141–5146.
17. Companik, J.; Bidstrup, S. *Polymer* **1994,** *35,* 4823–4840.
18. Bidstrup, S.; Sheppard, N.; Senturia, S. *ANTEC* **1987,** 987–993.
19. Bidstrup, W.; Bidstrup, S.; Senturia, S. *ANTEC* **1988,** 960–966.
20. MacKinnon, A.; Jenkins, S.; McGrail, P.; Pethrick, R. *Macromolecules* **1992,** *25,* 3492–3499.
21. Maistros, G.; Black, H.; Bucknall, C.; Partridge, I. *Polymer* **1992,** *33,* 4470–4478.
22. Hart, S.; Kranbuehl, D.; Hood, D.; Loos, A.; Koury, J.; Havery, J. *Int. SAMPE Symp. Exhib.* **1994,** *391,* 1641–1651.
23. Kranbuehl, D. *Matér. Tech.—Adv. Compos. Mater.* **1994,** *Nov.–Dec.* 18–22.
24. Kranbuehl, D.; Kingsley, P.; Hart, S.; Hasko, G.; Dexter, B.; Loos, A. C. *Polym. Compos.* **1994,** *15(4),* 297–305.
25. Hart, S.; Kranbuehl, D.; Loos, A.; Hinds, B.; Koury, J.; Harvey, J. *SAMPE* **1993,** *38,* 1009–1019.
26. Hart, S.; Kranbuehl, D.; Loos, A.; Hinds, B.; Koury, J. *SAMPE Symp.* **1992,** *37,* 225–230.
27. Loos, A. C.; Kranbuehl, D. E.; Freeman, W. T. *Intelligent Processing of Materials and Advanced Sensors;* Wadley, H. N. G.; et al., Eds.; Metallurgical Society: Warrendale, PA, 1987, pp 197–211.
28. Day, D. *Int. SAMPE Symp. Exhib.* **1990,** *35,* 1507–1516.
29. Kranbuehl, D.; Eichinger, D.; Hamilton, T.; Clark, R. *Polym. Eng. Sci.* **1991,** *31,* 56.
30. Ciriscioli, P.; Springer, G. "Smart Autoclave Cure"; Technomic: Lancaster, PA, 1990.
31. Kranbuehl, D. E.; Delos, S. E.; Jue, P. K. *Polymer* **1986,** *27,* 11–20.
32. Kranbuehl, D.; Delos, S.; Hoff, M.; Weller, W.; Haverty, P.; Seeley, J. In *Cross-Linked Polymers: Chemistry, Properties and Applications;* Dickie, R. A.; Labana, S. S.; Bauer, R. S., Eds.; ACS Symposium Series 367; American Chemical Society: Washington, DC, 1988; pp 100–112.

33. Sheppard, N.; Gavericle, S.; Day, D.; Senturia, S. *SAMPE Symp.* **1981,** 65–76.
34. *Cross-Linked Polymers: Chemistry, Properties and Applications;* Dickie, R. A.; Labana, S. S.; Bauer, R. S., Eds.; ACS Symposium Series 367; American Chemical Society: Washington, DC, 1988.
35. Kranbuehl, D.; *Plast Rubber Compos. Process Appl.* **1991,** *16,* 213–223.
36. Kranbuehl, D.; Eichinger, D.; Hamilton, T.; Clark, R. *Polym. Eng. Sci.* **1991,** *31,* 56–60.
37. Kranbuehl, D. *J. Non-Cryst. Solids* **1991,** *131–33,* 930–936.
38. Parthun, M. B.; Johari, G. *Macromolecules* **1992,** *25,* 3254–3263.
39. Mathieu, C.; Boiteux, G.; Seytre, G.; Villian, R.; Dublineau, P. *J. Non-Cryst. Solids* **1994,** *172–174,* 1012–1016.
40. Martin, G.; Tungare, A; Grotto, J. *Am. Chem. Soc., Polym. Charact.* **1990,** *227,* 205–215.
41. Deng, Y.; Martin, G. *ANTEC* **1994,** *94,* 1664–1670.
42. Kranbuehl, D.; Delos, S.; Hoff, M.; Weller, L.; Haverty, P.; Seeley, J.; Whitham, B. *Int. SAMPE Symp. Exhib.* **1987,** 338–348.
43. Kranbuehl, D. *Polym. Mater. Sci. Eng. Prepr.* **1993,** *68,* 224–225.
44. Kranbuehl, D. *Polym. Mater. Sci. Eng. Prepr.* **1990,** *68,* 90–93.
45. Kranbuehl, D.; Kim, T.; Liptak, S. C.; McGrath, J. E. *Polym. Prepr.* **1993,** *34,* 488–489.
46. Maistros, G.; Black, H.; Bucknall, C.; Partridge, I. *Polymer* **1992,** *33,* 4470–4478.
47. MacKinnon, A.; Jenkins, S.; McGrail, P.; Pethrick, R. *Macromolecules* **1992,** *25,* 3492–3499.

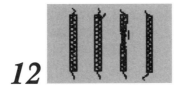

12

Dielectric Properties of Polymeric Liquid Crystals

George P. Simon

This chapter introduces polymeric liquid-crystalline polymers (PLCs) of both the side-chain (SCPLC) and main-chain (MCPLC) type and summarizes work done to date on the variable frequency dielectric relaxation properties of these materials. PLCs are macromolecules that demonstrate an anisotropic liquid state with an orientational order intermediate between that of three-dimensional lattices and isotropic liquids. In the case of SCPLCs, the nature of the mesogenic motion in the glassy, liquid crystalline, and isotropic state, the mathematical description of such motion, and its ability to be aligned in electrical fields are examined with particular reference to the influence of the connectivity of these groups with the polymeric main chain. The effect of different structural elements in the SCPLCs molecular architecture is also examined. In MCPLC, where considerably less work has been performed, the relationship among structure, morphology, and dielectric relaxation is reviewed, and emphasis is given to examination of some of the current commercial materials.

Molecular liquid crystals (MLCs) and polymeric liquid crystals (PLCs) have proven to be innovative and exciting areas of scientific research for a number of years, and the production of new materials (more than 50,000 liquid-crystalline (LC) products reported to date) (*1*) has continued unabated. Even though commercial application of PLCs has not kept pace with research, this situation is likely to improve in the near future.

The interest in PLCs is due to their ability to combine the properties of MLCs with the desirable properties of polymers. The LC state is intermediate between that of ordered three-dimensional crystalline arrays and isotropic disorder. In the LC, or mesophase, the molecules have an orientational order, and the direction along which the molecular long axes are roughly oriented is a vector quantity, the director, $\bar{n}$. The degree of this alignment is given by an order parameter, S (defined mathematically later). In thermotropic liquid crystals, the LC state is reached by heating the sample, and it occurs between the solid state (melting point temperature, T_m, in MLCs and the glass transition temperature, T_g, in PLCs) and the isotropic liquid state. The LC state is characterized by anisotropy of properties such as refractive index and dielectric constant whilst displaying liquid flow. This transition from LC to isotropic liquid is known as the isotropization or clearing temperature, T_{cl}. The chemical structures of molecules that show such behavior are often (although not exclusively) rigid and rod-like, and many consist predominantly of aromatic units. The orientational order of the LC state arises mainly from considerations of packing of molecules or *mesogens* with a sufficient aspect ratio. Mesogens need not be polar to demonstrate liquid crystallinity, although for a variety of reasons they are often synthesized with strong pendant or transverse dipoles, and this characteristic influences phase stability and other properties.

LC molecules may organize themselves in ways dependent on chemical structure and temperature, and the main classifications are nematic, smectic, and cholesteric (Figure 1), Some materials show more than one phase as a function of temperature. Nematic LC units (Figure 1a) share a common directional orientation, whereas the centers of mass are disordered. Smectic materials show greater order in that, in addition to orientational alignment, the centers of mass exist in spaced layers, although within a layer there is no regular arrangement of molecules. A variety of smectic phases are possible and are distinguished by the orientation of molecules within the layer to the layer normal, and smectic A and smectic C (S_C) phases are shown in Figures 1b and 1c, respectively. If the orientational direction (tilt angle) changes in a helical fashion between S_C layers, and strong dipoles are present, the S_C^* phase, which is often ferroelectric, may occur. If chiral centers are introduced into nematic materials a cholosteric phase may result that, as is clear from its depiction in Figure 1d, is also known as the twisted nematic phase. Orientational properties of MLCs and their ability to be readily switched by electric and magnetic fields have made them widely used in a range of applications such as high-speed displays, shutters, smart windows, and optical waveguiding and in a variety of nonlinear optical and fiber-optic applications.

By combining these mesogenic units with polymers, the possibility of a wide range of PLCs occurs (Figure 2). The predominant types of PLCs involve incorporation of the mesogen pendant to the main chain (side-chain polymer liquid crystals, SCPLCs) where the units are attached end-on (Figure 2a) or laterally (Figure 2b). The mesogens can also be incorporated within the polymer backbone (main-chain polymer liquid crystals, MCPLC)

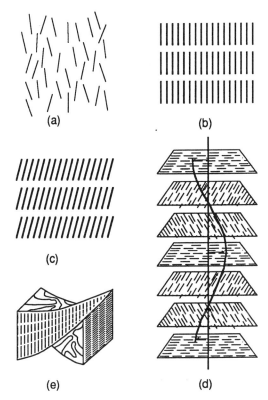

Figure 1. Classification of molecular organization in LC systems: (a) nematic, (b) smectic A, (c) smectic C, (d) cholosteric (twisted nematic), and (e) cholosteric structure in PLCs. (Redrawn from reference 6.)

(Figures 2e and 2f). Also, a range exists of other types of PLCs with intermediate or slightly different appearances as shown. In addition, instead of a rod or lathe-like moiety, disc-like groups can show LC behavior. Some rigid molecular materials show LC properties if dissolved in a nonmesogenic solvent at certain concentrations and are known as lyotropic PLCs. Only thermotropic materials will be considered in this review; dielectric relaxation behavior of lyotropics is summarized elsewhere (2).

A very large proportion of dielectric relaxation spectroscopy of PLCs reported thus far has involved SCPLCs where the mesogens are attached to polymer backbones via flexible spacers. This structure sets up a natural tension in the LC state between anisotropic, directionally ordered mesogens and random-coiling, entropy-driven polymer backbones. Whereas an LC state can be obtained via direct mesogen attachment to the polymer chain (3–5), spacers between mesogen and backbone tend to be used, allowing stabilization of the LC state, and the smectic phase is favored by longer

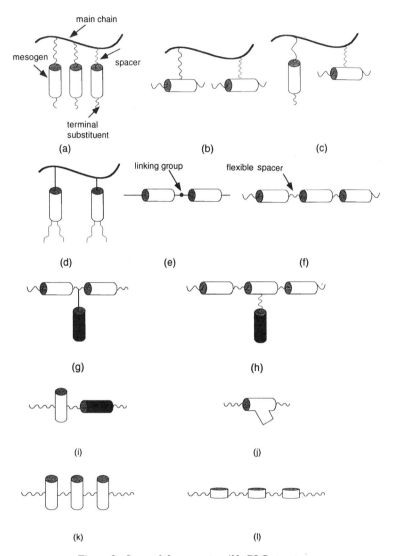

Figure 2. Some of the many possible PLC structures.

spacers. The various phases mentioned for MLCs also occur in SCPLCs. In
the case of cholosteric SCPLCs, the morphology is more of a twisted smectic
structure (6) (Figure 1e). Of much importance and interest is the degree of
coupling of motions between the backbone and mesogen. SCPLC materials
allow both the dielectric and optical properties of MLCs and their ability
to be aligned in electric and magnetic fields to be combined with the pro-
cessability and other properties of polymers. For example, the main chain
also allows the possibility of maintaining the directional alignment of the

mesogens caused by an applied voltage, whereas the electric field must be maintained for this process to occur in MLCs.

SCPLCs remain largely developmental materials although this has changed recently (7), and applications have been directed toward their use in optoelectronics such as in waveguides and as nonlinear optical media (predominantly second order (8)). Properties of these materials make them particularly attractive, as has been disclosed in particular for stereoregular SCPLCs (9). Equally promising is their use as materials for optical information storage. In the simplest form, this use involves production of a highly aligned monodomain onto which a laser can locally heat and disorder (write) information. Information thus stored must be stable and it may be erased by heat, realignment (electric–magnetic fields), or a combination of both (10, 11). Recent advances have focused in particular on SCPLC materials with azo units linking the aromatic groups in the mesogen because these functionalities can be switched between cis and trans isomers (writing and erasure) by application of either laser light or heat, causing manipulation of the birefringence and allowing holographic storage of large amounts of information (12–14). In most SCPLC applications, ease and stability of alignment (poling) and the nature of relaxation processes over a wide range of temperatures are important, and in this area dielectric analysis has an important role to play. This issue is a general one in many optoelectronic comb-shaped polymers, LC or otherwise (15). Comb-shaped polymers that do not exhibit an LC phase (perhaps because of short spacers units) have been suggested to be advantageous for some applications due to greater stability (16) or lower losses resulting from reduced domain scattering. To improve stability or *fix* alignment, strategies may include forming crosslinked networks or producing materials with very high T_g values (17). Recent reviews of the range of SCPLC materials (6, 18, 19) and details on general fields of their application (10, 11) can be found elsewhere.

MCPLCs have been commercially available since the mid-1980s and are used in engineering applications where they can be processed with conventional polymer processing machinery. To allow this processing to occur at reasonable temperatures, a range of strategies has been devised to lower the effective aspect ratio of the chains (reducing phase stability and usually crystallinity). This result has been achieved largely by either inclusion of flexible spacer units between the mesogenic groups or copolymerization of a range of monomeric units that in some way disrupt the linearity of the chain (20). MCPLC materials (especially those that are commercial) tend to be nematic, and advantages over other conventional engineering polymers include low viscosity because of low entanglement density and low die swell (in the molten state), high modulus and strength, and low thermal expansion (in the solid state). Applications include electrical components, fibers, optical fiber reinforcement, high strength housings, and flow modifiers and reinforcement in thermoplastic blends (21, 22).

Details of the range of molecular architectures that have been synthesized and their properties are given elsewhere (*23*).

Dielectric Relaxation of Side-Chain LC Polymers

Although the range of SCPLC materials synthesized and characterized by dielectric relaxation spectroscopy (DRS) to date is quite vast, the majority can usefully be described by a general structural formula based on schema developed elsewhere (*2, 24, 25*). Those materials based on an acrylate or methacrylate backbone can be represented as shown in **1**

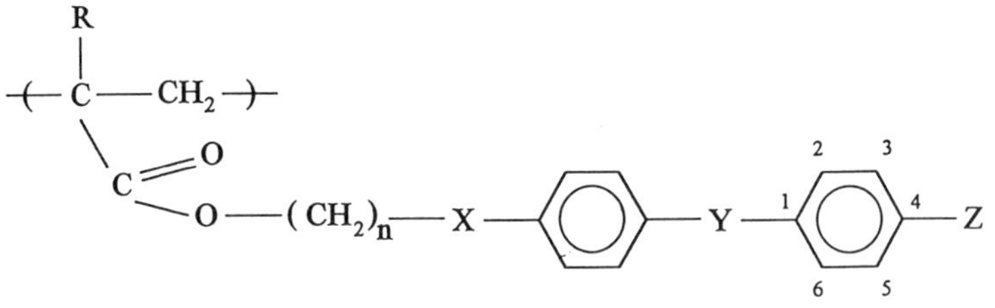

1

and labeled in shorthand as R/n/X/Y/Z/R"(i), where R differentiates between the varying main chain such as acrylate (H), methacrylate (CH$_3$), or chloroacrylate (Cl); n relates to the length of the aliphatic spacer; X is usually an ether linkage (O) although it may be a carbonyl (CO) moiety; Y is either nonexistent (biphenyl mesogen, Y shown as $-$) or an ester linkage (COO); and Z describes the terminal substituent, usually a cyano (CN) or alkoxy group (OCH$_3$, OC$_2$H$_5$, or OC$_4$H$_9$). R" allows for the possibility that there is one or more of either a CH$_3$, F, or CN substituent in various positions relative to the terminal group, and R"(i) is omitted if there are no adjacent functionalities. Control and characterization of average molecular weight (degree of polymerization, DP) and polydispersity are other important molecular parameters of these materials and can have a dramatic effect on thermal properties of SCPLCs including the LC phase observed (*6, 26*).

The other main class of SCPLC (siloxane-based materials) can be represented as shown in **2**

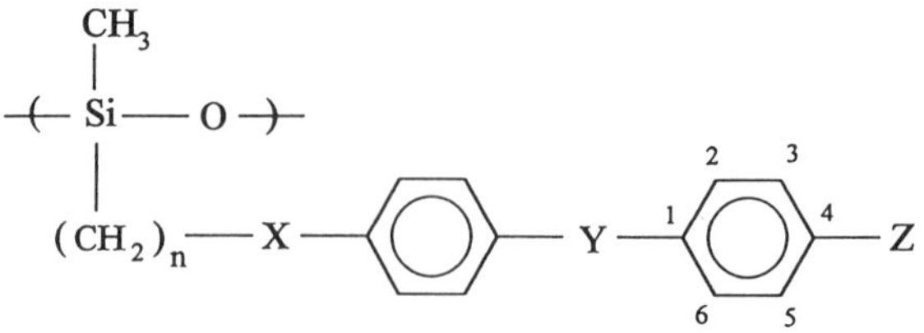

2

and represented by the notation Si/n/X/Y/Z/R''(i), where X is usually an ether or ester linkage, Y is an ester group or the aromatic units are a biphenyl moiety, and Z and R''(i) are as previously defined. DP will be mentioned where applicable and is more readily controlled than in acrylate systems, because the synthesis involves attachment of a side group to a preformed backbone, and DP tends to be between 35 and 50.

An important aspect of SCPLC architecture is the synthesis of copolymers (*5*, *27*, *28*) in both acrylate and siloxane materials. SCPLC copolymers can be divided into two main classes (*28*): backbone copolymers, where the comonomer of the LC monomer would, if homopolymerized, form a non-LC polymer and thus when copolymerized with the LC monomer effectively *dilutes* the liquid crystallinity of the copolymer; or side-chain copolymers, in which homopolymers of both constituents demonstrate an LC phase. In the case of acrylate materials, backbone copolymers mentioned herein involve copolymerization with nonmesogenic acrylate, methacrylate, or styrene monomer units. To obtain a similar result in siloxane systems, a poly(hydrogenmethyldimethyl siloxane) backbone is often used and replaces reactive sites with a concentration of unreactive methyl groups. Backbone copolymers are denoted co-(R/n/X/Y/Z)$_y$ or co-(Si/n/X/Y/Z)$_y$, where y is the mole fraction of the mesogen and that of the comonomer [the nature of which will be stated separately and whose mole fraction is $(1 - y)$]. In the case of side-chain copolymers, the formulas used are co-(R/n/X/Y/Z)$_y$(R'/n/X'/Y'/Z')$_{(1-y)}$ and co-(Si/n/X/Y/Z)$_y$-(Si/n/X'/Y'/Z')$_{(1-y)}$. Despite the wide variety of materials possible, there have been comparatively few synthetic characterization studies of copolymer properties across the composition range, and thus complete sets of dielectric data are limited. Structures of SCPLCs not conforming to the described notations will be shown separately if necessary.

Low-Temperature Relaxations in SCPLCs

The low-temperature, sub-T_g relaxations in side-chain LC polymers will now be discussed, and most data are for acrylate materials of structure 1 and resulting from local motions of side groups in the glassy state. Even though strong agreement exists on the origin of these motions in different systems (particularly where there is a central ester group between the aromatic rings in the mesogenic core), some uncertainty exists with regard to SCPLCs with biphenyl side chains (*2*) and will be discussed with reference to recent data. The relaxations are labeled β, γ_1, and γ_2, in order of decreasing temperature (increasing frequency) using the notation of Zentel et al. (*24*, *29*).

β-Relaxation

The β-relaxation has been ascribed to the internal motion of the central ester moiety residing between the aromatic groups. Zentel et al. (*24*, *29*) found it little affected by the nature and length of the spacer or the terminal substituent in a range of methacrylate and acrylate polymers of the form

$CH_3/n/O/COO/Z$ [Z = OCH_3 or $O(CH_2)_3CH_3$] and $H/n/O/COO/Z$ (Z = OCH_3 and CN) and n = 2 or n = 6 in both systems. The relaxation occurs at about -70 °C (100 Hz) and -35 °C (10^4 Hz), is Arrhenius in behavior, and has an activation energy (E_a) of approximately 50 kJ/mol. Bormuth et al. (30) reported an E_a of about 56 kJ/mol in similarly pendent-attached acrylate ($H/6/O/COO/OCH_3$), chloroacrylate ($Cl/6/O/COO/OC_4H_9$), and an acrylate polymer with laterally attached mesogenic units (31). As with other local motions, the relaxation is symmetric with a broad distribution of relaxation times (Cole–Cole exponent of 0.4) for $CH_3/6/O/COO/CN$ (32). Low-temperature data on siloxane materials of general structure 2 are sparse. Those data that were reported (33, 34) occur at a similar temperature location and a similar E_a(51 kJ/mol).

As discussed by Monnerie et al. (35), motion of the ester group adjacent to the chain can be ruled out due to the higher E_a values and lower frequencies found for poly(methyl methacrylate), where that secondary β-relaxation is due to a chain-adjacent rotational motion (relaxation occurs at about 80 °C at 10^3 Hz, E_a of between 80 and 90 kJ/mol). As for isotropic methacrylate materials with long alkyl side groups, a sub-T_g mobility of the adjacent unit as in SCPLCs tends not to be readily observed (36). ^2H NMR spectroscopy has confirmed that the local motion seen in SCPLCs is due to 180° flips around the long axis of the mesogen (37). Haase and Pfeiffer (38) demonstrated the nature of this motion by synthesizing and measuring the dielectric properties of methacrylate SCPLCs with varying numbers of lateral fluorine substituents on the end ring (which itself contains a pendant F group). The β-relaxation of these materials clearly involves the motion of the aromatic ring to which the lateral F group is attached because it is a higher E_a process (80 kJ/mol) than without it and in other SCPLC systems. Indeed, the β-relaxation of the disubstituted material merges with the lower frequency glass transition.

Recent work by two groups, however, has indicated a greater complexity and structural and environmental dependence of the β-relaxation than previously thought. Gedde et al. (39) synthesized and characterized polyether side-chain systems with long alkyl spacers with the structure shown in 3

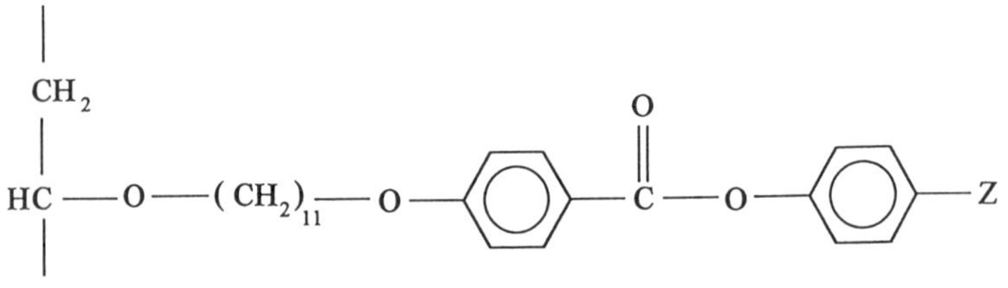

3

where Z = OC_2H_5 or CN. The resultant materials were smectic with high clearing temperatures, and a moderately wide range of β-relaxation E_a values were found depending on structure. The degree of molecular order was greater for the ethoxy-terminated material, and this material showed a greater E_a of the β-relaxation (74 to 105 kJ/mol) compared with approximately 60 kJ/mol for the cyano-terminated material. By calculating the intramolecular energy barrier for the required phenyl motion to be less than 30 kJ/mol, any E_a in excess of this was ascribed to intermolecular associations and variations in the degree of order. This result is borne out by the determination of the Kirkwood correlation parameter, g, to be considerably less than 1.

In other work (*40*), two β-relaxations (β$_1$ and β$_2$, in order of increasing frequency and decreasing E_a) are observed in novel poly(*N*-maleimides) SCPLC materials with two ester linkages and a range of mesogenic units as shown in structures **4–8**.

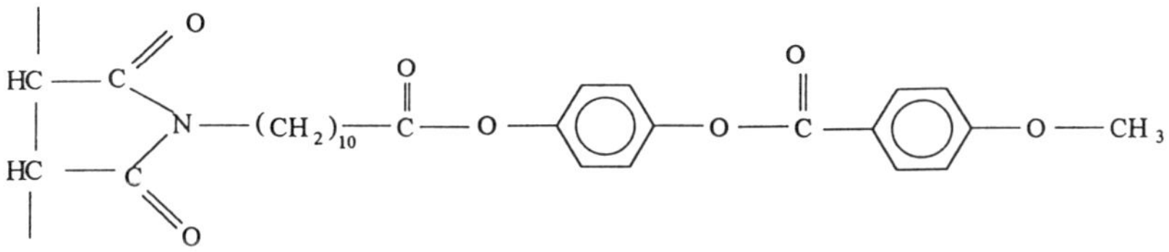

4

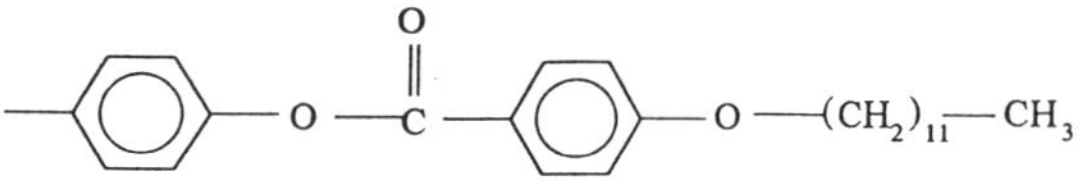

5

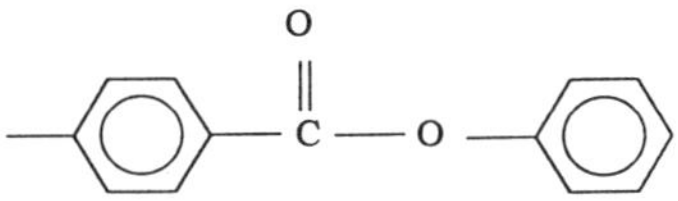

6

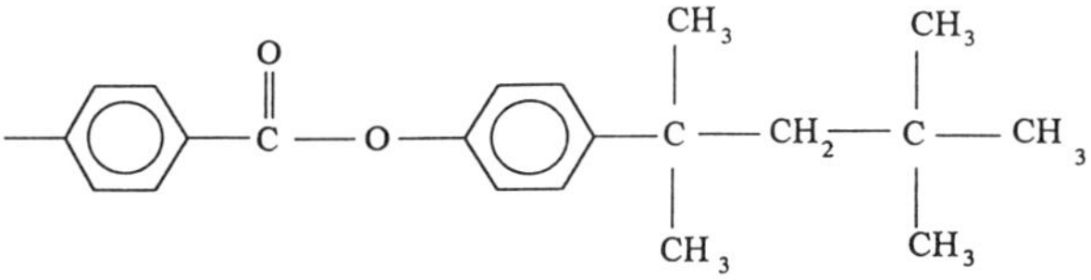

7

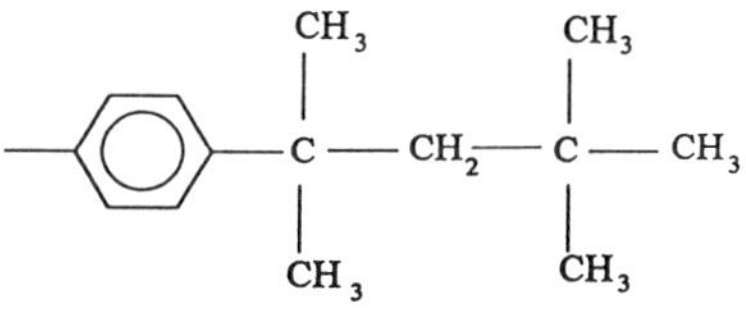

8

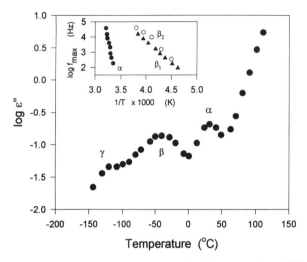

Figure 3. Dielectric loss spectrum of maleimide SCPLC structure 4 at 1 kHz with the frequency–temperature map for some of the relaxations (inset). (Redrawn from reference 40.)

These materials are very interesting because of the nature of fixation of the side chain to the main-chain backbone and because they contain both parallel and anti-parallel arrangement of the two ester dipoles, one of which replaces the more flexible ether group, which often links the alkyl spacer to the rigid mesogen. The nature of the linking ester group in structures 4 and 5 is uncommon in SCPLCs reported to date in dielectric studies and clearly has ramifications for properties such as longitudinal dipole moment. The strength of the β-relaxations correlates with the number of ester units and transverse dipole moments and is independent of the polarity (and hence strength) of the longitudinal dipole moment, basically confirming previous β-relaxation assignments. The dielectric curve and frequency–temperature location are shown in Figure 3 for structure 4, and the E_a values of all the materials are reproduced in Table I. The β-relaxations occur at

Table I. Activation Energies (kJ/mol) of Dielectric Relaxations of Poly(N-maleimides) of Structures 4–8

Structure	γ	β_1	β_2	α	δ
4		59	40	320	$-^a$
5	26	65	40		$-^a$
6		43	24	290	160
7			35	≈300	120
8		$-^a$			130

^{a} Relaxations not observed.

−38.7 °C and −28.4 °C, respectively, and clearly lie on the different E_a lines (Figure 3, inset).

The energy barrier to motion and the relaxation time of the ester group between the phenylene rings (β_1) are similar to those of the β-relaxation of the materials mentioned previously and are higher than those of the ester group between the alkyl spacer and phenylene ring (β_2). The frequency of motion of the relaxation at a given temperature is lower in the case of LC structures 4 and 5 compared with the other materials that are less ordered or amorphous. Clearly, the additional smectic potential barrier reduces motional speed. Even though this result was observed for motions of the mesogen in the LC state (as will be discussed later), it is not usually considered so important in motions in the glassy state. With sufficiently good data and the ability to test samples with a range of related structures, various β-relaxations can be a function of the dipolar orientation, order, and end group effects and not simply invariant to such parameters. No dielectric studies have been done on SCPLC materials of structures 1 and 2 where both X and Y are ester linkages, explaining why reports of more than one β-relaxation in the literature for most SCPLC materials are uncommon.

A dielectric study of a range of directly attached SCPLC materials (no flexible spacer) with multiple phenyl and carbonyl units was reported recently by Nikonorova et al. (*4*), and the materials had the structure shown in 9.

9

A multiplicity of dielectrically active local motions was observed in the glassy state and ascribed to local phenyl and ester mobility. The existence of such a range of local motions was used to explain the ability of the directly attached side chain to counteract the anisotropy of the random coiling main chain and form a mesophase.

γ_1-Relaxation

This relaxation has been ascribed to local motion in the glassy state of the spacer group (provided it is of sufficient length) and shows an Arrhenius behavior ($E_a = 35$ kJ/mol) (*24, 29*). Where the materials were of structures 1 and 2, the relaxation did not occur in side chains with two methylene unit spacers but only those with six spacers (H/6/O/COO/CN), and it was seen at approximately -90 °C at $10^{4.5}$ Hz. Because the alkyl spacer is not itself dielectrically active it implies local coupled motion of adjacent dipoles and appears to be independent of the terminal substituent or the nature of the main chain. In some systems (polymethacrylates) and at higher temperatures in others, it merges with the β-relaxation or has not always been observed (41), such as for (CH$_3$/6/COO/OC$_4$H$_9$).

Gedde et al. (39) found in polyether side-chain systems with a much longer alkyl spacer length of $n = 11$ (structure 3) a γ_1-relaxation with a similar E_a (34 kJ/mol) and temperature location (approximately -115 °C at 10^3 Hz) as discussed previously. As discussed in McCrum et al. (36) and references therein, local motions of higher order polyethers with a (CH$_2$)$_i$O repeat unit (where i is 4 or greater) are correctly predicted to show a local motion at -100 °C (10^2 Hz) and E_a of 40 kJ/mol, these results validating the mechanism assigned to the γ_1 motion. In the poly(N-maleimide) system mentioned previously (40) (structures 4–8), a relaxation at a temperature below that of the β-relaxation is only clearly visible in structure 5 occurring at -122 °C at 10^3 Hz, and the E_a is 26 kJ/mol. In the other materials it is either merged with the lower frequency β-relaxation or the higher frequency γ_2-relaxation (discussed shortly). On the basis of the temperature location of other γ_2-relaxations, this mixture is thought to be of both motions.

γ_2-Relaxation

This relaxation has been assigned to rotational motion of the terminal substituent (24, 29), where it was seen in materials of structure 1 if the terminal group was sufficiently long. It was observed if Z is a butoxy group (O(CH$_2$)$_3$CH$_3$) but not for CN or OCH$_3$. Its E_a is approximately 17–24 kJ/mol (29, 41) and occurs at about -130 °C (10^4 Hz).

To date, no mention has been made of the limited low-temperature data of motions of SCPLC biphenyl materials in the glassy state, such as for materials of structure 1 and R = H and directly linked phenylene rings. These materials should be useful in confirming the stated assignments because, without linking ester groups and its transverse dipole, the β-relaxation strength in particular should be reduced. Moscicki (2) outlined in some detail the confusion that exists in interpretation of the data for these materials. Much of this confusion arises from comparison with the data of Parneix et al. (25) who, with biphenyl SCPLCs, noted two low-temperature relaxations that, at -30 °C, were at 10^3 Hz and 10^7 Hz for H/6/O/–CN. Parneix et al. (25) ascribed the low-frequency motion to relaxation within the spacer, the high-frequency motion to the motion of the terminal CN group, and the lower frequency motion to that of the spacer. However, from E_a curves in Zentel et al. (29) of the β-, γ_1-, and γ_2-relaxations, at

-30 °C the β-relaxation (ester group motion) occurs at $10^{4.3}$ Hz and γ_1 (internal spacer motion) appears at $10^{6.5}$ Hz and γ_2 at $10^{8.7}$ Hz. Thus, the assignations of Parneix et al. (*25*) in their biphenyl systems apparently are at odds with those for acrylate–methacrylate systems. This difference is further compounded by the fact that their low frequency peak assigned to internal spacer motion occurs in the temperature–frequency location of the internal ester group rotation, despite the absence of a central ester group in their sample. Moscicki (*2*) suggested that the differences in motion in the biphenyl system to other materials could be due to different sample preparation or differing modes of motion in biphenyls. Very recent low-temperature data of Zhong et al. (*42*) largely substantiate the experimental results of Parneix et al. (*25*) with dielectric analysis of a biphenyl sample and a slightly different spacer as shown in **10**

$$10$$

where $n = 7$. Zhong et al. (*42*) found a relaxation at about -90 °C (for 1 kHz) and -40 °C (for 10 Hz, E_a of about 40 kJ/mol), not dissimilar to those values of Parneix et al. (*25*) and close to the γ_1- and β-relaxations of Zentel et al. (*29*). Therefore, motions of glassy biphenyl materials, as indeed in all side chains (as shown by the work of Gedde et al. (*39*) mentioned previously), may be more complex than originally thought and dependent on chemical structure and the order that results. Aspects such as antiparallel packing that are invoked in discussions of larger scale motion in the LC state may also have some validity in sub-T_g relaxations. These considerations may be important in application of these materials because any molecular motion that can occur in the glassy state may lead to loss of alignment and thus, for example, loss of information in an optical storage application.

Mathematical Description of Relaxation in LC State

This section will briefly outline the equations describing LC motion that have been reported, and an emphasis will be placed on their ability to aid understanding of the motions involved in SCPLCs and with a view to demonstrating how they can be applied. A more detailed exposition can be found elsewhere in the primary references (and those listed within) 43–47 and other relevant summaries (2, 48, 49).

To describe mathematically LC systems it is necessary to define the mesogenic, rod-like unit in a laboratory frame of reference {X, Y, Z}, where Z is the director (average direction) of the LC phase. Following standard notation (2, 46, 50), a molecular frame of reference {x, y, z} emanates from the mesogen with the z-direction along the mesogens long axis. The two frames of reference are linked by a set of Eulerian angles $\Omega = (\alpha, \beta, \gamma)$ (standard coordinate axis, as drawn for example, in reference 2).

In a monodomain of uniform director orientation the degree of alignment of the mesogen with respect to the director is quantified by an order parameter, S. This parameter quantifies the probability distribution function of the molecules around the director and generally requires a series of spherical harmonic functions to fully define the order (20), although generally the first even member of the harmonic series, a second-order Legendre polynomial, $P_2(\cos \beta')$, is sufficient where

$$S = \langle P_2 (\cos \beta') \rangle = \frac{1}{2} \langle 3 \cos^2 \beta' - 1 \rangle \qquad (1)$$

where β' is the angle between the molecule and director. Values of S are, conveniently, 1, -0.5, or 0 for orientations parallel, perpendicular, or random, respectively, to the director. Maier and Meier (51) extended the polar liquid theory of Onsager to predict the static dielectric permittivity of such a nematic LC monodomain assuming molecules of spherical shape and anisotropic polarizabilities, leading to equations

$$\varepsilon_\parallel^0 = \varepsilon_\parallel^\infty + \frac{G}{3kT} [\mu_\ell^2(1 + 2S) + \mu_t^2(1 - S)] \qquad (2)$$

$$\varepsilon_\perp^0 = \varepsilon_\perp^\infty + \frac{G}{3kT} \left[\mu_\ell^2(1 - S) + \mu_t^2 \left(1 + \frac{S}{2}\right) \right] \qquad (3)$$

where k is Boltzmann constant, T is temperature, and ε_i^0 and ε_i^∞ are the static and high-frequency values of permittivity for directions parallel ($i = \parallel$ and perpendicular ($i = \perp$) to the director, G is a field factor (constant and often approximately 1), and μ is the total dipole moment with the longitudinal component in the direction of the long axis ($\mu_l = \mu\cos \beta$) and the transverse in the direction of the short axis ($\mu_t = \mu\sin \beta$) and $\mu = (\mu_l^2 + \mu_t^2)^{1/2}$ and k and T are the Boltzmann constant and temperature, respectively.

Nordio et al. (52) extended these static equations to describe the *complex* relaxation function of fully aligned, LC monodomains: that is, the mathematical description of $\varepsilon_{\parallel}(\omega)$ and $\varepsilon_{\perp}(\omega)$. This approach assumed a rotational diffusion process that relaxed the dipole parallel and perpendicular to the molecular axis under the influence of a nematic potential.

To describe the situation where the order parameter, S, is other than 1, the autocorrelation function of the molecules $(\gamma_k(t))$, where k represents the kth component of the dipole moment, needs to be determined. The unnormalized autocorrelation function and complex dielectric permittivity are related by the equation

$$\varepsilon_k(\omega) - \varepsilon(\infty) = GL_{i\omega}[-d\gamma_k(t)/dt] \qquad (4)$$

where $L_{i\omega}$ is the Laplace transform, G is the local field factor, and $\varepsilon(\infty)$ is the static dielectric constant. Determination of the autocorrelation function requires knowledge of the probability that a molecule with a certain orientation (as defined by the Eulerian angles) at initial time, t_0, has a particular orientation at time t. The probability function, $P(\Omega, t)$, is assumed to be describable by a linear combination of functions of time and angle such that

$$P(\Omega,t) = \sum_{jlm} b^j_{lm}(t)D^j_{lm}(\Omega) \qquad (5)$$

where $D^j_{lm}(\Omega)$ is known as the Wigner rotation matrix element in the laboratory frame of reference, and b^j_{lm} is the coefficient used to expand the probability distribution function from the Wigner rotation matrix elements, where j refers to the order of the tensor (1 for an electric field), and l and m are $2j + 1$ tensor components. The analytical forms of the matrix elements are shown elsewhere (2). If the electric field is assumed to be in the Z direction (as is the director) then from these functions, the relevant dipole moment components and local order parameter, S, the correlation functions required in the Z and X directions $(\langle \mu_z^*(0)\mu_z(t)\rangle$ and $\langle \mu_x^*(0)\mu_x(t)\rangle$, respectively) can be determined where $\langle \rangle$ denotes a spatial average.

$$\langle \mu_z^*(0)\mu_z(t)\rangle = \frac{1}{3}[(1 + 2S)A_{00}(t)\,\mu_\ell^2 + (1 - S)A_{01}(t)\mu_t^2] \qquad (6)$$

$$\langle \mu_x^*(0)\mu_x(t)\rangle = \frac{1}{3}[(1 - S)A_{10}(t)\,\mu_\ell^2 + (1 + S/2)A_{11}(t)\mu_t^2] \qquad (7)$$

where $A_{ij}(t)$ are linear combinations of the normalized time correlation functions of the Wigner matrices (full equations are given in the appendices of reference 47). If these correlation functions are transformed to the frequency domain by using eq 4 the result is

$$\varepsilon_{\parallel}(\omega) = \varepsilon_{\parallel}(\infty) + \frac{G}{3kT}[\mu_{\ell}^2(1 + 2S)F_{\parallel}^{\ell}(\omega) + \mu_t^2(1 - S)F_{\parallel}^t(\omega)] \qquad (8)$$

$$\varepsilon_{\perp}(\omega) = \varepsilon_{\perp}(\infty) + \frac{G}{3kT}\left[\mu_t^2(1 - S)F_{\perp}^{\ell}(\omega) + \mu_t^2\left(1 + \frac{S}{2}\right)F_{\perp}^t(\omega)\right] \qquad (9)$$

where the form of the equation appears as a generalization of eqs 2 and 3 given earlier. The theory predicts that there are four main modes of motion, $F_{\parallel}^{\ell}(\omega)$, $F_{\parallel}^t(\omega)$, $F_{\perp}^{\ell}(\omega)$, and $F_{\perp}^t(\omega)$ (also known as 00, 01, 10, and 11) and are diagrammatically shown in Figure 4. The strengths of the various relaxations thus are (46): $[\mu_{\ell}^2(1 + 2S)]$, $[\mu_t^2(1 - S)]$, $[\mu_{\ell}^2(1 - S)]$ and $[\mu_t^2[1 + S/2)]$, respectively. In deriving these equations, Attard (43) generalized the theory of Nordio et al. (52), which assumed small-step rotational diffusion. Kozak et al. (45, 46) demonstrated that similar equations can be obtained based on the symmetry of the molecules without making assumptions about the motions. The nature of the four motions is described by the motion around particular angles described by the Wigner rotation matrices. In the case of the directing field parallel to the nematic director, two motions are possible. The variable $F_{\parallel}^{\ell}(\omega)$ involves relaxation of μ_l, which involves variation of the angle β (motion of the mesogen around the short molecular axis) (Figure 4a). Whereas in MLCs this is feasible, it is less easily understood in SCPLCs due to mesogen–polymer backbone connectivity. The motion in SCPLCs is thought to involve

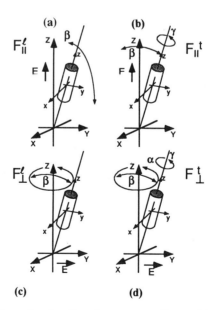

Figure 4. Four main modes of motion of mesogens as discussed in the text. (Redrawn from reference 2.)

limited angular motion that relaxes the longitudinal dipole and is tied in with some degree of chain cooperativity. The second motion, $F_{\parallel}^t(\omega)$, only occurs if the order parameter, S, is less than 1 and there can be a projection of the transverse dipole in the Z-direction (Figure 4b). This motion involves rotation around the mesogens long axis. If the electric field is in a direction perpendicular to the nematic director axis the other two relaxations occur. $F_{\perp}^{\ell}(\omega)$ involves motion or tumbling of the long molecular axis (motion of the β and λ angles), as in Figure 4c and is clearly also not observed if $S = 1$. $F_{\perp}^t(\omega)$ involves reorientations around the short axis of the mesogen (Figure 4d). The frequencies (or relaxation times) of these motions relate to the nature of the motion (52) and the effect of the nematic potential, which allows the speeds of motion to be ranked to some degree that

$$f_{00} < f_{01} \le f_{11} \le f_{10} \tag{10}$$

where f_{ij} represent the frequency locations of the maxima of the various modes. The natures of the motions are such that the motion around the short axis is that motion with the lower frequency and the other motions are somewhat grouped together at higher frequencies. The nematic potential barrier predominantly affects motion around the short axis and has an energy barrier of the order 10–25 kJ/mol (53), slowing down motions by a factor of 100 to the speed of motion if there was no such barrier. The speed of motion may increase slightly with improved alignment.

A further extension of this theory generalizes the mathematical model to the case where not only can the molecular order parameter, S, be incorporated but also a realization that on a macroscopic scale, polydomain samples exist in which the directors of local regions of S can themselves have a degree of domain order (characterized by domain order parameter, S_d, defined similarly to eq 1 although now the angle is between the domain director and electric field). The alignment of the LC domains (and hence the change in S_d) is influenced by application of an electric field. If the probing electric field is in the direction of the Z-axis of the laboratory frame of reference, a geometrical consideration of the relationship between the molecular and laboratory frame of reference 44 in terms of S_d leads to a value of the complex dielectric permittivity of

$$\langle \varepsilon_z(\omega) \rangle = \varepsilon_{\parallel}(\omega)(1 + 2S_d)/3 + \varepsilon_{\perp}(\omega)2(1 - S_d)/3 \tag{11}$$

Substitution of eqs 8 and 9 into 11 results in the following

$\langle \varepsilon_z(\omega) \rangle$

$$= \varepsilon_z^{\infty} + \frac{G}{3kT} \left[\frac{(1 + 2S_d)}{3} \{ \mu_{\ell}^2(1 + 2S)F_{\parallel}^{\ell}(\omega) + \mu_t^2(1 - S)F_{\parallel}^t(\omega) \} \right.$$
$$\left. + \frac{2}{3}(1 - S_d)\left\{ \mu_{\ell}^2(1 - S)F_{\perp}^{\ell}(\omega) + \mu_t^2\left(1 + \frac{S}{2}\right)F_{\perp}^t(\omega) \right\} \right] \tag{12}$$

where

$$\varepsilon_z^\infty = (\varepsilon_\parallel^\infty + 2\varepsilon_\perp^\infty)/3 + 2S_d(\varepsilon_\parallel^\infty - \varepsilon_\perp^\infty)/3 \tag{13}$$

The complex dielectric permittivity of a polydomain sample is made up of weighted components of the four basic motions described earlier where the strength of the relaxation is dependent on S, the two dipole moments, and S_d. Alignment of the sample and thus varying S_d and the mix of these modes results in the appearance of different dielectric loss spectra. In an unaligned polydomain sample ($S_d = 0$) all modes will be present: a lower frequency peak due to 00 motion and the 01, 10, and 11 relaxations grouped together as a broad relaxation in the higher frequency regime. In a homeo-tropically aligned sample ($S_d = 1$) two relaxations will appear (low frequency 00 and high frequency 01), the high-frequency relaxation arising because even though macroscopic domain director alignment in the direction of the Z-axis is perfect, a value of S (the molecular order parameter) may be less than 1 and an inclination of the molecule to the electric field still exists. Conversely, for planar or homogeneous alignment ($S_d = -0.5$) the 10 and 11 relaxations will be observed. As noted by Williams (49, 54), the theoretical analysis summarized previously does not take account of the Kirkwood g factors, which relate to cooperativity and correlation of motions of the dipoles and would influence the measured strength of the observed relaxations.

One of the motivations for this analysis was that it is possible to determine some properties of the materials by using dielectric data and this result has been demonstrated in a number of SCPLCs. For example, it is possible by measuring the intensities of various relaxations in aligned and unaligned relaxations (requiring peak deconvolution), parameters S and the ratio μ_t/μ_l (55, 56) can be determined. Also of interest is the ability of dielectric relaxation spectroscopy (DRS) to determine the degree of macroscopic director alignment induced by electrical fields, for example. This is given by expansion of complex permittivity into its real and imaginary components ($\langle \varepsilon_z(\omega) \rangle = \varepsilon_z(\omega) = \varepsilon_z'(\omega) - i\varepsilon_z''(\omega)$), and insertion into eq 11 leads to

$$S_d = \frac{3\varepsilon_z''(\omega) - (\varepsilon_\parallel''(\omega) + 2\varepsilon_\perp''(\omega))}{2(\varepsilon_\parallel''(\omega) - \varepsilon_\perp''(\omega))} \tag{14}$$

where $\varepsilon_z''(\omega)$ is the experimentally measured dielectric loss of the sample whose degree of director alignment is being determined, and $\varepsilon_\parallel''(\omega)$ and $\varepsilon_\perp''(\omega)$ are the dielectric losses homeotropically and planarly aligned samples, respectively. Clearly this can be done at a number of frequencies and the results averaged. Unlike determination of S, this method requires no curve fitting procedures. An interesting implication of the rearrangement of eq 14 in terms of $\varepsilon_z''(\omega)$ is that a frequency of intersection, ω_1, exists of the homeotropic and planar relaxation spectra $\varepsilon_\parallel''(\omega_1) = \varepsilon_\perp''(\omega_1) =$

$\varepsilon_z''(\omega_1)$, through which all other relaxation curves pass, independent of S_d. This existence of an isosbestic point is also true for $\varepsilon_z'(\omega_2)$, and the intersection occurs at a different frequency, ω_2. This frequency is the "cross-over frequency," important in electrical alignment of a SCPLC by using sinusoidal electrical fields, and it will be discussed more.

Because experimental SCPLC sample thickness is often not easy to determine precisely, a loss-related parameter $(G/\omega) = \varepsilon''C_0$ in picofarads (pF) is often reported in the PLC dielectric literature. Also reported is the capacitance (also in pF) $C_p = \varepsilon'C_0$, where in both cases C_0 is the empty capacitance (*36*) and G/ω is often substituted for ε'' in eq 14. Further experimental aspects of measurement of SCPLC properties are given by Haws et al. (*48*).

Experimentally Observed Dielectric Relaxations in LC State of SCPLCs

Much of the dielectric relaxational work in SCPLC dielectric measurement has concentrated on the motion of the mesogenic units at temperatures when the sample is within the LC or isotropic state. The analysis and understanding of the experimental results were strongly supported by the theoretical models and the description of the modes of motion, discussed previously. The main features of the dielectric loss spectra of the SCPLC materials that will be compared in this review look at the general shape and deconvolution and assignment of motions, the speed of motion (position of the loss frequency maxima or relaxation time at a given temperature), activation energy, and strength of the relaxation. These motions will be examined as a function of chemical architecture (including copolymer structure), temperature, and degree of alignment. Because the molecular mobility observed by DRS is related to that motion that allows alignment in electrical fields as well as loss of alignment, this aspect (dc and ac alignment) will also be discussed.

However, when comparing different SCPLCs, comparison of peak maxima from different samples at the same absolute temperature is often not sufficient. The changing of the nature of the mesogen, for example, usually alters phase stability by varying the glass transition (or melting point) and the clearing temperature. Past work has mainly compared mobility data of different systems either at constant temperature (*57–59*) or at a constant reduced temperature compared to the clearing temperature ($T_{red(cl)} = T/T_{cl}$), where temperatures are in absolute kelvin (*60, 61*). Haws et al. (*48*) stated that in general, relaxation properties such as τ (relaxation time) or f_{max} (position of frequency maxima) should be scaled by the T_g: that is, comparisons between different samples should be made at constant reduced temperature, $T_{red(g)} = T/T_g$, where temperature is in kelvin and measurement temperatures are just above T_g. Static properties related to the orientational order of the LC phase (such as permittivity) should scale better with $T_{red(cl)}$. If scaling of relaxational properties with the glass transition results in overlaying data from different systems, this result may be a good indication that the backbone is dominating such properties. In addition to scaling with temperature, different samples compared at the same reduced temperature may exist in different mesophases (i.e., nematic vs. smectic) resulting in further complications, and this aspect will be discussed in more detail later.

As indicated when introducing structures 1 and 2, most DRS measurements have been performed on SCPLCs with either acrylate (62–91) or siloxane backbones (92–130).

SCPLCs with Acrylate Backbones

Some of the seminal work in this area was performed by Kresse et al. (78, 79, 81), who measured properties of unaligned biphenyl materials, H/5/O/–CN and CH_3/5/O/–CN above the glass transition in the LC state where dielectric relaxations in the 10–10^5 Hz frequency region were observed and ascribed to motion of the polar CN dipole around the short axis of the side chain. The E_a of this motion was greater in the LC than isotropic state, and a slight increase in frequency was observed in going from the LC to isotropic state. Zentel et al. (29) examined a range of unaligned methacrylate and acrylate polymers of structure 1 (R = CH_3 or H) with combinations of varying spacer lengths (n = 2 and 6) and substituent endgroups (Z = CN, OCH_3, O$(CH_2)_3CH_3$). Most materials showed a nematic mesophase although samples with n = 6 or with the longer n-butoxy endgroup also showed an S_A phase. In the acrylates they found two relaxations above the glass transition, a lower frequency relaxation labeled δ, and a higher frequency relaxation (some 10 °C above the static T_g measured by thermal methods) labeled α. The α relaxation was assigned to the glass transition motion and thought to originate largely from the ester group adjacent to the main chain because it demonstrated strong curvature similar to Williams–Landel–Ferry (WLF) curvature indicative of mobility and free volume effects. This motion yielded E_a values of between 150 and 350 kJ/mol. The lower frequency δ relaxation was more visible in the acrylates with a strong longitudinal dipole moment (such as CN), although it was observed in varying degrees in the others and had a lower E_a than the α-relaxation. It was found to occur in the LC and isotropic phase. In the methacrylate materials, although the α-relaxation was observed, the δ-relaxation was only seen at higher temperatures associated with the phase transition and even then was less distinct than in acrylate materials. This relaxation was assigned to 180° motions of the mesogen around its short axis. As indicated by the strong dependence of the δ-relaxation on the nature of the chain backbone (R = H or CH_3, structure 1), its motion is clearly coupled and thus, large-scale, 180° reorientation motions are probably not as possible as they are in MLCs.

SCPLCs with Siloxane Backbones

The other most widely studied class of material has been that of the siloxane SCPLCs (structure **2**). A main structural difference of these materials compared with acrylate–methacrylates is that usually no ester group is adjacent to the main chain. Unaligned spectra of a nematic polymer (*106*) Si/5/O/COO/CN/CH$_3$(6) and a smectic polymer (*55*) Si/8/O/COO/CN/CH$_3$(2) measured in the frequency domain also showed very broad spectra that could be deconvoluted by using Fuoss–Kirkwood functions into two peaks: a narrower, lower frequency peak of a width at half height (Δ) of log f = 1.2 (cf. 1.14 for single relaxation time motion) and a broader high frequency relaxation with Δ approximately log f = 3.2. The high and low frequency relaxations were assigned α and δ, respectively. As with the acrylate materials, the δ-relaxation was assigned to relaxation of the longitudinal dipole moment by a relaxation of the mesogen around its long axis. Because no ester group was adjacent to the siloxane chain (which is relatively nonpolar compared to the side chain), the high-frequency motion was assigned to the motion of the bridging ester group on rotation of the mesogen around its long axis. The connectivity of the mesogen to the main chain is clearly demonstrated by the higher E_a of both motions compared to that of MLCs (approximately double) and a dramatically higher relaxation time (lower frequency of motion) (*131*, *132*).

General Comments

The understanding of the various motions involved in these relaxations is aided by reference to the extension of the Nordio theory outlined in the previous section (*43–49*). The δ process has been identified with the 00 rotation of the mesogenic unit relaxing the longitudinal dipole. It generally has a value for β (Fuoss–Kirkwood parameter) of about 0.8, whereas the α-process is of about 0.4 (*55*). The broad α-relaxation has been associated with motion of the transverse dipole and is a mixture of the 01, 10, and 11 (Figure 4b–d) relaxations, which occur at similar frequencies (*95*). This multiplicity of motions and the range of molecular environments are possibly responsible for the broadness of this peak (*55*). Often broadness is taken as meaning great interchain cooperativity, and indeed, if the α-relaxation is associated with main-chain motion, such cooperativity and broadness would be likely. Williams et al. (*127*) pointed out, however, that the fact that the δ-relaxation is somewhat narrow and close to a single relaxation time width is not *necessarily* indicative of the absence of cooperative motions. Many highly associated liquids such as water, despite moving cooperatively, show close to single relaxation time behavior. Further understanding of the processes involved is aided by the ability of SCPLCs to be aligned in magnetic or electric fields. With reference to eq 12, by aligning materials in either a homeotropic or planar manner and thus altering S_d, the mix of modes present can be changed and has proven useful in siloxane (*33, 47, 55, 56, 95, 97, 102, 119, 127*) and acrylate materials (*31, 66*). Magnetic fields are similarly useful in both siloxane (*34*) and methacrylate systems (*25, 41, 70, 76, 90*). Deconvolution using Fuoss–Kirkwood functions of the dielectric relaxation spectra for $Si/6/0/COO/-CN/CH_3(6)$ material (*47*) demonstrates the various relaxations (Figure 5). Full homeotropic alignment results in a strong 00 peak at low frequencies with a weaker 01 loss peak broadening the relaxation at high frequencies (Figure 5a). Full planar alignment results in the disappearance of the 00 and 01 modes, and only the combination of 10 (lower frequency) and 11 (higher frequency) relaxation should remain (Figure 5b). The 00, 01, 10, and 11 relaxations occur at $\log f_{max}$ of 1.72, 3.68, 2.62, and 3.17, respectively. The strength of the δ-relaxation (00) mode in the homeotropic sample is clearly the strongest of all modes due to the strong longitudinal dipole due to the CN substituent. Determination of the frequency positions of the 10 and 11 motions in planarly aligned material is much harder than discriminating between the 00 and 01 relaxations in a homeotropically aligned sample, because in the latter case, the peaks are well separated in frequency. Intermediate alignments and unaligned samples will produce spectra that are a mixture of the various modes; the 00 mode is the δ-relaxation, and the α-relaxation is comprised of a mixture of 01, 10, and 11 motions. The degree of director order, S_d, can be readily determined from eq 14, provided the dielectric loss is known for the planar and homeotropic alignments at the same frequency. Both the relaxation frequency maximum and the broadness of the various modes should and do remain relatively unchanged by alignment in these systems (*55*).

Temperature–frequency maps of the δ and α peak maxima were reported and had the appearance of that shown in Figure 6 (*43*) for $Si/5/ O/COO/CN/CH_3(2)$, although often the α-relaxation shows a more pronounced WLF-like downward curvature. The linearity of the δ-relaxation may be due to the limits of frequency often used. For example, in a study where a very wide frequency range was employed (10^{-4} to 10^6 Hz) (*100*), the δ-relaxation showed a similar curved dependency to that often seen in α motion. A deviation of the peak position is seen as the sample moves through the biphasic region (Figure 6), and the δ-relaxation becomes faster on being released from the LC potential field. The E_a for δ-process is usually between 120–180 kJ/mol, whereas that of the α-relaxation is approximately 150–350 kJ/mol. The E_a of material in the isotropic state is between 80–120 kJ/mol. This lower value is usually explained by the greater ease of motion

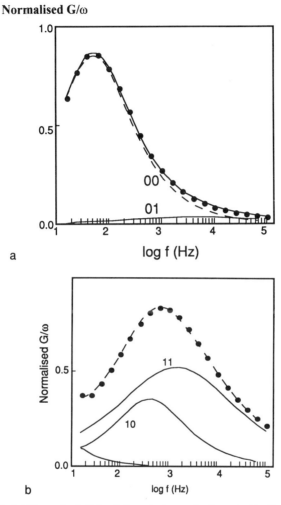

Figure 5. (a) Normalized dielectric loss spectrum of a homeotropically aligned SCPLC deconvoluted into 00 and 01 modes of motion at 34 °C. (b) Normalized dielectric loss spectrum of a planarly aligned SCPLC deconvoluted into 10 and 11 modes of motion at 34 °C. The broken line is 00 peak, low solid line is 01, ● is data, and solid line through ● is the sum of 00 and 01, showing that their sum is similar to data. (Redrawn from reference 47.)

in the isotropic state due to the absence of the nematic potential mentioned previously (*76*).

The motion of laterally attached SCPLCs (Figure 2b) should also give insight into the nature of the relaxation because the motion of the mesogenic unit around its long axis (which forms part of the α motion) would clearly be hindered. Bormuth and Haase (*31*) demonstrated that this result

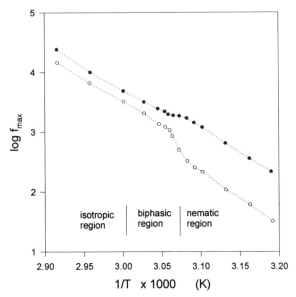

Figure 6. Position of frequency maxima vs. reciprocal temperature for δ (○) and α (●) relaxations of a SCPLC. (Redrawn from reference 43.)

indeed was the case with a laterally attached, triple-phenyl ring mesogen. These materials demonstrated a relatively high value of E_a for the α-relaxation (350 kJ/mol) and a lower relaxation frequency compared with most end-attached comb-shaped PLC materials. Interestingly, the δ-relaxation did not show a markedly different E_a or frequency of motion compared to mesogens attached end-on.

Of relevance to this discussion is the effect of chemical structure on the degree of orientational order and the subsequent effect on dielectric mobility; for example, the effect of smectic versus nematic potential fields on molecular relaxation. Comparing this effect between differing samples is often difficult because a change in phase type usually involves a change in other properties (nature of backbone, spacer length, or molecular weight in addition to the nature of the phase). In this sense, if both phases are accessible in the same material by changing temperature, this effect is the most useful. Pranoto et al. (*34*) found that the smectic E_a is greater than that of the nematic phase relaxation in a copolymer sample that showed both phases, and this result is usually ascribed to the effect of the higher degree of order and the fact that the polymer chains are confined between the smectic layers, increasing the potential barriers. However, Parneix et al. (*25*) had a sample H/6/O/–CN and found E_a values of 126, 193, and 83 kJ/mol for the smectic, nematic, and isotropic phases, respectively. These results indicate that other factors are involved. In a number of MLCs (*133, 134*), the E_a can be lower in the smectic state than the nematic state and has been related to the interaction of the dipoles. Even within the S_A meso-phase for MLCs with strong longitudinal dipoles (such as a terminal CN group), dipoles can align in a number of different ways (monolayer, bilayer, and interdigitated layers (*135*)) and the nature of the alignment is depen-dent on the precise chemical structure of the mesogen (for example, whether or not the liquid crystal possessed a biphenyl or phenyl benzoate core mesogen). This result indicates that the effect of the smectic type may also be important in mobility of PLCs, but detailed studies have not been carried out.

Despite differences in ascribed mechanism of the α-relaxation in silox-anes and acrylates (*136*), what is clear is that motion of the main chain is involved. A clear indication of this involvement is in the case of acrylates where an α-relaxation is observed in biphenyl material (*57*) (H/6/O/–CN) where no lateral dipole exists within the phenyl benzoate group. This result emphasizes that the ester adjacent to the main chain (and hence motion of the polymer chain itself) contributes to the α-relaxation in those materi-als. The additional contribution of the transverse dipole to the α-relaxation is demonstrated by the influence on relaxation time when comparing (H/6/O/–CN) to a sample (H/6/O/COO/CN) where the phenyl rings are bridged by an ester group (*57*). Bormuth and Haase (*67, 68*) sought to quantify the effect of polar groups located on or near the main chain in eqs 8 and 9 by adding them in as an additional contribution. A further indicator of side-chain–main-chain connectivity is demonstrated by the ef-fect of pressure on the δ- and α-relaxations (*77, 118*) where the effect is similar on both relaxations. The lower frequency, higher temperature δ-relaxation, which involves larger scale motion, may have been expected to have a different dependence on free volume than the α-relaxation, but this result was not found.

Aspects of Electric Field Alignment of SCPLC Materials

SCPLC materials are of technological interest because they retain the prop-erties of polymers whilst also having the ability to be readily aligned in the LC state. The anisotropic LC order provides some opposition to thermal randomization, both allowing achievement and maintenance of alignment, even above the glass transition for a period of years (*107, 124*). Alignment can result from application of dc voltage, and the alignment is usually un-changed until a certain threshold voltage (U_0) is achieved, after which the director rotates. If a voltage U is applied, both a rise time and a decay time (after turning off the voltage) can be measured by monitoring the variation in intensity of the birefringent light (*137*), and the rise time (t_r) and decay time (t_d) are the time to change from 90% to 10% of initial intensity or vice versa, depending on which value is being measured. The relevant equations (for rise time) are

$$\frac{1}{t_r} = \frac{1}{t_d^0}\left[\left(\frac{U}{U_0}\right)^2 - 1\right]$$ (15)

$$\frac{1}{t_d^0} = \left(\frac{\pi}{d}\right)^2 \frac{k_{ii}}{\eta_i}$$ (16)

$$U_0 = \pi \left(\frac{k_{ii}}{\varepsilon_0 \Delta\varepsilon}\right)^{1/2}$$ (17)

where t_d^0 is the passive decay time, which is a function of cell thickness, d (usually 5–10 μm), and LC elastic constant and viscosity (k_{ii} and η_i, respectively). The variable k_{ii} represents splay (k_{11}), bend (k_{22}), and twist (k_{33}) elastic constants that quantify the curvature deformations occurring in LC systems that are being aligned. Rather than absolute values, usually a low ratio of k_{33}/k_{11} is desired in switching cells. The threshold voltage is a function of both the elastic constant and the dielectric anisotropy ($\Delta\varepsilon = \varepsilon_\parallel - \varepsilon_\perp$). Even though rise times are of the order of 100 ms for MLCs, they are often in the order of between 0.1 to 10 s, dependent on system and temperature in SCPLCs (86, 108), and better values are close to the clearing point. The value of t_d^0 is found to be significantly greater for acrylate than siloxane-backbone material (measured at the same temperature distance below the clearing point). Because elastic constants are similar for both MLCs and SCPLCs (138), the polymer backbone viscosity is significantly greater and reduces response time (depending on the distance the test temperature is above T_g). Threshold voltages should be of the same order as for MLC because, as seen from eq 17, they depend on k_{ii} and $\Delta\varepsilon$, which can be manipulated by altering mesogen chemistry. As with MLCs, these voltages tend to be of the order of 5 to 50 V, and the threshold voltage is affected, for example, by alkyl spacer length (86). Ujiie et al. (137) illustrated these effects in some detail for a range of spacer lengths and backbones at constant reduced temperature, $T_{red(cl)}$.

A relationship between coupled backbone and mesogen motion from alignment measurements and variable frequency dielectric studies was demonstrated by Birenheide et al. (66), who showed that the E_a of the temperature dependence of viscosity, η_i, was similar to that of the δ-relaxation, motion around the mesogenic groups short axis. Indeed, η_i from switching measurements is similar to that of viscosity from rheometric shear measurements (66, 130). As with MLCs, electrohydrodynamic instabilities occur, especially with conductive impurities, and this result may affect the efficacy of alignment with dc voltages and low-frequency ac fields (136, 139, 140).

Two-Frequency Addressing

As mentioned earlier, much use has been made of the ability to align SCPLC materials by ac voltages. Most of this work has been done by the group of Williams, Attard, Araki, and co-workers (*33, 54, 119*) and others (*73, 124*). Much of the interest lies in the ability to alter these materials by two-frequency addressing. That is, by altering the frequency of the applied voltage (approximately within the 20–100-kHz range), alignment of the mesogens perpendicular or parallel to the capacitive cell plates can be induced. The critical frequency, f_c, above and below which the orientation induced differs, is known as the cross-over frequency. This phenomenon, also known as the dual-frequency effect, is well known in MLCs (*141*), often for its ability to improve switching response times by allowing changes of frequency, rather than voltage. However, f_c usually occurs at much higher frequencies (in the mega-Hertz region for cyano biphenyls, for example (*142*)) and cannot always be readily accessed. (*Note*: some MLC mixtures can have quite low cross-over frequencies (*143*).)

The effect in both MLCs and SCPLCs arises because of the change in dielectric anisotropy, $\Delta\varepsilon$, with frequency. If $\Delta\varepsilon$ is positive at low frequencies (as it is for many MLCs and PLCs with a strong longitudinal dipole, such as due to a terminal CN group), the long molecular axis will align with the electric field (homeotropic alignment). At high frequencies $\Delta\varepsilon$ becomes negative and the long axis orients perpendicularly to the field (planar alignment). The cross-over frequency, then, is determined simply by determining the point at which the real part of the complex dielectric permittivity of a planarly aligned and a homeotropically aligned sample intersect, for a given measuring temperature. Figure 7a shows this phenomenon for sample Si/6/O/COO/CN/CH$_3$(2) at 34 °C where the cross-over frequency is $\log f_c = 1.8$. SCPLCs with somewhat different degrees of alignment are necessary if only the position of the cross-over frequency is sought. As mentioned previously, the corresponding isosbestic point for the loss curves at the same temperature (Figure 7b) is different, and for this sample and temperature it is about $\log f_c = 2.8$. Even if the transverse dipole itself is weak there still may be a small cross-over effect, as demonstrated by Zhong et al. (*42*) for an SCPLC material with biphenyl side chain (structure **10**).

Because of the high viscosity of the main chains to which the mesogens are coupled, alignment generally has to occur when the viscosity is low. This alignment most commonly involves cooling the sample very slowly from the isotropic to the LC state through the biphasic region with the voltage on at the frequency required. The biphasic region can be broad due to contaminants, impurities, and the molecular weight distribution in SCPLCs (*101, 112, 144*). The alignment process occurs as discrete regions of LC material within an isotropic matrix become aligned with the amount of aligned material increasing as the sample is cooled. Evidence obtained

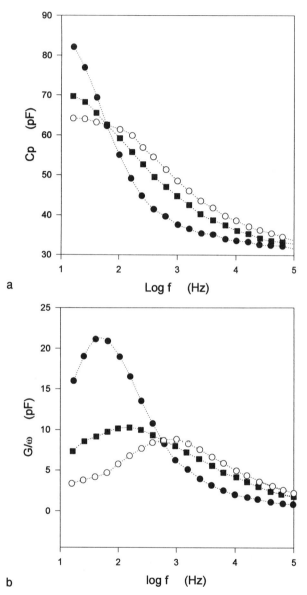

Figure 7. (a) Capacitance as a function of log frequency in the LC state for a homeotropically aligned (●), planarly aligned (○), and unaligned (■) SCPLC at 34 °C. (b) Dielectric loss as a function of log frequency in the LC state for a homeotropically aligned (●), planarly aligned (○), and unaligned (■) SCPLC at 34 °C. (Redrawn from reference 98.)

by monitoring the dielectric relaxation spectra suggests that the regions of material can retain their homeotropic or planar nature if an aligned sample is partially raised into the biphasic region (*92*). This sample acts as a template and encourages the previous alignment on recooling, which is the degree of recovery depending on the number of thermal recycles, cooling rate, and how far into the biphasic region the sample is heated (*93*).

Alignment in LC State

Also, some samples can be aligned whilst in the LC state, sometimes relatively far below the clearing point. This result has been achieved in a range of materials such as copolymers (*128*), a malonate polymer (*113, 129*), and a laterally attached siloxane SCPLC (*127*). In the case of the copolymer and laterally attached polymers, the easier alignability is thought to be due to poorer packing than that in homopolymers (*128*). In some of these materials (*127, 128*), samples thus aligned may slowly disalign with time. Generally, alignment in the LC state is not readily achieved in homopolymers, although some commercial SCPLCs (*7*) such as $CH_3/10/COO/-CN$ and $H/5/COO/-CN$ align at temperatures approximately 10 °C below the clearing point, although they must be rapidly quenched to room temperature (below their glass transitions) to maintain this alignment.

The ability to align materials in the LC state allows for the possibility of in situ dielectric monitoring of dc alignment at least, because many dielectric bridges allow a dc biasing voltage to be placed on a sample concurrent with obtaining its dielectric spectra. If a measurement of dielectric loss is made at a given frequency (or a full spectrum is obtained) as a function of time with the voltage on and the spectra of the homeotropic and planarly aligned materials are known, eq 14 allows S_d to be determined as a function of time and voltage. Because of problems of conductivity and potential electrohydrodynamic instabilities, low degrees of homeotropic alignment usually result compared with ac alignment. In the main kinetic electrical alignment study reported to date (*114*), alignment achievable by the dc field was only slightly lower than that from ac fields. The material used was a malonate side-chain material with a chiral group in the backbone. The kinetic alignment curves demonstrate two stages: an initial rise followed by a plateau, the magnitude of the plateau dependent on the voltage applied. Kozak et al. (*114*) applied the equations of Martins et al. (*145*), an extension of the Leslie–Ericksen continuum theory of the LC state, to the data. These equations contain parameters such as the electric torque applied by the dc field, conductivity, elastic forces of the LC state (primarily bend and splay), and rotational viscosity. The competition among elastic forces, conductivity, and the electric torque determines the magnitude of the plateau (level of alignment achievable). The viscosity relates to the time to attain the plateau. On removal of the dc voltage, some recovery of the chiral structure (and hence of the relaxation loss peak) was observed.

The temperature dependence of the cross-over frequency and the frequency location of the δ-relaxation are parallel with similar E_a values (*119*), demonstrating the connection between the mesogenic motion that leads to alignment and that related to the dielectric relaxation of the mesogen. Chemical structure may also affect the cross-over frequency, although little work has been done on this aspect. A recent study has shown that increasing the flexible spacer length has the same effect as increasing temperature, that is, increasing the cross-over frequency (*111*). A range of other practical issues relating to alignment such as dc conductivity and instabilities (at low alignment frequencies) and dielectric heating (especially if samples are being aligned near the cross-over frequency and the relaxations are prominent) was reviewed elsewhere (*119, 120*).

Effect of Polymer Architecture on SCPLC Dielectric Relaxation

To date, a number of the results presented have indicated that the motion of the decoupled mesogen is indeed linked to the motion of the polymer backbone. The theme of connectivity in both the α- and δ-relaxations is further apparent if the effect of chemical structure on dielectric relaxation of SCPLCs is examined.

Molecular Weight

A number of studies have examined the effect of molecular weight on motions of the side chain (*57, 60, 78*). In general, the effect is not great with regards to the E_a except at the low molecular weight end where there is a decrease from 144 to 140 to 126 kJ/mol for the δ-relaxation in H/6/ O/COO/CN for degrees of polymerization of 150, 21, and 14, respectively. The increase in speed of the motion with decreasing molecular weight is dramatic for the δ-relaxation (3.8, 28, and 32 kHz) and less so for the α motion (0.63, 2, and 2.5 MHz). Direct comparison is straightforward in these systems because the glass transition varies little with increase in molecular weight. An increase in motional speed with decreasing molecular weights in methacrylate systems over a wide temperature range, including within the isotropic phase, has also been observed (*60*).

Polymer Backbone

The effect of the type of polymer backbone on mesogenic motion was mentioned previously by Zentel et al. (*29*), who observed a much more separated α- and δ-relaxation in acrylate, as opposed to methacrylate backbone samples. Bormuth and Haase (*68*) examined the motions of copolymers of acrylate and methacrylate mesogens (R/6/O/COO/CN). Greater asymmetry (broadening and high frequency skewing) was found with greater methacrylate content, and this broadening was monatomic with even a small dilution of methacrylate comonomer changing mobility. Zentel et al. (*70, 130*) compared the effects of an acrylate with the polar, bulky chloroacrylate backbone with a similar terminal substituent (H/6/O/COO/OCH$_3$ and Cl/6/O/COO/OC$_4$H$_9$). The E_a value is greater for the δ-relaxation of the chloroacrylate than the acrylate (220 kJ/mol vs. 140 kJ/mol, respectively). The strength of the δ-relaxation was increased and that of the α-relaxation decreased due to the orientation of the dipole moment from the Cl group. As with the methacrylate backbone, the chloroacrylate material shows a broadened δ-relaxation, indicative of a stiffer backbone. This interpretation is reinforced by rheological results with all three systems (acrylate, methacrylate, and chloroacrylate) and the ascending E_a values, respectively (*130*).

Spacer Length

Spacer length is one of the attributes of SCPLC materials that are often varied to influence LC properties. In general, very short spacers do not allow sufficient decoupling for liquid crystallinity to occur, although it is still possible (*5*). On increasing the length of the spacer group, the LC phase generally becomes stabilized with the appearance of smectic phase, an increased clearing point, and a slightly lower T_g. Dielectric studies with varying spacer lengths have been performed on both acrylates (*25, 29, 30, 41, 63, 64, 86, 91*) and siloxane materials (*58, 61, 111, 125*). The general effect is that of speeding up of the molecular motion with increased spacer length. This effect is demonstrated by analysis of the data of Kresse et al. (*58*), who compared Si/n/COO/–CN where n = 4 and n = 10. At similar reduced temperatures in the LC state the relaxation times are 0.013 and 0.008 s for n = 4 and n = 10, respectively, and the n = 10 sample has a lower E_a, approximately two-thirds that of n = 4. Similar results were observed in other siloxane materials with spacers of quite different chemistry (*111*). Parneix et al. (*25*) found that E_a in the isotropic state decreased monotonically (as did relaxation time at the same $T_{red(cl)}$) with spacer length in an acrylate system with a biphenyl mesogen H/n/O/–CN and n = 3–6. Activation in the nematic state was more complex, decreasing from n = 4 to n = 5 and increasing for n = 6. Such a complexity with length is also seen in the work of Kim et al. (*111*) and Simon and Coles (*125*) and has been variously ascribed to effects such as increased decoupling, greater resistance to motion for much longer spacer chains, and the odd–even effect. The odd–even effect in LC polymers relates to the fact that spacers with even numbers of bonds tend to be able to take up rather more linear conformations (and hence show greater LC order) than those with an odd number.

The shape of the relaxation peaks has also been seen to change as a function of spacer length, and the δ-relaxation becomes broader for shorter spacers (30). The use of a novel vinylacetic acid spacer type by Kim et al. (111, 112, 146, 147) of the general form shown in 11

$$\text{H}_3\text{C} \longrightarrow \underset{\underset{\text{O}}{|}}{\overset{|}{\text{Si}}} \longrightarrow (\text{CH}_2)_3\text{COO}(\text{CH}_2)_n\text{O} \langle\bigcirc\rangle \longrightarrow \text{COO} \longrightarrow \langle\bigcirc\rangle \longrightarrow \text{CN}$$

11

was also found to change the appearance of the relaxation peaks of siloxanes with structure (n = 3, 6, and 8). The relaxation spectra showed clear separation between the δ and α peaks (compared with broad peaks if an alkyl spacer was used (111)), even in unaligned samples. This result indicated that the very flexible spacer resulted in decoupling of the motion of the longitudinal and perpendicular dipolar motions. This separation was accentuated for longer spacer lengths (n = 8 and greater). Zhong et al. (91) noted anomalous dielectric behavior generally in samples with very long spacer lengths (structure 10, n = 11) and suggested that in such cases phase separation may influence SCPLC phase morphology and mobility. The length of the spacer chain can also affect alignment properties. The length of spacer units can affect rise times and threshold voltages in dc alignment applications (86, 88), as well as onset of turbulence (electrohydrodynamic instabilities) (124) with shorter spacer lengths requiring higher threshold voltages and voltages at which turbulence in the sample commences.

Nature of Bridging Group

Most SCPLCs characterized by dielectric relaxation spectroscopy (DRS) have contained mesogen with biphenyl moieties or the aromatic units are linked by an ester group. Parneix et al. (25) compared two such materials (H/6/O/–CN and H/6/O/COO/CN) and found that the second sample with the phenyl benzoate moiety had a lower glass transition and clearing point and demonstrated only a nematic phase, whereas the biphenyl showed a nematic, smectic, and reentrant nematic phase. The E_a of the δ-relaxation was higher in the nematic state for the biphenyl material than the phenyl benzoate (193 vs. 149 kJ/mol), although this order was reversed for motions in the isotropic state (83 vs. 110 kJ/mol). This result can possibly be explained by better alignment, higher order, and thus a greater nematic potential to overcome in the nematic state for the biphenyl, whereas in the isotropic regime, the bulkier phenyl benzoate unit has a greater E_a barrier to overcome to move. Other data on a similar system showed a similar (albeit less strong) trend (*66*) or the trend was not observed at all (*76*). Haws et al. (*61*) found the reverse effect with siloxane backbone material.

Kresse et al. (*117*) recently reported on the effect of reversing the bridging ester (Si/4/COO/COO/OCH$_3$ and Si/4/COO/OOC/OCH$_3$), which has implications for the stability, motion, strength of relaxation, and the nature of the resulting mesophase (*5*) (in this case, the second material has a slightly higher clearing point). The difference between E_a of the δ- and α-relaxations in the LC state of the two was not found to be strong (although this difference is complicated because the material with the OOC bridging group showed a more ordered smectic mesophase and a higher clearing point). Mainly, local motions in the glassy state seem more affected, and faster motion of the OOC than the COO group leads to its phase stabilization.

Other forms of isomerism of bridging groups have been examined. Zhong et al. (*91*) indicated a significant influence of constitutional isomerization of a bridging methoxy–hydroxy-α-methylstilbene on the dielectric relaxation functions, and this influence results in differing appearances of the spectra due to different degrees of homeotropic and planar alignment and due to such isomerism affecting the dipole moments. Recent work by Wendorff and Fuhrmann (*89*) involved in situ dielectric measurement of the motion of an SCPLC mesogen of a trans–cis isomerization of bridging azo groups by illuminating films with the appropriate wavelength light. The cis isomer is more bulky, decreases the perfection of packing, lowers the clearing point, and speeds up the δ motion, although the E_a is not significantly altered.

Terminal Substituent

The nature of the terminal substituent may influence the dielectric spectra in the LC state in a number of ways. Clearly, greater polarity may increase the relaxation strength and magnitude of loss peaks. However, greater immobility due to greater order in the mesophase could also result, as could a decrease in the effective dipole moment should antiparallel arrangements such as CN groups cancel each other. Haws et al. (*61*) found a lower E_a for the δ motion of a mesogen with a terminal CN group compared to an OCH$_3$ moiety, although this result was reversed for the α-relaxation. Parneix et al. (25) also found a greater E_a for the δ motion in the LC region of acrylate SCPLCs with CN terminal substituents, although the order was reversed in the isotropic state. Vallerien et al. (41) and Haws et al. (61) both found faster relaxation for side chains with CN terminal groups than for those without. It seems that considerations of size and resultant steric hindrance dominate the effect of the terminal substituent, as opposed to effects of altering phase stability and order.

Copolymerization

Thermal and mesophase dependence of copolymer composition across the whole copolymer range is scarce and reports of dielectric relaxation of these materials are even more so. The dielectric properties of the first type of copolymers, *backbone copolymers*, have been studied for siloxane copolymers by Haws et al. (61). They measured properties of homopolymer Si/8/O/COO/CN/CH$_3$(6) and a copolymer, co-(Si/8/O/COO/CN/CH$_3$(6))$_{0.5}$. For a constant $T_{red(cl)}$ = 0.92 as calculated by Haws et al. (61), the α-relaxation had a faster speed of motion in the copolymer than the homopolymer, whereas the reverse was true for the δ-relaxation. However, by reanalyzing other data of Attard et al. (95) in conjunction with the work of Haws et al. (61), the copolymers can be compared at constant $T_{red(g)}$, which is the favored way of comparing such relaxational properties. If this process is done at $T_{red(g)}$ = 1.16, the speed of motion is greater in both instances for the copolymer, as expected: the δ-relaxation occurs at log f = 2.75 and 3.3 for the homopolymer and copolymer, respectively. The E_a is lower for the copolymer, also indicating the effect of a diluted LC phase and a more flexible main chain (T_g of the homopolymer is about 1 °C compared with − 15 °C for the copolymer), the effect of which is transmitted to the motion of the side groups. Carr et al. (72) and Al-Ammar and Mitchell (62) also noted that in methacrylate and acrylate dilution copolymers, the loss in order from non-LC copolymer units due to disruption to LC order (and also reduced antiparallel packing) allows easier poling and lower threshold voltages in electronic applications.

In a number of dielectric studies of siloxane-based side-chain copolymers, where only one composition is reported, usually only one δ-relaxation is observed, a result indicating that the different copolymers randomly placed on the chain are moving cooperatively (rather than separately) (33, 110, 119). Detailed work has also been done on side-chain copolymers with siloxane backbones (59, 121). In the work of Scheuermann et al. (59) the copolymers were of the form co-(Si/6/O/COO/CN/CH$_3$(4))$_y$(Si/4/O/COO/OCH$_3$)$_{(1-y)}$ where y equals 0, 0.25, 0.5, 0.75, and 1. The T_g is constant across the composition range (≈11 °C), and the clearing point monotonically decreases with increasing content of the cyano-terminated comonomer. Interestingly, the strength of the dielectric relaxation at $T_{red(cl)}$ =

0.97 shows a maximum with $y = 0.75$ due to disruption of the antiparallel packing of the CN-terminated mesogen.

Novotna et al. (*121*) examined the effect of copolymerization of a siloxane copolymer with quite different side chains shown in **12** and **13**.

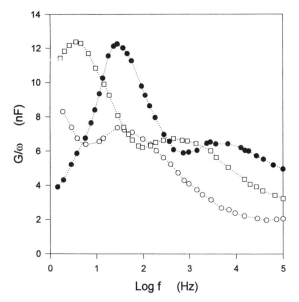

In this work two relaxations were observed and ascribed to the motion of the different constituent side chains, and these relaxations are shown in Figure 8 for the copolymer with 20 mol% of **13** units. The homopolymer of group **12** does not show liquid crystallinity, whereas that with structure

Figure 8. Dielectric loss maxima as a function of frequency of a siloxane-backbone SCPLC based on side-chain structures S12 (80 mol%) and S13 (20 mol%) at 36 °C (○), 45.4 °C (●), and 55.3 °C (□). (Redrawn from reference 121.)

13 is nematic with a high clearing point. The copolymer is liquid crystalline, provided that there is greater than 10% mesogen **13** content. Even in the isotropic state, a broadened relaxation indicative of multiple processes is observed where (surprisingly) it is clear that the higher frequency motion relates to mesogen **13**, the larger, bulkier of the two. In the smectic phase, a homeotropically aligned two-frequency spectrum is observed, and the lower relaxation is clearly that of the more polar, but shorter, mesogen **12**. Unfortunately, data are limited at higher copolymer compositions due to problems with dc conductivity.

The complexity that underlies these results has been very recently emphasized in dielectric characterization (*148*) of a range of novel methacrylate copolymer SCPLC materials with siloxane moieties in the spacer and copolymers of double and triple phenyl group side chains. The synthesis and physical characterization of these materials were reported elsewhere (*149, 150*) and samples have been copolymerized and tested across the entire composition range. The homopolymer of the triple-core mesogen material has an LC phase, whereas that of double-core mesogen is isotropic above its glass transition (the flexible siloxane spacer units requiring a very high degree of rigidity for the LC state to occur). Liquid crystallinity is observed in copolymers containing approximately 50 mol% or less of the double-core mesogen. The dielectric spectra of the LC copolymers with low contents of the double-core mesogen show very complex, multi-peak spectra that are indicative of a combination of molecular motions, and this complexity disappears at higher double-core mesogen dilution. The E_a of the δ-relaxation of this copolymer series across the composition range shows a positive deviation of E_a from a rule of mixtures line between that of the homopolymers, a result indicating that in this system, the mixture of different mesogens actually inhibits their motion. The area of dielectric mobility of side-chain copolymer clearly warrants further research.

Other Side LC Polymer Systems Examined by DRS

Other interesting and often novel SCPLC systems have been examined by DRS. These systems include discotic SCPLCs (*151, 152*), crossing mesogens (*153–155*), swallow-tailed mesogens (*156–159*), main-chain–side-chain mesogens (*160–163*), Y-shaped mesogens (*164*), blends of MLCs with SCPLCs (*165–168*), cross-linked thermoset SCPLCs (*169–171*), and polymer-dispersed LC systems (*172*). Further discussion of these systems is beyond the scope of this chapter and will be presented elsewhere (*173*). In most cases, dielectric relaxation is able to give useful structural information about the mobility of the mesogens, which can be important in helping design and modify existing structures (i.e., when synthesizing main-chain–side-chain materials, as well as the many molecular variables discussed previously in the sections on molecular architecture, the choice of attaching the side chain to either the flexible or rigid section of the main chain also arises). In the case of blends of SCPLC–MLC materials (often done to improve response times of the SCPLC materials by lowering viscosity) there is the additional possibility of influencing the miscibility and mobility of the blend (*174*), for example, by changing chemical structure of both MLC and SCPLC, changing the SCPLC spacer length and glass transition, or mixing MLCs with dilution SCPLC copolymers.

Dielectric Relaxation of Main-Chain Polymeric Liquid Crystals

Comparatively less DRS has been performed on MCPLC materials. The dielectric properties of the various types of materials that have been examined are divided into two main categories for the purpose of this review: rigid MCPLCs (*175–187*) and semiflexible MCPLCs (*188–197*). The common link is that the vast majority of these materials are copolyesters, usually containing a combination of two or more units. Some of the most often used groups are shown in Figure 9, and they are used in the majority of commercial or patented MCPLC systems.

Rigid MCPLCs

The most dielectric work of materials in this class has been done on HBA/HNA copolymers (HBA is *P*-hydroxybenzoic acid and HNA is 2-6-hydroxynaphthoic acid): One of which, HBA73/HNA27 (73 mol% HBA units, copolymerized with 27% HNA units) represents the most widely commercial and studied resin, Vectra A950 (Hoechst-Celanese, United States). This material has a relatively weak T_g of about 100 °C and a small melting point at about 275 °C ($\Delta H_m \approx 1$ J/g) (no clearing point or decomposition point occurring before the isotropic region). Although early variable-temperature DRS work with this material only reported two relaxations (*177, 182*), more recently workers have observed three relaxations (*176, 179, 180*). These relaxations are listed in order of decreasing temperature with approximate values of position and E_a values: α (170 kJ/mol, 100 °C at 10 Hz), β (25 kJ/mol, 20 °C at 1 Hz), γ (12 kJ/mol, −50 °C at 1 Hz). These relaxations are shown from the results of Alhaj-Mohammed et al. (*176*) for a number of frequencies in Figure 10 (all three relaxations are seen best at 10 Hz).

The most prominent relaxation is that of the β motion. Because an increase in the amount of HNA (HBA30/HNA70) increases the strength of the β-relaxation and moves it to higher temperatures (lower frequencies) (*176, 181*), the relaxation has been assigned to the motion of carbonyl

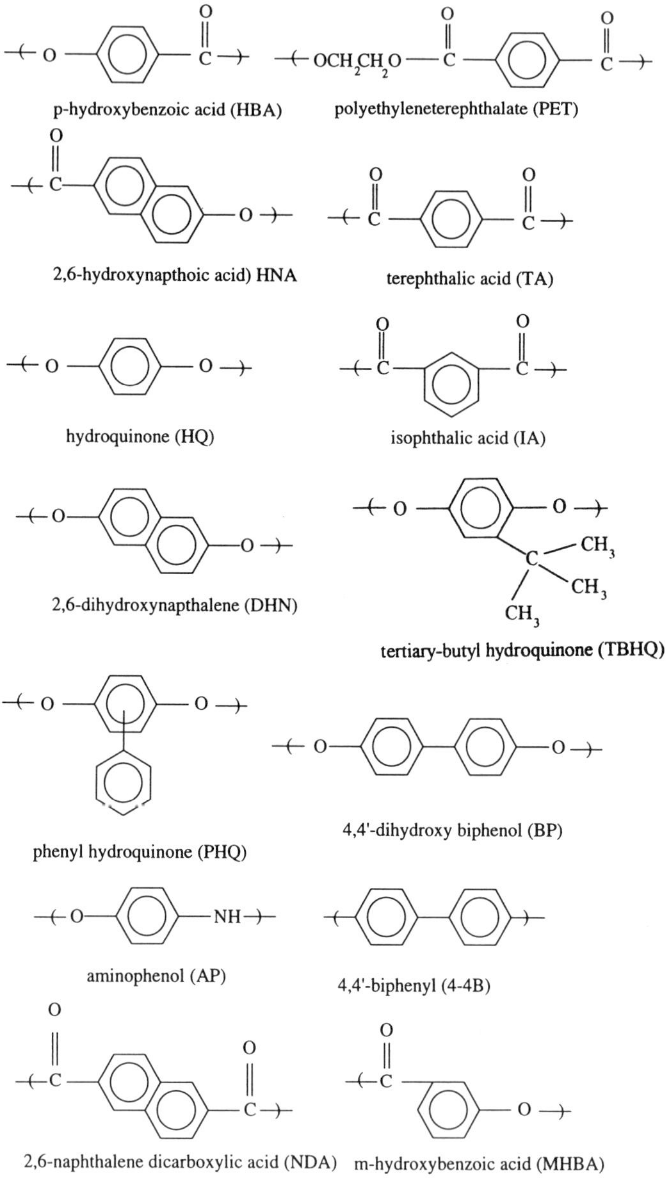

p-hydroxybenzoic acid (HBA)

polyethyleneterephthalate (PET)

2,6-hydroxynapthoic acid) HNA

terephthalic acid (TA)

hydroquinone (HQ)

isophthalic acid (IA)

2,6-dihydroxynapthalene (DHN)

tertiary-butyl hydroquinone (TBHQ)

phenyl hydroquinone (PHQ)

4,4'-dihydroxy biphenol (BP)

aminophenol (AP)

4,4'-biphenyl (4-4B)

2,6-naphthalene dicarboxylic acid (NDA)   m-hydroxybenzoic acid (MHBA)

Figure 9. Some of the chemical units found in commercial MCPLC copolyesters.

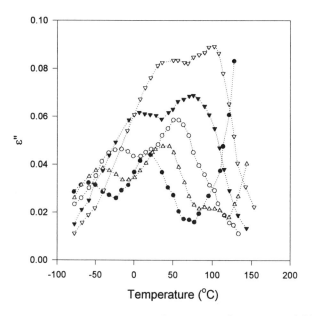

Figure 10. Dielectric loss as a function of temperature for a commercial MCPLC, HBA73/HNA27, for 1 Hz (●), 10 Hz (△), 100 Hz (O), 1 kHz (▼), and 10 kHz (▽). (Redrawn from reference 176.)

groups associated with the HNA moiety. (This result is shown in the temperature–frequency map in Figure 11. Note that the α- and γ-relaxation positions remain unchanged with different HNA content.) The assignation as the β-relaxation has also been confirmed by its absence in related systems with no HNA groups (*180*). Blundell and Buckingham (*177*) proposed that the β motion involves rotation around the oxygen aromatic bonds that are in the same direction as the main chain, this being the lowest energy barrier to such a motion. Green et al. (*180*) demonstrated by comparison between dynamic mechanical and dielectric data that the motion of the HNA unit was dominated by the naphthalene unit itself, rather than the connecting linkages. This result was shown by comparing the dynamic mechanical spectrum of a copolymer with 2,6-dihydroxynaphthalene (DHN) (as opposed to HNA), which is not directly attached to carbonyl units. The DHN unit demonstrates similar mechanical relaxational energetics to those of HNA. In work in which the relaxational properties of HBA/HNA have been determined at the same frequency by using dielectric and mechanical techniques (*184*), the β-relaxation occurs at temperatures some 15 °C greater than that of the mechanical peak, perhaps indicating the slightly different kinetic units probed by each technique.

The precise nature of the α-relaxation and the distinction between it and the β-relaxation in rigid aromatic PLCs such as Vectra A950 are still

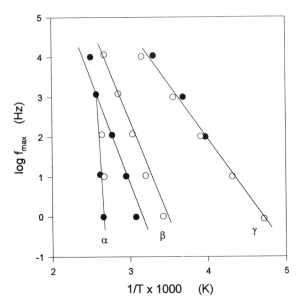

Figure 11. Dielectric loss maxima as a function of frequency for MCPLC HBA73/ HNA27 (●) and HBA30/HNA70 (○) for α-, β-, and γ-relaxations. (Redrawn from reference 176.)

not fully understood. As mentioned, the α-relaxation is generally weak in HBA73/HNA27 because most of the dipoles transverse to the chain (in particular, the carbonyls) have been relaxed by secondary relaxations before the larger scale segmental motions associated with the α-relaxation. Indeed, the very nature of motion at the glass transition in these anisotropic materials differs from that of nonmesogenic polymers, as indicated by the greater strength of the MCPLC T_g when measured in mechanical shear, rather than tension (*180, 198–200*). The rotation of the dipoles around the long axis of the chain, which results in the α and β motions, leads to greater loss peaks (higher relaxation strengths) for samples that have been mechanically aligned (*176*). Given that relaxations of most dipoles are mobilized in the β- and γ-relaxations, it is not clear precisely which additional motions occur in the α-relaxation. Whereas early results for these materials seemed to indicate an Arrhenius-like behavior for the β-relaxation (*176*), more recent work (*179, 181*) noted a degree of WLF-like curvature in the β-relaxation and indicated a relationship between these two motions, and both the α- and β-relaxations demonstrated a degree of translational and larger scale mobility.

The dipolar relaxation strength of these units is also lower than expected, and only a fraction of these units is mobile (*176, 181, 183*). This characteristic is probably due to conformational constraints and the nature of the motion that results in cancellation of dipoles. Interestingly, increased crystallinity due to annealing has little effect on dielectric properties (*180, 186*), and a high degree of similarity between crystalline and amorphous regions is indicated in these ordered MCPLC materials.

The lowest temperature motion, the γ-relaxation, on the other hand, does not change position with composition although it is more intense with greater HBA concentrations and has thus been ascribed to relaxation of carbonyl dipoles associated with motion of the HBA moiety. It is quite weak due to the ease of rotation of this unit without the need to disturb neighboring chains and dipoles (*179*). This relaxation is suppressed in copolymers where there is greater hydrogen bonding between chains (*184*), such as the inclusion of the AP unit (commercial Vectra B950 is a TA-AP–HNA copolymer). A further indication of the strength of the hydrogen bonding in this system resulting from the NH moiety is that it is one of the few MCPLCs with a greater α-relaxation strength than that of the β motion, and this characteristic demonstrates that quite high temperatures are required in this instance for the secondary bonding forces to be overcome (*175, 184*). The assignments of the increased motion of the various HBA and HNA units with increasing temperature have also been confirmed by NMR measurements in oriented and unoriented samples (*202, 203*).

Semi-Flexible MCPLCs

The most widely studied polymer in this class is HBA60/poly(ethylene terephthalate) (PET)40, sold originally as XG7 (Eastman, United States) and now as Rodrun 3000 (Unitika, Japan). This polymer is essentially a nonrandom copolymer resulting in a two-phase material with both mesogenic HBA-rich and flexible PET-rich phases. Despite its complex morphological nature and problems with repeatable thermal characterization (*201*), the PET phase has a T_g of around 55–75 °C and melting point around 250 °C. The PHB phase has a broad T_g of around 140–175 °C and possibly a clearing point–degradation temperature of 450 °C (*204–206*).

Dielectric properties of these materials are also quite complex (*182, 190, 191*). Gedde et al. (*190*) examined both HBA60/PET40 and HBA80/PET20. The HBA60/PET40 material showed an HBA T_g (seen by differential scanning calorimetry at 150 °C) and two α-relaxations due to different PET phases at 64 °C and 91 °C. The reason for the existence of the two DRS-observed T_g values is not fully understood, nor is the reason for their appearance at lower temperatures than that of nonmesogenic PET homopolymer, reported by Coburn et al. (*207*). Recent dielectric and dynamic mechanical work by Brostow et al. (*208*) indicated that the higher T_g region may be due to HBA units present in the PET-rich regions. (The presence of multiple glass transitions has been observed in other MCPLC systems such as MCPLC with three aromatic groups in the repeat unit and flexible polyether linkages, as reported by Laupretre et al. (*194*). In the work of Laupretre et al., however, multiple relaxations were assigned to increased motion of the flexible polyether segments in the amorphous isotropic regions and the smectic regions, which occur at a higher temperature.)

The observed β-relaxation in HBA40/PET60 is probably due to local motions of the PET groups and occurs with a similar position (–50 °C at 50 Hz) and E_a (56 kJ/mol) to that of PET homopolymer (*207*). The complexity of the motions involved and the environments in which the dipoles find themselves are demonstrated by the fact that it is influenced by HBA content. A greater mole fraction of HBA increases the temperature of the β-relaxation. The miscibility of each component in the other phase and the influence of crystallinity on local chain motion (*190*) are important, as is the crystallinity in non-LC homopolymers, for example PET (*207*) or HBA. Because semiflexible MCPLCs have a greater degree of crystallinity than rigid HBA/HNA materials, this difference influences mobility, although crystallinity does not necessarily affect dielectric properties strongly. Kalika et al. (*192*) showed in measurements on a semirigid PLC composed of a diethylene glycol spacer and a rigid triad mesogenic unit that dielectric properties change only mildly as materials pass from the crystalline into the smectic LC state. For semiflexible MCPLCs, the length and flexibility of the spacer have an effect on E_a and speed of the various motions (*188, 191, 193*). The frequency location of the α-relaxation increases with greater spacer length (flexibility).

Very few studies of frequency sweeps of MCPLCs at isothermal temperatures above the glass transition have been reported (as compared to SCPLCs), largely due to the high levels of conductivity at these temperatures (*192*) and the relative weakness of the relaxation. Given the nature of the motion below and at the glass transition being related to rotation around the long axis of the molecule, mainly the longitudinal dipoles (if they exist) remain to be relaxed at and above the α-relaxation. Araki et al. (*188*) reported isothermal, frequency-scan data for two mesogenic, semiflexible polycarbonate MCPLCs with differing polyether spacer lengths. Scans were performed at temperatures above the T_g and were broad (half widths of $\log f = 6.3$) and demonstrated a strongly curved, non-Arrhenius temperature dependence. The α-relaxation is ascribed to motion of the perpendicular azoxy dipole with some possible contribution from the longitudinal dipole motion. In both cases, the nematic potential would affect the mobility, although as with SCPLCs a full flip–flop motion is not required to relax dipoles, and fluctuations relative to the director are sufficient. Mucha (*195*) recently reported observation of a δ-relaxation (above the α-relaxation) in some main-chain semiflexible polyesters and showed that the E_a is higher in the mesomorphic state under the influence of the nematic potential. As discussed by Araki et al. (*188*) greater understanding could be achieved if the MCPLC samples could be homeotropically or planarly aligned in electric fields, but this process proved difficult. Some success has been obtained with such electrical alignment in certain geometries recently (*209, 210*) and may hold some promise although high voltages are required. Increased understanding of the nature of MCPLC relaxation in LC state is aided, as in SCPLCs, by good theoretical expositions of MCPLC dynamics. Gotlib and co-workers attempted to do this for semiflexible (*211*) and increasingly flexible MCPLCs (*212, 213*), and such work should help shed light on the modes of motion of these polymers as influenced by their chemical structures, the LC potential field, and order parameter.

Conclusions

DRS has been widely used as a characterization method for PLCs, most of which are polar. This use is particularly true for SCPLCs in which many of the potential electrooptic applications for this class of material (nonlinear optics, optical information storage, and waveguide materials) are reliant on such polarity, as well as mobility and ease or stability of alignment. By contrast, MCPLCs are primarily used as engineering plastics. Because they are generally copolyesters (at least those available commercially), they tend to contain carbonyl and other polar groups, and DRS is also a potent method of characterization. Primary and secondary relaxations in such materials are of some importance with regards to defining the thermal behavior of such materials. DRS is a useful relaxation technique because of the wide frequency/temperature relaxation map that can be produced. In particular, the wide frequency range that can be measured allows a more fundamental understanding and modeling of the motions than either temperature scanning or dynamic mechanical analysis alone can provide.

DRS has mainly been applied to SCPLCs and the motion of the LC unit pendant to the polymer chain is predominantly monitored. This may be limited motion in the glassy state or larger scale motions of the unit and chain as a whole above T_g or T_m (should it occur). Whereas secondary sub-T_g relaxations tend to be due to rotations within the side chain itself, it appears that this motion is sufficiently cooperative to be influenced by its environment. The bases of the larger-scale motions above T_g or T_m are related to motions of the unit or the main chain or a mixture of both. Whereas this is still the basis of some ongoing discussion, it appears clear that the lower frequency (highest temperature) δ-relaxation is mainly due to the longitudinal dipole, and the higher frequency α-relaxation has been ascribed variously to motion of the perpendicular dipole on the side group and the main chain. The nature of the α-relaxation may be system-dependent, and careful synthesis and relaxation studies of homologous series of materials are necessary to clarify this point. Dielectric monitoring of these higher temperature side-group motions is useful in that they allow determination of the various orders of the LC regions. Whereas order within domains is determined by chemical structure and temperature, the larger scale orientation or order of the domains is influenced by surface forces or alignment by electric and magnetic fields. Indeed, to better assign dielectric relaxations in these materials, the ability to be able to align sample parallel or perpendicular (and thus intermediate) to the capacitive plates between which the sample is contained is of great value from a characterization as well as a technological point of view. Despite much work done to date, opportunity remains to further explore DRS structure–property relationships in these and associated materials. The fields of copolymers, blends, networks, gels, as well as other more complex non-combshaped architectures currently are being pursued.

MCPLC DRS analysis to date has not progressed to the same level as with SCPLCs, and frequency scanning of relaxations (particularly the α-relaxation) is hindered by conductivity. Most commercial MCPLC materials have been studied to some degree, and because most are copolymers, the influence of copolymer content on the relaxations has been elucidated. Even though further work on these and noncommercial materials continues, most is in the form of temperature scans. Because blends of MCPLCs with other thermoplastics continue to be an active area of research, the DRS techniques used as sensitive determinations of miscibility may be very valuable. This technique is used particularly if MCPLCs and other plastic materials are compatibilized to improve blend properties, because any relaxational changes due to interfacial agents may be small.

References

1. Vill, V. *Adv. Mater.* **1994,** *6,* 527.
2. Moscicki, J. K. In *Liquid Crystal Polymers: From Structures to Applications;* Collyer, A. A., Ed.; Elsevier Applied Science: London, 1992; Chapter 4.
3. Nikonorova, N. A.; Malinovskaya, V. P.; Borisova, T. I.; Burshtein, L. L.; Korshun, A. M.; Skorokhodov, S. S. *Polym. Sci. USSR* **1987,** *29,* 612.
4. Nikonorova, N. A.; Stepanova, T. P.; Malinovskaya, V. P.; Lija, L.; Burshtein, L. L.; Borisova, T. I.; Korshun, A. M.; Skorokhodov, S. S. *Makromol. Chem.* **1992,** *193,* 2771.
5. Plate, N. A.; Shibaev, V. P. *Comb Shaped Polymers and Liquid Crystals;* Plenum: New York, 1987.
6. Shibaev, V. *Mol. Cryst. Liq. Cryst.* **1994,** *243,* 201.
7. *Merck Liquid Crystals;* Liquid Crystal Polymer Catalogue; Merck & Co., Inc.: Rahway, NJ, 1994.
8. Dubois, J-C.; Le Barny, P.; Robin, P.; Lemoine, V.; Rajbenbach, H. *Liq. Cryst.* **1993,** *14,* 197.
9. Warner, M. *Mater. Res. Soc. Symp. Proc.* **1989,** *134,* 61.
10. Attard, G. S. In *High Value Polymers;* Fawcett, A. H., Ed.; Special Publication 87; Royal Society of Chemistry: London, 1991: pp 131–150.

11. Attard, G. S. *Trends Polym. Sci.* **1993**, *1*, 79.
12. Eich, M.; Wendorff; J. H. *Makromol. Chem., Rapid Commun.* **1986**, *8*, 467.
13. Shiabev, V. P.; Yakovlev, I. V.; Kostromin, S. G. *Polym. Sci. USSR* **1990**, *32*, 1478.
14. Anderle, K.; Wendorff, J. H. *Mol. Cryst. Liq. Cryst.* **1994**, *243*, 51.
15. Baer, Ch.; Glusen, B.; Wendorff, J. H. *Macromol. Chem. Rapid Commun.* **1994**, *15*, 327.
16. Xie, S.; Nantansohn, A.; Rochan, P. *Macromolecules* **1994**, *27*, 1489.
17. Becker, M. W.; Sapochak, L. S.; Ghosen, R. *Chem. Mater.* **1994**, *6*, 104.
18. Finkelmann, H. In *Liquid Crystallinity in Polymers;* Ciferri, A., Ed.; VCH: New York, 1991; Chapter 8.
19. Shibaev, V. P.; Freidzon, Ya. S.; Kostromin, S. G. In *Liquid Crystalline and Mesomorphic Polymers;* Shibaev, V. P.; Lam, L., Eds.; Springer-Verlag, Berlin, Germany, 1994; Chapter 3.
20. Donald, A. M.; Windle, A. H. *Liquid Crystalline Polymers;* Cambridge University: Cambridge, England, 1992.
21. Jansson, J-F. In *Liquid Crystal Polymers: From Structures to Applications;* Collyer, A. A., Ed.; Elsevier Applied Science: London, 1992; Chapter 9.
22. *Liquid Crystalline Polymers;* National Materials Advisory Board, Commission on Engineering and Technical Systems, National Research Council, National Academy Press: Washington, DC, 1990.
23. MacDonald, W. A. In *High Value Polymers;* Fawcett, A. H., Ed.; Special Publication 87; Royal Society of Chemistry: London, 1991; pp 428–454.
24. Zentel, R.; Strobl, G.; Ringsdorf, H. In *Recent Advances in Liquid Crystalline Polymers;* Chapoy, L. L., Ed.; Elsevier: London, 1985; Chapter 17.
25. Parneix, J. P.; Njeumo, R.; Legrand, C.; Le Barny, P.; Dubois, J. C. *Liq. Cryst.* **1987**, *2*, 167.
26. Shibaev, V. P. *Mol. Cryst. Liq. Cryst.* **1988**, *155*, 189.
27. Shibaev, V. P.; Plate, N. A. *Adv. Polym. Sci.* **1984**, *60/61*, 173.
28. Gray, G. W. In *Side Chain Liquid Crystal Polymers;* McArdle, C. B., Ed.; Blackie: Glasgow, Scotland, 1989; Chapter 3.
29. Zentel, R.; Strobl, G. R.; Ringsdorf, H. *Macromolecules* **1985**, *18*, 960.
30. Bormuth, F. J.; Biradar, A. M.; Quotschalla, U.; Haase, W. *Liq. Cryst.* **1989**, *5*, 1549.
31. Bormuth, F. J.; Haase, W. *Liq. Cryst.* **1988**, *3*, 881.
32. Gotz, S.; Stille, W.; Strobl, G.; Scheuermann, H. *Macromolecules* **1993**, *26*, 1520.
33. Araki, K. *Polym. J.* **1990**, *22*, 546.
34. Pranoto, H.; Bormuth, F-J.; Haase, W.; Kiechle, U.; Finkelmann, H. *Makromol. Chem.* **1986**, *187*, 2453.
35. Monnerie, L.; Laupretre, F.; Noël, C. *Liq. Cryst.* **1988**, *3*, 1013.
36. McCrum, N. G.; Read, B. E.; Williams, G. *Anelastic and Dielectric Effects in Polymeric Solids;* Wiley: New York, 1967.
37. Pschorn, U.; Spiess, H. W.; Hisgen, B.; Ringsdorf, H. *Makromol. Chem.* **1986**, *187*, 2711.
38. Haase, W.; Pfeiffer, M. *Mat. Res. Soc., Symp. Proc.* **1990**, *175*, 257.
39. Gedde, U. W.; Liu, F.; Hult, A.; Sahlen, F.; Boyd, R. H. *Polymer* **1994**, *35*, 2056.
40. Romero Colomer, F.; Meseguer Duenas, J. M.; Gomez Ribelles, J. L.; Barrales-Rienda, J. M.; Bautista de Ojeda, J. M. *Macromolecules* **1993**, *26*, 155.
41. Vallerien, S. U.; Kremer, F.; Boeffel, C. *Liq. Cryst.* **1989**, *4*, 79.
42. Zhong, Z. Z.; Gordon, W. L.; Schuele, D. E. *Liq. Cryst.* **1994**, *17*, 199.
43. Attard, G. S. *Mol. Phys.* **1986**, *58*, 1087.
44. Attard, G. S.; Araki, K.; Williams, G. *Br. Polym. J.* **1987**, *19*, 119.
45. Kozak, A.; Moscicki, J. K.; Williams, G. *Mol. Cryst. Liq. Cryst.* **1991**, *201*, 1.

46. Kozak, A.; Moscicki, J. K. *Liq. Cryst.* **1992**, *12*, 377.
47. Araki, K.; Attard, G. S.; Kozak, A.; Williams, G.; Gray, G. W.; Lacey, D.; Nestor, G. *J. Chem. Soc., Faraday Trans.* **1988**, *84*, 1067.
48. Haws, C. M.; Clark, M. G.; Attard, G. S. In *Side Chain Liquid Crystal Polymers;* McArdle, C. B., Ed.; Blackie: Glasgow, Scotland, 1989; Chapter 7.
49. Williams, G. In *Comprehensive Polymer Science. Volume 2. Polymer Properties;* Allen, G.; Bevington, J. C., Eds.; Pergamon: New York, 1989; Chapter 18.
50. Clark, M. G. *Mol. Cryst. Liq. Cryst.* **1985**, *127*, 1.
51. Maier, W.; Meier, G. *Z. Naturforsch.* **1961**, *16a*, 262.
52. Nordio, P. L.; Giorgio, R.; Segre, U. *Mol. Phys.* **1973**, *25*, 129.
53. Birenheide, R.; Budesheim, K. W.; Wendorff, J. H. *Angew. Makromol. Chem.* **1991**, *185/186*, 319.
54. Williams, G. In *Structures and Properties of Polymers;* Cahn, R. W.; Haasen, P.; Kramer, E. J., Eds.; Materials Science and Technology, Vol. 12; VCH: Basel, Switzerland, 1992; Chapter 11.
55. Attard, G. S.; Williams, G. *Liq. Cryst.* **1986**, *1*, 253.
56. Araki, K.; Attard, G. S. *Liq. Cryst.*, **1986**, *1*, 301.
57. Bormuth, F. J.; Muhlberger, B.; Haase, W. *Makromol. Chem.* **1989**, *10*, 231.
58. Kresse, H.; Wiegeleben, A.; Krucke, B. *Acta Polymer.* **1988**, *39*, 583.
59. Scheuermann, H.; Tsukruk, V.; Finkelmann, H. *Liq. Cryst.* **1993**, *14*, 889.
60. Kresse, H.; Stettin, H.; Tennstedt, E.; Kostromin, S. *Mol. Cryst. Liq. Cryst.* **1990**, *191*, 135.
61. Haws, C. M.; Clark, M. G.; McArdle, C. B. *Mol. Cryst. Liq. Cryst.* **1987**, *153*, 537.
62. Al-Ammar, K. H.; Mitchell, G. R. *Polymer* **1992**, *33*, 11.
63. Borisova, T. I.; Burshtein, L. L.; Stepanova, T. P.; Freidzon, Ya. S.; Kharitonov, A. V.; Shibayev, V. P. *Polym. Sci. USSR* **1982**, *24*, 1664.
64. Borisova, T. I.; Burshtein, L. L.; Nikonorova, N. A.; Tal'roze, R. V.; Shibayev, V. P. *Polym. Sci. USSR*, **1986**, *28*, 2594.
65. Borisova, T. I.; Burshtein, L. L.; Stepanova, T. P.; Kostromin, S. G.; Shibayev, V. P. *Polym. Sci. USSR* **1986**, *28*, 1150.
66. Birenheide, R.; Eich, D. A.; Jungbauer, O.; Hermann-Schonherr, O.; Stoll, O.; Wendorff, J. H. *Mol. Cryst. Liq. Cryst.* **1989**, *177*, 13.
67. Bormuth, F. J.; Haase, W. *Liquid Crystal Chemistry, Physics and Applications;* SPIE Vol. 1080; The International Society for Optical Engineering: Bellingham, WA, 1989; p 232
68. Bormuth, F. J.; Haase, W. *Liq. Cryst.* **1989**, *5*, 1849.
69. Bormuth, F. J.; Haase, W. *Mol. Cryst., Liq. Cryst.* **1987**, *153*, 207.
70. Bormuth, F. J.; Haase, W.; Zentel, R. *Mol. Cryst., Liq. Cryst.* **1987**, *148*, 1.
71. Canessa, G.; Reck, B.; Reckert, G.; Zentel, R. *Makromol. Chem., Macromol. Symp.* **1986**, *4*, 91.
72. Carr, P. L.; Davies, G. R.; Ward, I. M. *Polymer* **1993**, *34*, 5.
73. Findlay, R. B.; Windle, A. H. *Mol. Cryst. Liq. Cryst.* **1991**, *206*, 55.
74. Finkelmann, H.; Naegele, D.; Ringsdorf, H. *Makromol. Chem.* **1979**, *180*, 803.
75. Haase, W.; Pranoto, H. In *Polymeric Liquid Crystals;* Blumstein, F., Ed.; Plenum: New York, 1985; pp 313–329
76. Haase, W.; Pranoto, H.; Bormuth, F. J. *Ber. Bunsenges. Phys. Chem.* **1985**, *89*, 1229.
77. Heinrich, W.; Stoll, B. *Colloid Polym. Sci.* **1985**, *263*, 895.
78. Kresse, H.; Kostromin, S.; Shibaev, V. P. *Makromol. Chem., Rapid Commun.* **1982**, *3*, 509.
79. Kresse, H.; Shibaev, V. P. *Z. Phys. Chemie (Leipzig)* **1983**, *1*, 161.
80. Kresse, H.; Shibaev, V. P. *Makromol. Chem., Rapid Commun.* **1984**, *5*, 63.

81. Kresse, H.; Talrose, R. V. *Makromol. Chem., Rapid Commun.* **1981**, *2*, 369.
82. Tennstedt, E.; Kresse, H.; Zentel, R. *Acta Polymer.* **1986**, *37*, 685.
83. Kresse, H.; Tennstedt, E.; Zentel, R. *Makromol. Chem., Rapid Commun.* **1985**, *6*, 261.
84. Nikonorova, N. A.; Borisova, T. I.; Burshtein, L. L.; Freidzon, Ya. S.; Shibaev, V. P. *Polym. Sci. USSR* **1990**, *32*, 82.
85. Nikonorova, N. A.; Borisova, T. I.; Burshtein, L. L.; Freidzon, Ya. S.; Shibaev, V. P. *Polym. Sci. USSR* **1990**, *32*, 970.
86. Ringsdorf, H.; Zentel, R. *Makromol. Chem.* **1982**, *183*, 1245.
87. Talroze, R. V.; Kostromin, S. G.; Shibaev, V. P.; Plate, N. A.; Kresse, H.; Sauer, K.; Demus, D. *Makromol. Chem., Rapid Commun.* **1981**, *2*, 305.
88. Ujiie, S.; Koide, N.; Iimura, K. *Mol. Cryst. Liq. Cryst.* **1987**, *153*, 191.
89. Wendorff, J. H.; Fuhrmann, Th. Dielectrics Newsletter (NOVOCONTROL GmbH), 1994, July 1.
90. Zhong, Z. Z.; Gordon, W. L.; Schuele, D. E.; Akins, R. B.; Percec, V. *Mol. Cryst. Liq. Cryst.* **1994**, *238*, 129.
91. Zhong, Z. Z.; Schuele, D. E.; Gordon, W. L. *Macromolecules,* **1993**, *26*, 6403.
92. Araki, K.; Attard, G. S.; Williams, G. *Polymer* **1989**, *30*, 432.
93. Attard, G. S. *Polymer* **1989**, *30*, 438.
94. Attard, G. S.; Araki, K. *Mol. Cryst. Liq. Cryst.* **1986**, *141*, 69.
95. Attard, G. S.; Araki, K.; Moura-Ramos, J. J.; Williams, G. *Liq. Cryst.* **1986**, *3*, 861.
96. Goodwin, A. A.; Beevers, M. S. *Macromol. Chem. Phys.* **1995**, *196*, 1465.
97. Attard, G. S.; Araki, K.; Williams, G. *J. Molec. Electron.* **1986**, *2*, 107.
98. Attard, G. S.; Araki, K.; Williams, G. *J. Molec. Electron.* **1987**, *3*, 1.
99. Attard, G. S.; Araki, K.; Moura-Ramos, J. J.; Williams, G.; Griffin, A. C.; Bhatti, A. M.; Hung, R. S. L. In P*olymer Association Structures, Microemulsions and Liquid Crystals;* El-Noakly, M. A., Ed.; ACS Symposium Series 384; American Chemical Society: Washington, DC, 1989; pp 255–264.
100. Attard, G. S.; Moura-Ramos, J. J.; Williams, G. *J. Polym. Sci., Polym. Phys.* **1987**, *25*, 1099.
101. Attard, G. S.; Moura-Ramos, J. J.; Williams, G.; Nestor, G.; White, M. S.; Gray, G. W.; Lacey, D.; Toyne, K. *J. Makromol. Chem.* **1987**, *188*, 2769.
102. Attard, G. S.; Williams, G. *Polym. Commun.* **1986**, *27*, 2.
103. Attard, G. S.; Williams, G. *Polym. Commun.* **1986**, *27*, 66.
104. Attard, G. S.; Williams, G. *Nature (London)* **1987**, *326*, 544.
105. Attard, G. S.; Williams, G.; Fawcett, A. H. *Polymer* **1990**, *31*, 928.
106. Attard, G. S.; Williams, G.; Gray, G. W.; Lacey, D.; Gemmel, P. A. *Polymer* **1986**, *27*, 185.
107. Coles, H. J.; Simon, R. *Polymer* **1985**, *26*, 1801.
108. Haase, W.; Pranoto, H. *Progr. Colloid, Polym. Sci.* **1984**, *69*, 139.
109. Simon, G. P. *Polymer* **1989**, *30*, 2227.
110. Keller, E. J. C.; Williams, G.; Krongauz, V.; Yitzchaik, S. *J. Mater. Chem.* **1991**, *1*, 331.
111. Kim, H. J.; Jackson, W. R.; Simon, G. P. *Eur. Polym. J.* **1994**, *30*, 1201.
112. Kim, H. J.; Jackson, W. R.; Simon, G. P. *J. Mater. Chem.* **1993**, *3*, 537.
113. Kozak, A.; Moura-Ramos, J. J.; Simon, G. P.; Williams, G. *Makromol. Chem.* **1989**, *190*, 2463.
114. Kozak, A.; Simon, G. P.; Williams, G. *Polym. Commun.* **1989**, *30*, 102.
115. Kozak, A.; Simon, G. P.; Moscicki, J. F.; Williams, G. *Mol. Cryst. Liq. Cryst.* **1990**, *193*, 149.
116. Kozak, A.; Simon, G. P.; Moscicki, J. F.; Williams, G. *Mol. Cryst. Liq. Cryst.* **1990**, *193*, 155.

117. Kresse, H.; Ernst, S.; Krucke, B.; Kremer, F.; Vallerien, S. U. *Liq. Cryst.* **1992**, *11*, 439.
118. Moura-Ramos, J. J.; Williams, G. *Polymer* **1991**, *32*, 909.
119. Nazemi, A.; Kellar, E. J. C.; Williams, G.; Karasz, F. E.; Hill, J. S.; Lacey, D.; Gray, G. W. *Liq. Cryst.* **1991**, *9*, 307.
120. Nazemi, A.; Williams, G.; Attard, G. S.; Karasz, F. E. *Polym. Adv. Technol.* **1992**, *3*, 157.
121. Novotna, E.; Kresse, H.; Krucke, B. *Acta Polym.* **1992**, *43*, 279.
122. Simon, G. P.; Williams, G. *Polymer* **1993**, *34*, 2038.
123. Simon, G. P.; Kozak, A.; Williams, G. W.; Wetton, R. E. *Mater. Forum* **1991**, *15*, 71.
124. Simon, R. J.; Coles, H. J. *Polymer* **1986**, *27*, 811.
125. Simon, R. J.; Coles, H. J. *J. Polym. Sci., Phys. Ed.* **1989**, *27*, 1823.
126. Williams, G. *Polymer* **1994**, *35*, 1915.
127. Williams, G.; Nazemi, A.; Karasz, F. E.; Hill, J. S.; Lacey, D.; Gray, G. W. *Macromolecules* **1991**, *24*, 5134.
128. Williams, G.; Nazemi, A.; Karasz, F. E. In *Multifunctional Materials;* Buckley, A.; Gallagher-Daggit, G.; Karasz, F. E.; Ulrich, D. R., Eds.; Materials Research Society: Pittsburgh, PA, 1990; pp 227–239.
129. Kozak, A.; Moura-Ramos, J. J.; Simon, G. P.; Williams, G., unpublished results.
130. Zentel, R.; Wu. J. *Makromol. Chem.* **1986**, *187*, 1727.
131. Davies, M.; Moutran, R.; Price, A. H.; Beevers, M. S.; Williams, G. *J. Chem. Soc., Faraday Trans. II* **1973**, *69*, 1486.
132. Maurel, P.; Price, A. H. *J. Chem. Soc., Faraday Trans.* II **1973**, *69*, 1486.
133. Druon, C.; Wacrenier, J. M. *Mol. Cryst. Liq. Cryst.* **1983**, *98*, 201.
134. Madhusama, N. V.; Srikanta, B. S.; Subramanya, M.; Raj Urs. *Mol. Cryst. Liq. Cryst.* **1984**, *108*, 19.
135. Noel, C. In *Liquid Crystal Polymers: From Structures to Applications;* Collyer, A. A., Ed.; Elsvier Applied Science: London, 1992; Chapter 2.
136. Findlay, R. B. *Mol. Cryst. Liq. Cryst.* **1993**, *231*, 137.
137. Ujiie, S.; Koide, N.; Iimura, K. *Mol. Cryst. Liq. Cryst.* **1991**, *153*, 191.
138. Finkelmann, H.; Rehage, G. *Adv. Polym. Sci.* **1984**, *60/61*, 99.
139. Blinov, L. M.; Barnik, M. I.; Trufanov, A. N. *Mol. Cryst. Liq. Cryst.* **1982**, *89*, 47.
140. Kriegbaum, W. R. *J. Appl. Polym. Sci., Polym. Symp.* **1985**, *41*, 149
141. Bucher, H. K.; Klingbiel, R. T.; Van Meter, J. P. *Appl. Phys. Lett.* **1974**, *5*, 186.
142. Bose, T. K.; Campbell, B.; Yagihara, S.; Thoen, J. *J. Phys. Rev. A.* **1987**, *A36*, 5767.
143. Schadt, M. *Mol. Cryst. Liq. Cryst.*, **1982**, *89*, 77.
144. Nestor, G.; White, M. S.; Gray, G. W.; Lacey, D.; Toyne, K. *Makromol. Chem.* **1987**, *186*, 2759.
145. Martins, A. F.; Esnault, P.; Volino, F. *Phys. Rev. Lett.* **1986**, *56*, 1745.
146. Jacobs, H. A.; Day, G. M.; Jackson, W. R.; Simon, G. P.; Watson, K. G.; Zheng, S. *Aust. J. Chem.* **1992**, *45*, 695.
147. Day, G. M.; Jackson, W. R.; Jacobs, H. A.; Kim, H. J.; Simon, G. P.; Sarna, R.; Watson, K. G. *Polym. Bull.* **1992**, *29*, 21.
148. Gaff, K.; Simon, G. P.; Akiyama, E.; Takamura, Y.; Nagase, Y., unpublished results.
149. Nagase, Y.; Takamura, Y.; Abe, H.; Ono, K.; Saito, T.; Akiyama, E. *Makromol. Chem.* **1993**, *194*, 2517.
150. Nagase, Y.; Saito, T.; Abe, H.; Takamura, Y. *Macromol. Chem. Phys.* **1994**, *195*, 263.
151. Vallerien, S. U.; Kremer, F.; Huser, B.; Spiess, H. W. *Colloid Polym. Sci.* **1989**, *267*, 583.

152. Huser, B.; Spiess, H. W. *Makromol. Chem., Rapid Commun.* **1988,** *9,* 337.
153. Rotz, U.; Lindau, J.; Weissflog, W.; Reinhold, G.; Unseld, W.; Kuschel, F. *Mol. Cryst. Liq. Cryst.* **1989,** *170,* 185.
154. Kresse, H.; Ernst, S.; Rotz, U.; Lindau, J.; Kuschel, F. *Mol. Cryst. Liq. Cryst.* **1990,** *193,* 211.
155. Kresse, H. S.; Rotz, U.; Lindau, J.; Kuschel, F. *Makromol. Chem.* **1989,** *190,* 2953.
156. Weissflog, W.; Wiegeleben, A.; Diele, S.; Demus, D. *Cryst. Res. Technol.* **1984,** *19,* 583.
157. Kresse, H.; Rabenstein, P.; Stettin, H.; Diele, S.; Demus, D.; Weissflog, W. *Res. Cryst. Technol.* **1988,** *23,* 135.
158. Heinemann, S.; Pirwitz, G.; Kresse, H. *Mol. Cryst. Liq. Cryst.* **1993,** *237,* 277.
159. Haddawi, S.; Diele, S.; Kresse, H.; Pelzl, G.; Weissflog, W.; Wiegeleben, A. *Liq. Cryst.* **1994,** *17,* 191.
160. Reck, B.; Ringsdorf, H. *Makromol. Chem., Rapid Commun.* **1985,** *6,* 291.
161. Kremer, F.; Vallerien, S. U.; Zentel, R.; Kapitza, H. *Macromolecules* **1989,** *22,* 4040.
162. Enderes, B. W.; Wendorff, J. H.; Reck, B.; Ringsdorf, H. *Makromol. Chem.* **1987,** *188,* 1501.
163. Enderes, B. W.; Wendorff, J. H.; Reck, B.; Ringsdorf, H. *Liq. Cryst.* **1990,** *7,* 217.
164. Kresse, H.; Novotna, E.; Rotz, U.; Bohme, A.; Lindau, J.; Kuschel, F. *Acta Polym.* **1992,** *43,* 183.
165. Araki, K.; Namiki, M.; Usami, T.; Iimura, K. *Mol. Cryst. Liq. Cryst.* **1991,** *105,* 1.
166. Chen, F-L.; Jamieson, A. M. *Macromolecules* **1993,** *26,* 6576.
167. Ringsdorf, H.; Schmidt, H.-W.; Schneller, A. *Makromol. Chem., Rapid Commun.* **1982,** *3,* 745.
168. Sierble, H.; Stille, W.; Strobl, G. *Macromolecules* **1990,** *23,* 2008.
169. Hikmet, R. A. M.; Broer, D. J. *Polymer* **1991,** *32,* 1627.
170. Hikmet, R. A. M.; Zwerver, B. H. *Liq. Cryst.* **1991,** *10,* 835.
171. Hikmet, R. A. M.; Lub, J.; Maassen vd Brink, P. *Macromolecules* **1992,** *25,* 4194.
172. Zhong, Z. Z.; Scheule, D. E.; Gordon, W. L.; Adamic, K. J.; Akins, R. B. *J. Polym. Sci., Polym. Phys.* **1992,** *30,* 1443.
173. Simon, G. P., in preparation.
174. Hardouin, F.; Sigaud, G.; Achard, M. F. In *Liquid Crystalline and Mesomorphic Polymers;* Shibaev, V. P.; Lam, L., Eds.; Springer-Verlag: Berlin, Germany, 1994; Chapter 4.
175. Abdul Jawad, S.; Ahmad, M. S. *Mater. Lett.* **1993,** *17,* 91.
176. Alhaj-Mohammed, M. H.; Davies, G. R.; Abdul Jawad, S.; Ward, I. M. *J. Polym. Sci., Polym. Phys. Ed.* **1988,** *26,* 1751.
177. Blundell, D. J.; Buckingham, K. A. *Polymer* **1985,** *26,* 1623.
178. Davies, G. R. *Makromol. Chem., Makromol. Symp.* **1988,** *20/21,* 293.
179. Davies, G. R.; Ward, I. M. In *High Modulus Polymers—Approaches to Design and Development;* Zachariades A. E.; Porter, R. S., Eds.; Marcel Dekker: New York, 1988; Chapter 2.
180. Green, D. I.; Davies, G. R.; Ward, I. M.; Alhaj-Mohammed, M. H.; Abdul Jawad, S. *Polym. Adv. Technol.* **1990,** *1,* 41.
181. Kalika, D. S.; Yoon, D. Y. *Macromolecules* **1991,** *24,* 3404.
182. Takase, Y.; Mitchell, G. R.; Odajima, A. *Polym. Comm.* **1986,** *27,* 76.
183. Ward, I. M. *Macromol. Chem., Makromol. Symp.* **1993,** *69,* 75.
184. Abdul Jawad, S.; Alhaj-Mohammad, M. H. *Mater. Lett.* **1992,** *13,* 312.
185. Damman, S. B.; Buijs, J. A. H. M.; van Turnhout, J. *Polymer* **1994,** *35,* 2364.

186. Kalika, D. S.; Yoon, D. Y.; Iannelli, P.; Parrsih, W. *Macromolecules* **1991**, *24*, 3413.
187. Sato, T.; Tsujji, Y.; Kita, Y.; Fukuda, T.; Miyamoto, T. *Macromolecules* **1991**, *24*, 4691.
188. Araki, K.; Aoshima, M.; Namiki, N.; Ujiie, S.; Koide, N.; Iimura, K.; Imamura, Y.; Williams, G. *Makromol. Chem., Rapid Commun.* **1989**, *10*, 265.
189. Borisova, T.; Fridrikh, S.; Gotlib, Y.; Medvedev, G.; Nikonorova, N.; Skorokhodov, S.; Zuev, V. *Macromol. Chem., Macromol. Symp.* **1993**, *72*, 67.
190. Gedde, U. W.; Buerger, D.; Boyd, R. H. *Macromolecules* **1987**, *20*, 988.
191. Gedde, U. W.; Liu, F.; Hult, A. *Polymer* **1991**, *32*, 1219.
192. Kalika, D. S.; Shen, M.-R.; Yu, X.-M.; Denn, M. M.; Iannelli, P.; Masciocchi, N.; Yoon, D. Y.; Parrsih, W.; Friedrich, C.; Noel, C. *Macromolecules* **1990**, *23*, 5192.
193. Kresse, H.; Hempel, E.; Kuschel, F.; Kremer, F. *Polym. Bull.* **1990**, *24*, 93.
194. Laupretre, F.; Noël, C.; Jenkins, W. N.; Williams, G. *Faraday Discuss. Chem. Soc.* **1985**, *79*, 191.
195. Mucha, M. In *Liquid Crystalline Polymers;* Carfagna, C., Ed.; Pergamon: London, 1994.
196. Pu'ertolas, J. A.; Alonso, P. J.; Oriol, L.; Serrano, J. L. *Macromolecules* **1992**, *25*, 6018.
197. Zhukov, S. V.; Malinovskaya, V. P.; Burshtein, L. L.; Koshun, A.; Skorokhodov, S. S. *Polym. Sci.* **1993**, *35*, 28.
198. Wissbrun, K. F.; Yoon, H. N. *Polymer* **1989**, *30*, 2193.
199. Troughton, M. J.; Unwin, A. P.; Davies, G. R.; Ward, I. M. *Polymer* **1988**, *29*, 1389.
200. Troughton, M. J.; Davies, G. R.; Ward, I. M. *Polymer* **1989**, *30*, 58.
201. Brostow, W.; Hess, M.; Lopez, B. L. *Macromolecules* **1994**, *27*, 2262.
202. Clements, J.; Humphreys, J.; Ward, I. *J. Polym. Sci., Polym. Phys. Ed.* **1986**, *24*, 2293.
203. Allen, R. A.; Ward, I. M. *Polymer* **1991**, *32*, 202.
204. Kodama, M. *Polym. Eng. Sci.* **1992**, *32*, 267.
205. Zheng, J. Q.; Kyu, T. *Polym. Eng. Sci.* **1992**, *32*, 1004.
206. Carfagna, C.; Amendola, E.; Nicolais, L. *Int. J. Mater. Prod. Technol.* **1992**, *7*, 205.
207. Coburn, J. C.; Boyd, R. H. *Macromolecules* **1986**, *19*, 2238.
208. Brostow, W.; Samatowicz, D. *Polym. Eng. Sci.* **1993**, *33*, 581.
209. Tsvetkov, V. N.; Tsvetkov, N. V.; Andreeva, L. N.; Bilibin, A. Yu.; Skorokhodov, S. S. *Eur. Polym. J.* **1993**, *29*, 1003.
210. Tsvetkov, V. N.; Tsvetkov, N. V.; Andreeva, L. N.; Bilibin, A. Yu.; Skorokhodov, S. S.; *Polym. Sci.* **1993**, *35*, 335.
211. Maksimov, A. V.; Gotlib, Y. Y. *Vsokmol. Soedin.* **1989**, *A31*, 1013.
212. Gotlib, Y. Y. *Prog. Coll. Polym. Sci.* **1989**, *80*, 245.
213. Gotlib, Y. Y.; Medvedev, G. A.; Fridrikh, S. V. *Makromol. Chem., Makromol. Symp.* **1991**, *52*, 209.

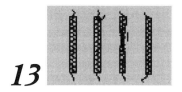

13

Dielectric Properties of Fluoropolymers

Peter Avakian, Howard W. Starkweather, Jr., John J. Fontanella, and Mary C. Wintersgill

Dielectric properties have been measured for polytetrafluoroethylene (PTFE) and a number of copolymers of TFE as well as poly(perfluoropropylene oxide). Of the three dynamic mechanical relaxations in PTFE, only the low temperature γ-relaxation is dielectrically active. However, the α-relaxation is seen in certain copolymers. Polymers having side groups exhibit an additional relaxation near 94 K. Small guest molecules absorbed in PTFE also produce a relaxation at low temperatures. Other studies included the effect of pressure, which caused the γ-relaxation to shift to higher temperatures and lower frequencies.

In this chapter, we review our work on the dielectric relaxations in fluoropolymers based on tetrafluoroethylene, including some high pressure measurements, in poly(perfluoropropylene oxide), and in polytetrafluoroethylene (PTFE) containing guest molecules. Carbon–fluorine bonds are highly polar. However, in perfluoropolymers such as PTFE, the dipoles associated with these bonds are largely balanced. As a result, the dielectric losses due to viscoelastic relaxations are very small. To study these effects, one must be able to measure dissipation factors at least as small as 10^{-5}. One instrument that can do this measurement is the C. Andeen & Associates impedance bridge (Model CGA-85), which was used in the studies reported here.

Some fluoropolymers, such as poly(vinylidene fluoride) and poly-(chloro trifluoroethylene), are much more polar. Even though these mate-

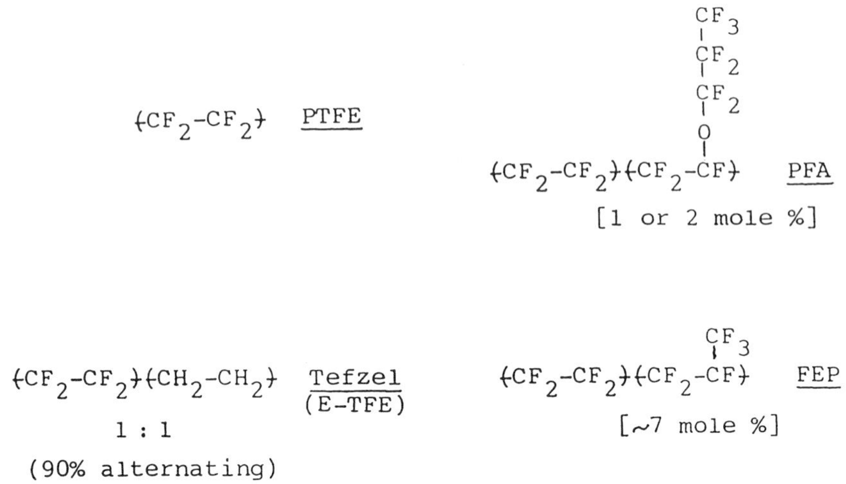

$$\text{+CF}_2\text{-CF}_2\text{+} \quad \underline{\text{PTFE}}$$

$$
\begin{array}{l}
\text{CF}_3 \\
|\\
\text{CF}_2 \\
|\\
\text{CF}_2 \\
|\\
\text{O} \\
|\\
\text{+CF}_2\text{-CF}_2\text{+}\text{+CF}_2\text{-CF}\text{+} \quad \underline{\text{PFA}}
\end{array}
$$

[1 or 2 mole %]

$$\text{+CF}_2\text{-CF}_2\text{+}\text{+CH}_2\text{-CH}_2\text{+} \quad \underline{\text{Tefzel}}$$
$$\text{(E-TFE)}$$
1 : 1
(90% alternating)

$$
\begin{array}{l}
\text{CF}_3 \\
|\\
\text{+CF}_2\text{-CF}_2\text{+}\text{+CF}_2\text{-CF}\text{+} \quad \underline{\text{FEP}}
\end{array}
$$

[~7 mole %]

Figure 1. Semicrystalline polymers based on TFE.

rials have been studied extensively, they are outside the scope of the present discussion.

Semicrystalline Polymers Based on Tetrafluoroethylene

In PTFE, three viscoelastic relaxations are seen in dynamic mechanical experiments at temperatures below the melting point (*1–3*), designated α, β, and γ in the order of decreasing temperature. In early work (*3–5*), all of these processes were reported to be dielectrically active. However, subsequent studies have found that only the low temperature γ-relaxation exhibits a significant dielectric loss (*6–8*). The α- and β-relaxations are thought to be seen dielectrically only in samples that have been decorated with polar groups and are not characteristic of the basic structure of PTFE.

FEP and PFA resins are copolymers of tetrafluoroethylene with hexafluoropropylene and perfluoropropyl vinyl ether, respectively. Thus, they contain CF_3 and $n\text{-}C_3F_7O$ side groups. These polymers have an additional dielectric relaxation at temperatures below the γ-relaxation (*6, 8*). For FEP, a small loss peak also is associated with the high temperature α-relaxation (*6, 7*). ETFE is a largely alternating copolymer of ethylene and TFE. Compared with the isomeric polymer, polyvinylidene fluoride, its melting point is more than 100 °C higher, and its dielectric loss is much lower (*9*). Dielectric peaks have been observed for both the α- and γ-relaxations (*7, 9*).

The structures of these four polymers are shown in Figure 1, and their dissipation factors at a frequency of 1 kHz are plotted against temperature

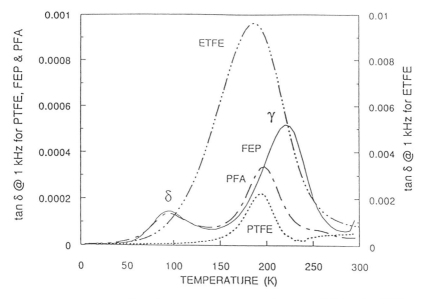

Figure 2. Temperature dependence of the dissipation factors (at 1 kHz) for PTFE, PFA, FEP, and ETFE. (Adapted from reference 8. Copyright 1991 American Chemical Society.)

up to 300 K in Figure 2. Each of the polymers exhibits the γ-relaxation that is attributed to motions of short chain segments in the amorphous regions. At this frequency, the peak occurs at 194 K for PTFE, 220 K for FEP, 198 K for PFA, and 186 K for ETFE. The height of the loss peak is smallest for PTFE and much larger for ETFE than for the perfluorocarbon polymers.

Frequency–Temperature Relationships

The Arrhenius activation energy (E_a) is defined by the following expression:

$$E_a = -R \frac{d \ln f_{max}}{d(1/T)} \tag{1}$$

where R is the gas constant, f_{max} is frequency of the loss maximum, and T is temperature. As shown in Figure 3, the activation energies for the γ-relaxations derived from Arrhenius plots of log frequency versus the reciprocal of the absolute temperature are about 16 kcal/mol for PTFE, FEP, and PFA and 10.7 kcal/mol for ETFE (8, 9). The higher temperature of the γ-relaxation in FEP than in the other polymers and the larger peak than in PTFE and PFA may be due to an overlapping β-relaxation (8).

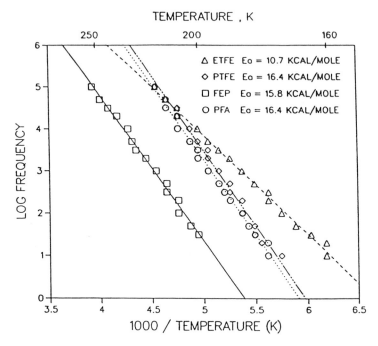

Figure 3. Arrhenius plot of the γ-relaxations in PTFE, FEP, PFA, and ETFE. (Reproduced from reference 8. Copyright 1991 American Chemical Society.)

The β-relaxation is attributed to internal motions in the crystalline regions (*1–3*).

Another formulation for the dependence of the frequency of the loss maximum in a relaxation on temperature follows Eyring:

$$f_{max} = \frac{kT}{2\pi h} \, e^{-\Delta H^{\neq}/RT} \, e^{\Delta S^{\neq}/RT} \tag{2}$$

where k is the Boltzmann constant, h is the Planck constant, and $\Delta H^{\neq}$ and $\Delta S^{\neq}$ are the activation enthalpy and entropy, respectively. Equation 2 can be rearranged in terms of the activation free energy, $\Delta F^{\neq}$.

$$\Delta F^{\neq} = \Delta H^{\neq} - T\Delta S^{\neq} = \Delta E^{\neq} + P\Delta V^{\neq} - T\Delta S^{\neq}$$

$$\Delta F^{\neq} = RT[\ln(k/2\pi h) + \ln(T/f_{max})] \tag{3}$$

Each observation of a combination of frequency and temperature for a loss maximum defines a value of $\Delta F^{\neq}$. From the dependence of $\Delta F^{\neq}$ on T and pressure (P), one obtains $\Delta S^{\neq}$ and the activation volume, $\Delta V^{\neq}$,

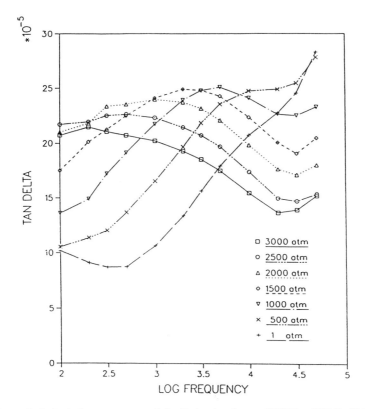

Figure 4. Isobaric frequency scans of the dissipation factor of PTFE at 212 K. (Reproduced from reference 10. Copyright 1992 American Chemical Society.)

$$\Delta S^{\neq} = - \left(\frac{\partial \Delta F^{\neq}}{\partial T} \right)_{P} \qquad (4)$$

$$\Delta V^{\neq} = - RT \left(\frac{\partial \ln f_{max}}{\partial P} \right)_{T} \qquad (5)$$

Effects of Pressure

With increasing pressure, internal motions are restricted due to decreased volume. These effects were investigated for the γ-relaxations in PTFE and FEP at pressures up to 3000 atm (*10*). As is shown in Figure 4, for isobaric frequency scans on PTFE at 212 K, increasing the pressure causes the peak to shift to lower frequencies. At a defined frequency, the relaxation shifts to higher temperatures as the pressure is increased. This result is illustrated in Figure 5 by using isobaric temperature scans at 1 kHz for FEP.

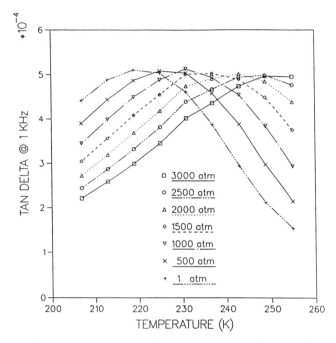

Figure 5. Isobaric temperature scans of the dissipation factor of FEP at 1 kHz. (Reproduced from reference 10. Copyright 1992 American Chemical Society.)

With increasing pressure, the activation energy increases from 16.4 kcal/mol at 1 atm to 21.1 kcal/mol at 2500 atm for PTFE as shown in Table I, but it remains essentially constant at 14 kcal/mol for FEP. In the same vein, the activation volume defined by eq 5 decreases with increasing temperature for PTFE as shown in Table II, but remains almost constant for FEP. The response of the γ-relaxation in PTFE to pressure is similar to that of the analogous relaxation in polyethylene (*11*). The different behavior in the closely related polymer, FEP, is consistent with the suggestion that the observed dielectric relaxation in FEP is actually a combination of the β- and γ-relaxations. For the fluoropolymers, the activation volumes for the γ-relaxations are equivalent to the volume of 1 to 2 CF_2 units in the amorphous regions compared to the volume of a little less than one CH_2 unit for the corresponding relaxation in polyethylene.

Other Relaxations

Figure 2 shows that FEP and PFA exhibit an additional loss peak at 94 K, the δ-relaxation. It is attributed to motions involving the side groups that are not present in PTFE and ETFE (*6, 8*). The Arrhenius plot in Figure 6

Table I. Effect of Pressure on Activation Energy

P *(atm)*	E_a *(kcal/mol)*	
	PTFE	*FEP*
1	16.4	13.5
500	16.7	14.3
1000	17.4	14.2
1500	18.2	14.1
2000	19.2	14.2
2500	21.1	13.9
av		14.0
σ		0.3

NOTE: av is average; σ is standard deviation.
SOURCE: Reproduced from reference 10. Copyright 1992 American Chemical Society.

Table II. Effect of Temperature on Activation Volume

T *(K)*	$\Delta V^{\neq}$ *(L/mol)*	
	PTFE	*FEP*
188	0.0455	
194	0.0460	
200	0.0472	
206	0.0449	
212	0.0426	
213		0.0225
218	0.0414	
219		0.0272
225	0.0362	0.0255
231	0.0337	0.0262
237	0.0313	0.0228
243		0.0220
249		0.0222
av		0.0241
σ		0.0022

NOTE: av is average; σ is standard deviation.
SOURCE: Reproduced from reference 10. Copyright 1992 American Chemical Society.

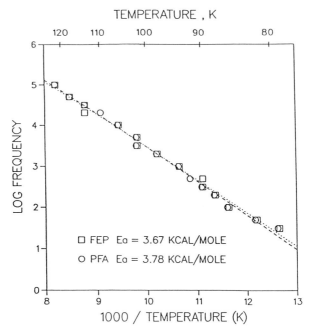

Figure 6. Arrhenius plot of the low temperature δ-relaxations in FEP and PFA. (Reproduced from reference 8. Copyright 1991 American Chemical Society.)

corresponds to an activation energy of 3.7 kcal/mol. This relaxation belongs to the category that has been called simple, noncooperative relaxations for which the activation entropy is close to 0 (*12, 13*). This fact is demonstrated by the Eyring activation free energy (from eq 3), which is essentially independent of temperature. This result means that the groups responsible for these relaxations move independently of each other.

In Figure 7, the dissipation factor at 1 kHz in a higher temperature range is shown for FEP and ETFE. In addition to the γ-relaxation, the higher temperature α-relaxation is seen at 103 °C (376 K) for FEP and 113 °C (386 K) for ETFE. This relaxation is associated with the glass transition and is attributed to the onset of motions of longer chain segments in the amorphous regions.

Teflon AF

Teflon AF is a family of amorphous copolymers of tetrafluoroethylene and 2,2-bis(trifluoromethyl)-4,5-difluoro-1,3-dioxole (*8, 14*). Teflon AF 1600 contains 66 mol% dioxole and has a glass transition temperature of 160 °C.

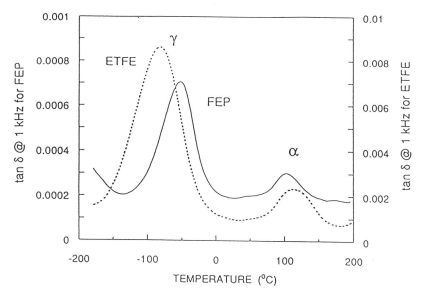

Figure 7. Dissipation factors of FEP and ETFE at 1 kHz extended to higher temperatures than in Figure 2.

The dissipation factor of Teflon AF at three frequencies is plotted against temperature in Figure 8. The large peak representing the α-relaxation or glass transition occurs at temperatures ranging from 188 °C at 100 Hz to 215 °C at 100 kHz. Its apparent activation energy is 115 kcal/mol. The β-relaxation, which is attributed to a local motion of the dioxole units, is observed at 88 °C and 111 °C at 100 Hz and 1 kHz, respectively, corresponding to an activation energy of 27 kcal/mol.

The γ-relaxation in polymers is attributed to local motions of aliphatic segments involving at least four chain atoms (*3*). Because Teflon AF contains only 34 mol% tetrafluoroethylene, the concentration of sequences of two or more TFE units is probably quite low. The γ-relaxation appears as a broad maximum near 220 K and has an activation energy of about 10 kcal/mol. This result is shown in Figure 9 in comparison with PTFE.

Teflon AF also exhibits a very weak relaxation that appears as a shoulder near 70 K. This relaxation probably reflects the motion of the CF_3 groups of the dioxole units and would be related to the δ-relaxation at 94 K in FEP and PFA.

Poly (perfluoropropylene oxide)

Krytox oils are low molecular weight polymers of hexafluoropropylene epoxide, and they are used in lubrication and related applications. The chemi-

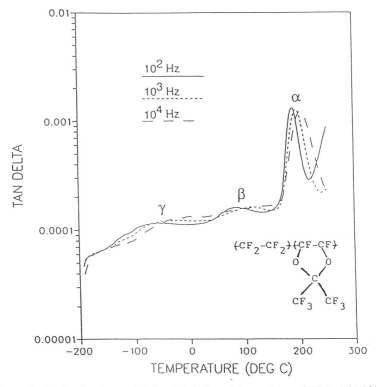

Figure 8. Dissipation factor of Teflon AF. (Adapted from reference 8. Copyright 1991 American Chemical Society.)

cal structure is $F(CF(CF_3)CF_2O)_nCF_2CF_3$. Dielectric measurements were made on samples of molecular weight 1850 and 8250, corresponding to values of n of about 10 and 49 (*15*). Figure 10 is a plot of the dissipation factor at 1 kHz versus temperature from 50 to 300 K. A logarithmic scale is used to present both large and small peaks clearly.

In scans by differential scanning calorimetry at 10 °C/min, glass transitions were observed at 195 and 219 K for the lower and higher molecular weight oils, respectively. At 1 kHz, the α-relaxation, which corresponds to the glass transition, occurred at 213 and 232 K. In work by Alper and co-workers (*16*) on a sample of molecular weight 3700 (n = 21), the glass transition was observed at an intermediate temperature. Arrhenius plots for the α-relaxations were curved, a familiar characteristic of glass transitions.

For both materials, a β-relaxation occurred near 100 K, about 6 K higher than the δ-relaxations that are associated with motions of the side groups of FEP and PFA. The activation energy for the β-relaxation in Krytox is 4.4 kcal/mol, slightly higher than the value of 3.7 kcal/mol for the δ-relaxations in FEP and PFA. All of these relaxations have activation entro-

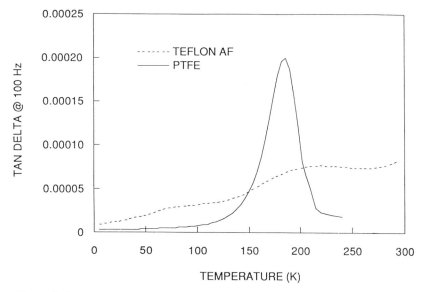

Figure 9. Dissipation factors for Teflon AF and PTFE at low temperatures. (Adapted from reference 8. Copyright 1991 American Chemical Society.)

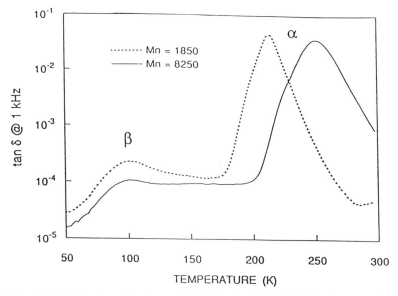

Figure 10. Temperature dependence of the dissipation factors of poly(perfluoropropylene oxide) with two different molecular weights. (Adapted from reference 15. Copyright 1992 American Chemical Society.)

pies close to 0 and are attributed to noncooperative motions involving fluorinated side groups.

Even though the peak heights for the α-relaxations are similar, that for the β-relaxation is about twice as large for the lower molecular weight sample. This result is probably related to the fact that it contains more end groups that, like the side groups, are perfluoromethyls.

Effect of Guest Molecules in PTFE

Even though PTFE is resistant to many chemicals, it can absorb limited amounts of compounds that contain little or no hydrogen (*17*). Three liquids, carbon tetrachloride, chloroform, and fluorocarbon-113 (CCl_2FCClF_2), were selected for dielectric measurements to study the differences between polar and nonpolar additives (*18*). Samples of PTFE were immersed in each liquid until the weight no longer increased. The increases in weight were 2.4% for carbon tetrachloride, 1.4% for chloroform, and 5.6% for F-113.

The temperature dependencies of the dissipation factor at 1 kHz for samples saturated with chloroform and F-113 as well as a PTFE control are shown in Figure 11. The data for the sample saturated with carbon tetrachloride were similar to those for the control and are not shown.

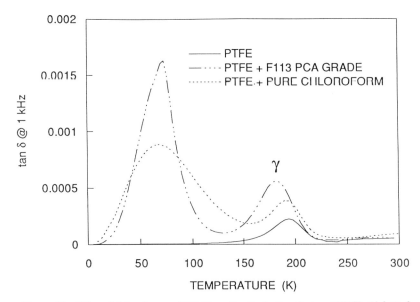

Figure 11. Effect of chloroform and F-113 on the dissipation factor of PTFE. (Adapted from reference 18. Copyright 1992 American Chemical Society.)

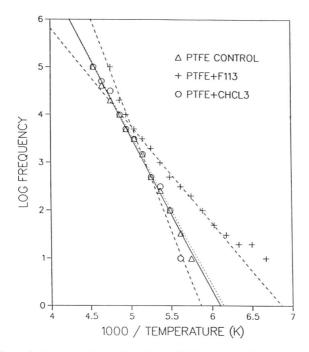

Figure 12. Arrhenius plot of the γ-relaxation in PTFE showing the effects of chloroform and F-113. (Reproduced from reference 18. Copyright 1992 American Chemical Society.)

Both of the polar additives increased the maximum for the γ-relaxation and introduced a new, large loss peak at lower temperatures. For the sample containing F-113, the γ-relaxation was also shifted to a lower temperature. Figure 12 is an Arrhenius plot for the γ-relaxations based on the maxima in isothermal frequency scans. The relationships for the control and the sample saturated with chloroform are very similar, and activation energies of 14 kcal/mol based on isothermal frequency scans and 16 kcal/mol based on isochronal temperature scans were obtained.

In the sample saturated with F-113, the γ-relaxation occurs at lower temperatures and the Arrhenius plot exhibits a "dog leg." The activation energy is only 9 kcal/mol for frequencies up to 1 kHz. At higher frequencies, it is 15 kcal/mol, and the data are closer to those for the other samples. The data in the low frequency region correspond to an activation entropy close to 0. This result suggests that at low temperatures and frequencies, the presence of F-113 facilitates the noncooperative motion of short segments in the amorphous regions of PTFE.

The low temperature relaxations shown in Figure 11 occur at 67 and 73 K for the samples saturated with chloroform and F-113, respectively.

Because the peak heights in isochronal temperature scans increase with increasing frequency, the maxima occur at lower temperatures in isothermal frequency scans. On that basis, the maxima for a frequency of 1 kHz occur at 49 K with chloroform and 61 K with F-113. The activation energies are ~2–3 kcal/mol, and the activation entropies are essentially 0. These relaxations are attributed to the reorientation of the absorbed molecules moving independently of each other. Subsequent experiments using hydrofluorocarbon additives have produced similar results (*19*).

Conclusions

A requirement for a relaxation in a fluoropolymer to be dielectrically active is that it involve the motion of a molecular conformation in which the dipolar vectors of the chemical bonds are not fully balanced. In PTFE, this requirement pretty much limits the dielectric loss to the γ-relaxation, which is attributed to motions of short chain segments in the amorphous regions. Increasing the pressure restricts internal motions because of decreasing volume. This result causes the γ-relaxation to shift to lower frequencies and higher temperatures. The glass transition or α-relaxation, which is attributed to motions of longer segments in the amorphous regions, is observed dielectrically in the amorphous polymers, Teflon AF and poly(perfluoropropylene oxide), as well as the semicrystalline polymers, FEP and ETFE.

Fluoropolymers such as FEP, PFA, poly(perfluoropropylene oxide), and Teflon AF exhibit a δ-relaxation at a temperature below 100 K. These relaxations have activation entropies close to 0, a characteristic of local, noncooperative motions. However, the fact that they are dielectrically active means that they must involve more than a simple rotation of a CF_3 group.

When small, dipolar molecules are absorbed in PTFE, a new relaxation appears at a very low temperature. This result is attributed to the noncooperative reorientation of the guest molecule. When the absorbed chemical is a nonpolar molecule such as carbon tetrachloride, this phenomenon does not occur.

References

1. McCrum, N. G. *J. Polym. Sci.* **1959**, *34*, 355.
2. Sperati, C. A.; Starkweather, H. W. *Adv. Polym. Sci.* **1961**, *2*, 465.
3. McCrum, N. G.; Read, B. E.; Williams, G. *Anelastic and Dielectric Effects in Polymeric Solids;* Wiley: New York, 1967.
4. Kabin, S. P. *Sov. Phys.-Tech. Phys.* **1956**, *1*, 2542.
5. Krum, F.; Muller, F. H. *Kolloid Z.* **1959**, *164*, 8.
6. Eby, R. K.; Wilson, F. C. *J. Appl. Phys.* **1962**, *33*, 2951.
7. Sacher, E. *J. Macromol. Sci.* **1981**, *B19*, 109.

8. Starkweather, H. W.; Avakian, P.; Matheson, R. R.; Fontanella, J. J.; Wintersgill, M. C. *Macromolecules* **1991,** *24,* 3853.

9. Starkweather, H. W. *J. Polym. Sci., Polym. Phys. Ed.* **1973,** *11,* 587.

10. Starkweather, H. W.; Avakian, P; Fontanella, J. J.; Wintersgill, M. C. *Macromolecules* **1992,** *25,* 7145.

11. Sayer, J. A.; Swanson, S. R.; Boyd, R. H. *J. Polym. Sci., Polym. Phys. Ed.* **1978,** *16,* 1739.

12. Starkweather, H. W. *Macromolecules* **1981,** *14,* 1277.

13. Starkweather, H. W. *Polymer* **1991,** *32,* 2443.

14. Resnick, P. R. *Polym. Prepr. (Am. Chem. Soc. Div. Polym. Chem.)* **1990,** *31(1),* 312.

15. Starkweather, H. W.; Avakian, P.; Fontanella, J. J.; Wintersgill, M. C. *Macromolecules* **1992,** *25,* 3815.

16. Alper, T.; Barlow, A. J.; Gray, R. W.; Kim, M. G.; McLachlan, R. J.; Lamb, J. *J. Chem. Soc., Faraday Trans. 2* **1980,** *76,* 205.

17. Starkweather, H. W. *Macromolecules* **1977,** *10,* 1161.

18. Starkweather, H. W.; Avakian, P.; Matheson, R. R.; Fontanella, J. J.; Wintersgill, M. C. *Macromolecules* **1992,** *25,* 1475.

19. Starkweather, H. W.; Avakian, P.; Fontanella, J. J.; Wintersgill, M. C. *Macromolecules* **1994,** *27,* 610.

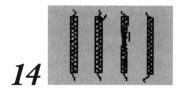

14

Dielectric Studies of Polymeric Nonlinear Optical Materials

C. Y. Stacey Fu, Mark H. Ostrowski, and Hilary S. Lackritz*

Dielectric relaxation and second-harmonic generation, a second-order nonlinear optical technique, can be used to study the rotational Brownian dynamics of the nonlinear optical chromophores doped or functionalized into a polymer matrix as a function of time or frequency and temperature. By combining these two techniques, one can determine more quantitatively how the thermal and temporal stability of chromophore orientation relates to specific polymer motions. Furthermore, the relationships between the intra- and intermolecular cooperativity and polymer structures and properties can be better understood. This information is critical for designing novel nonlinear optical polymeric materials for optical device applications.

Glassy polymer relaxations, particularly at temperatures well below the glass transition, have been difficult to study at the molecular level. Studies involving second-order nonlinear optical (NLO) techniques are sensitive in probing relaxation phenomena in doped and functionalized glassy polymers at room temperature and above at the molecular level (*1–11*). The rotational Brownian dynamics (*10*) of the NLO chromophores as a function of time and temperature have been studied using second-order NLO techniques such as second-harmonic generation (SHG), or frequency doubling. By monitoring the chromophore orientational dynamics, information con-

* Corresponding author.

cerning local mobility in the polymer matrices can be obtained. To determine more quantitatively how the relaxations obtained from SHG relate to specific polymer motions, and to traditionally measured relaxations, thermally and temporally dependent SHG studies are compared to dielectric relaxation studies. Dielectric relaxation measures both polymer relaxations and chromophore rotations in the polymer systems. By combining information obtained using these two techniques, a better understanding of the cooperative relaxations of both the chromophores and polymers can be achieved.

Organic NLO systems are of interest in the optoelectronic and photonic industries and have applications such as optical switching, coupling devices, and spatial light modulators (12). The polymer secondary properties, such as physical and optical stability and consistency, ease of preparation, compatibility with microelectronic processing methods, adhesion, and mechanical properties, are factors that determine whether the materials are technologically practical.

Second-harmonic generation is sensitive to small degrees of chromophore rotational mobility. The initial local free volume surrounding the dopants changes with time in the local glassy microenvironment, thus affecting the second-harmonic intensity. Electric field-induced dopant orientation occurs in regions of sufficient local free volume and segmental mobility. The disorientation of the dopants is due to mobility of the polymer chains and local free volume present in the vicinity of the dopants; the relaxations of the polymer chains and the presence of local free volume prevent the "freezing in" of the imposed orientation. Dopant orientation can thus be examined over a wide range of time and temperature scales as a function of the local free volume and segmental mobility in the glassy polymer matrix. Dielectric relaxation has been shown to be a useful technique for studying the transitions, relaxations, and intra- and intermolecular interactions in polymeric materials (13–15). In this study, dielectric relaxation is employed to examine the segmental relaxation in the homo-, doped, and functionalized polymer systems.

The two techniques do not necessarily measure the identical relaxations in polymers; dielectric relaxation is most sensitive to changes in bulk polarization (such as the side-group motion in poly(methyl methacrylate) (PMMA)), whereas second-order nonlinear optical studies are sensitive to rotational motion of the NLO active chromophore: this motion may or may not be identical.

The experiments discussed in this chapter evaluate the effects of dopant–polymer interactions, polymer backbone structures, and chromophore functionalization on dopant orientational dynamics and polymer relaxations in a special class of thermally stable polymers that was recently developed. By examining the second-order NLO properties of the doped or functionalized polymeric material as a function of time and temperature and the dielectric relaxation phenomena as a function of frequency and temperature, information concerning the local mobility and relaxation phenomena of the polymer microenvironment surrounding the NLO dopants can be obtained.

Background

This section outlines the general theory of dielectric relaxation and second-order nonlinear optics. Some of the more extensive efforts of dielectric relaxation and NLO studies on the dynamic behavior of various NLO polymer systems are also reviewed.

Dielectric Relaxation

The basic theory of dielectric relaxation has been extensively covered elsewhere (*16, 17*). Here, the dielectric dispersion phenomena and the Havriliak–Negami (HN) equation and the Schonhals and Schlosser model that have been used to describe the loss curves in polymers are briefly outlined. Dielectric dispersion or dissipation of energy takes place in a particular frequency region for a given polymer because of the dynamics of chain motions (*16, 17*). At any temperature, a frequency or characteristic band of frequency is associated with the motion of dipoles. Dielectric constant, ε', is observed to decrease with frequency in a relaxation region, and dielectric loss factor, ε'', passes through a maximum when the dispersion occurs.

Two types of dielectric experiments have been performed (*16, 17*). A sample can be held at a constant temperature while the frequency is scanned, or it is held at constant frequency while the temperature is varied. In both constant temperature and constant frequency experiments, dispersion is depicted by a peak in ε'' or $\tan\delta_\varepsilon$ (dissipation), which indicates that some forms of dipolar molecular motions are taking place. These motions can be associated to the micro-Brownian motion of the polymer main chain (α-relaxation) or motions of the flexible side groups along the chain (β-relaxation).

The HN empirical function has been extensively used to describe the breadth and asymmetry of the dielectric loss curves in polymers (*18*). Havriliak and Negami found that the relaxation process can be represented as the sum of two dispersions, and they proposed an empirical expression to represent the data

$$\frac{\varepsilon^*(\omega) - \varepsilon_\infty}{\varepsilon_0 - \varepsilon_\infty} = \frac{1}{[1 + (i\omega\tau_0)^\alpha]^\beta} \tag{1}$$

where ε_0 and ε_∞ are the relaxed and unrelaxed relative permittivities, respectively, and τ_0 is the relaxation time. The parameter α represents the symmetric broadening of the dielectric loss curve, and β is correlated to the asymmetric nature of the dispersion curve. If $\alpha = 1$ and $\beta = 1$ in eq 1, it becomes the Debye function (*19*), which predicted that a molecule exhibits only a single relaxation time. If $\beta = 1$ and α is allowed to vary between $0 < a \leq 1$, it becomes the Cole–Cole function (*20*), which modeled materials that exhibit a broad but symmetrical distribution of relaxation times. If $\alpha = 1$ and β is allowed to vary between $0 < \beta \leq 1$, the expression becomes the Davidson–Cole equation (*21*), which was able to describe materials with an asymmetrical distribution of relaxation times. These three empirical models, however, cannot successfully describe the breadth and asymmetrical behavior of the relaxation curves in polymers. By combining the Cole–Cole and Davidson–Cole models, Havriliak and Negami were able to adequately describe the shape of dielectric dispersion curves in many polymers. In this model, α and β are able to vary between $0 < \alpha \leq 1$ and $0 < \beta \leq 1$.

Because the HN function is a phenomenological relationship, a model was later developed by Schonhals and Schlosser to relate the fitting parameters α and β to the structures of polymers (*22, 23*). In this model, the frequency dependence of dielectric loss factor in the low-frequency limit was scaled as $\varepsilon''(\omega) \propto \omega^{\alpha}$ for $\omega \ll \tau_-^{-1}$ and in the high-frequency limit, $\varepsilon''(\omega) \propto \omega^{-\alpha\beta}$ for $\omega \gg \tau_-^{-1}$. Thus, on a log–log plot of the dielectric loss factors as a function of frequency, the slope of the dielectric loss peak on the low-frequency side ($\omega\tau_- \ll 1$) is proportional to α, whereas it is proportional to $-\alpha\beta$ on the high-frequency side ($\omega\tau_- \gg 1$). The parameter α is correlated with the intermolecular dynamics, whereas the product $\alpha\beta$ describes the local intramolecular dynamics of the polymer. Furthermore, α decreases in the range $1 > \alpha > 0$ with increasing large-scale correlation of the segments of different chains, and $\alpha\beta$ decreases in the range $0.5 > \alpha\beta > 0$ with an increase of hindrance of the local segments. In this research, the HN equation and the Schonhals and Schlosser model are employed to determine the values of α and β and to evaluate the extent of intermolecular coupling arising from various steric interferences of the polymer backbones. We make the assumption that this model, developed for homopolymer systems, will hold equally well for plasticized systems. Schonhals (private communication) indicates that this assumption is appropriate. Because this is a phenomenological model, it gives us not a direct measure of exact properties, but allows a relative comparison of the effects of different perturbations or inputs on the observed relaxations.

Second-Order Nonlinear Optics

The principles of second-order nonlinear optics are briefly outlined here. A complete development is available elsewhere (*12, 24, 25*). Dopant orientation in the polymer matrix is related to the observed NLO intensity. It has been shown that second-harmonic generation (SHG), a second-order NLO effect, is a sensitive, novel technique for examining the rotation of NLO dopants and studying the local microenvironment surrounding these chromophores. Second-harmonic generation is a special case of frequency mixing occurring when light waves of frequency ω passing through an array of molecules interact with them in such a way as to produce light waves at double the original frequency. Another second-order NLO effect is the linear electrooptic (LEO) effect, which involves electric-field-induced bire-fringence in the optical material (*24, 25*).

Nonlinear optical effects occur at the molecular level because of a devia-tion in the electronic potential energy from a simple harmonic dependence on the field (*24, 25*). When an intense field is applied, the polarization response becomes nonlinear. The microscopic polarization, **p**, in a power series of the field strength, **E**, can be expressed as (*26*)

$$\mathbf{p} = \alpha\cdot\mathbf{E} + \beta:\mathbf{EE} + \gamma:\mathbf{EEE} + \cdots \tag{2}$$

where α, β, and γ are the linear, second-, and third-order hyperpolarizabili-ties relating the vector quantities, **p** and **E**. The magnitude of β depends on the charge-transfer-induced delocalization giving rise to the NLO re-sponse, and can be independently measured by using SHG induced by a dc field (*26, 27*).

Second-order NLO properties including SHG arise from the second-order NLO susceptibility $\chi^{(2)}$ tensor found in the expansion of bulk polari-zation, P (*26*)

$$P = \chi^{(1)}\cdot\mathbf{E} + \chi^{(2)}:\mathbf{EE} + \chi^{(3)}:\mathbf{EEE} + \cdots \tag{3}$$

where $\chi^{(1)}$, $\chi^{(2)}$, and $\chi^{(3)}$ are linear, second-, and third-order susceptibility tensors relating the bulk polarization, P, to the field strength, **E**. $\chi^{(1)}$ and $\chi^{(3)}$ are nonzero in all materials. $\chi^{(1)}$ is responsible for linear optical proper-ties of the material such as absorbance and refraction. $\chi^{(3)}$ is responsible for third-harmonic generation, optical bistability, and other third-order NLO effects (*26, 27*). When both dc (or low-frequency ac) electric field, $\mathbf{E}(0)$, and optical field, $\mathbf{E}(\omega)$, are applied, the total field (as in eqs 2 and 3) the medium is subject to is $\mathbf{E} = \mathbf{E}(0) + \mathbf{E}(\omega)$ (*12*).

For a polymeric material to be second-order nonlinear optically active, NLO dopants are generally polar and have delocalized π-electron systems with an electron donor and an electron acceptor on either end and provid-ing a charge-transfer resonance dominant along one direction (*28, 29*). Because the second-order polarization is proportional to the field squared, SHG (and $\chi^{(2)}$) is zero in a centrosymmetric or randomly oriented medium due to cancellation of the net dipole moment. A material with inversion symmetry obeys the property that a reversal of all of the coordinate axes must leave the interrelationships between physical quantities unaltered. To make a material capable of SHG, the NLO dopants must be oriented noncentrosymmetrically in the polymer matrix (26, 27). This result is achieved by electric field poling as described subsequently.

The relationship between $\chi^{(2)}$ and β for poled polymer films was developed (30, 31). For SHG, $\chi^{(2)}_{333}$ and $\chi^{(2)}_{311}$ are the only two nonzero susceptibility tensors. The expressions for $\chi^{(2)}_{333}$p and $\chi^{(2)}_{311}$, ignoring the contributions of electric-field-induced third-order effects, are

$$\chi^{(2)}_{333} \approx Nf_3^2(\omega)f_3(2\omega)\beta_{zzz}\langle\cos^3\theta\rangle \tag{4}$$

and

$$\chi^{(2)}_{311} \approx Nf_1^2(\omega)f_3(2\omega)\beta_{zzz}(\langle\cos\theta\rangle - \langle\cos^3\theta\rangle)/2 \tag{5}$$

where N is the number of molecules per unit volume, and $f_i(\Omega)$ is the local field factor at frequency Ω for fields polarized along the i direction (32). The subscript 3 refers to the direction of the orienting (poling) field and z denotes the axis of the molecule parallel to the molecular dipole moment. The angle between 3 and z directions is represented by θ. The variable β_{zzz} is the second-order hyperpolarizability along the longest axis of the cigar-shaped molecule and is assumed to be roughly equal to β of the molecule ($\beta = \beta_{zxx} + \beta_{zyy} + \beta_{zzz}$). The variables $\langle\cos^3\theta\rangle$ and $\langle\cos\theta\rangle$ represent the orientational average values for $\cos^3\theta$ and $\cos\theta$, respectively. In the limit of low poling fields, eqs 4 and 5 become

$$\chi^{(2)}_{333} \approx \frac{Nf_3^2(\omega)f_3(2\omega)\beta_{zzz}\mu E}{5kT} \tag{6}$$

and

$$\chi^{(2)}_{311} \approx \frac{Nf_1^2(\omega)f_3(2\omega)\beta_{zzz}\mu E}{15kT} \tag{7}$$

where k is the Boltzmann constant, T is temperature, and μ is the dipole moment of the molecule, and thus

$$\chi^{(2)}_{311} \approx (1/3)\chi^{(2)}_{333} \tag{8}$$

The relationships between the experimentally measured second-harmonic intensity and relative values of $\chi^{(2)}$ have been developed (12, 25)

$$I_{2\omega} = (l^2 I_\omega^2(\chi^{(2)})^2/n_1^2 n_2)(\sin^2(\pi l/2l_c)/(\pi l/2l_c)^2) \tag{9}$$

where $I_{2\omega}$ is the second-harmonic intensity, I_ω is the fundamental intensity, n_1 and n_2 are the indices of refraction of the film at frequency ω and 2ω, respectively, l is the film thickness, and l_c is the coherence length of the film that characterizes the phase mismatch of magnitude π of the nonlinear polarization wave to the free propagating wave at 2ω. After further simplification and assuming no dispersion ($n_2 = n_1$), the $\chi^{(2)}$ of the polymer film, $\chi_{\text{film}}^{(2)}$, is

$$\chi_{\text{film}}^{(2)} \approx (I_{\text{film}}/I_{\text{ref}})^{1/2}(2/\pi)(n^3/l^2)_{\text{film}}^{1/2}(l_c^2/n^3)_{\text{ref}}^{1/2}\chi_{\text{ref}}^{(2)} \qquad (10)$$

where $I_{\text{film}}/I_{\text{ref}}$ is the ratio of the experimentally measured second-harmonic intensity of the film to that of a reference, and l_c is the coherence length of the reference. Quartz is a common reference and its l_c is about 20 μm (*33*). The square root of the experimentally measured second-harmonic intensity decay ratio is thus directly proportional to $\chi_{\text{film}}^{(2)}$, which is related to the NLO dopant orientation in the polymer matrix.

Residual surface and space charges persist in the films following corona poling (*34, 35*), which complicates the analyses of relaxation data. Polymer relaxations and residual charge effects must both be considered when evaluating the dopant orientational dynamics, and no *absolute* measure of the relaxation time of the chromophores can be determined. Several researchers have assumed the effects were negligible at low corona fields, whereas others, including our work, have tried to measure these effects directly (*35, 36*). The electric field effects in this case are poorly understood and a topic of much ongoing research (*37–39*). A surface voltage probe has been used to measure the surface voltage decay following poling in the doped PMMA systems. However, because the probe of the electrostatic voltmeter that we are using cannot withstand temperatures greater than 50 °C, surface voltage measurements could not be performed on these polymers. A high temperature stable probe is under design. This work thus focuses on aspects of polymer relaxations and all relaxation data reported here using second-harmonic generation include electric field effects, as is done by many researchers in the field. Near the glass transition temperature (T_g) these effects are relatively short lived (in PMMA doped with 4-dimethylamino-4'-nitrostilbene (DANS) at T_g, the residual field persisted for less than an hour), but they are still significant. The SHG decay data thus obtained, however, allow one to qualitatively interpret trends in the temporal stability of dopant orientation following poling.

Literature Review

Dielectric relaxation has been used in conjunction with NLO techniques to study the dynamic behavior in various guest–host, side-chain, and main-chain functionalized NLO polymer systems. Some of the more extensive efforts are summarized.

Singer and co-workers at AT&T were one of the first groups to report the observation of SHG in electric field poled polymer glasses, specifically the 4-[N-ethyl-N-(2-hydroxyethyl)]amino-4'-nitroazobenzene (Disperse Red 1, DR1, an NLO chromophore) doped PMMA system (1). In addition, isothermal decay of the second-order nonlinear susceptibility of the 10 wt% DR1 doped PMMA was studied by using SHG (39). The decay of the second-harmonic signal was consistent with the distribution of relaxation rates of an Arrhenius relaxation. Runt and co-workers (14) later investigated the dielectric behavior of unoriented DR1-PMMA mixtures and found that the dielectric loss curves became broader as the DR1 concentration increased from 0 to 10 wt%. This result was because the coupling between the dopants and polymer became stronger as the dopant concentration increased, broadening the loss curves. The average activation energy for α-relaxation process in the 10 wt% DR1 doped PMMA system was approximately 58 kcal/mol. A β-transition was also observed at around 60 °C in mixtures with low DR1 concentration. However, Runt and co-workers found that above 20 wt% DR1 content, the relaxation time distribution became independent of dye concentration and the β-relaxation disappeared. This result may have been caused by phase separation. Ren and co-workers (40) also performed the dielectric relaxation studies in a 12 wt% DR1 doped PMMA system. An activation energy of 50.7 kcal/mol was determined, a result that agreed fairly well with the data reported by Lei et al. (14). It was noted that the shape of the dipoles influences the magnitude of the activation energies and that larger dipoles correspond to higher activation energies.

Eich and co-workers (41) at IBM performed the dielectric relaxation measurements of the low-molecular-weight organic compounds, (S)-2-N-α-(methylbenzylamino)-5-nitropyridine and 2-N-(cyclooctylamino)-5-nitropyridine, to assess the mobility of NLO-active groups. These organic compounds have several advantages over doped or functionalized polymers, in that higher densities of nonlinear active groups and thus possibly higher optical nonlinearities can be achieved because no polymer backbone and spacer groups are incorporated. In both glasses, no low-temperature relaxation below T_g down to − 150 °C was observed at frequencies between 100 Hz and 4 MHz. This result illustrates that below T_g only electronic and no orientational contributions to the dielectric displacement were present. Also, the loss maxima plotted in an activation diagram showed no linear Arrhenius-type dependence. From these results, it was concluded that the only relaxation of the ground-state dipole occurred at the glass transition of the bulk, and thus a given orientational state of the ensemble should be stable at temperatures well below T_g. Second-harmonic experiments were performed on these systems as well. The dielectric dipole relaxation and the SHG relaxation were coupled to the glass transition of the bulk. The relaxation time, τ, obtained from SHG agreed well to extrapolated relaxation times from the dielectric results. This agreement suggested that both phenomena have the same relaxation mechanism. In addition, no sub-T_g relaxation was observed in the dielectric relaxation and SHG experiments; therefore, these systems had a stable noncentrosymmetrical orientational state below T_g.

The dielectric relaxation behavior of a side-chain functionalized polymer, poly-p-nitroaniline, was examined (5). Again, no low-temperature relaxation below T_g down to 20 °C was observed in the frequency range between 100 Hz and 100 kHz, a result that indicated that only electronic and no orientational contributions to the dielectric displacement were present below T_g. Furthermore, the loss maxima plotted in an activation diagram showed no linear Arrhenius-type dependence. This nonlinear dependence in the vicinity of T_g was interpreted as a temperature-dependent change of the activation energy for molecular rearrangements. The main orientational relaxation of the ground-state dipole, which was strongly coupled to the axes of the nonlinear polarizability tensor, was concluded to occur near T_g. In addition, the SHG measurements on this polymer were performed. Analysis of their data revealed a multiple exponential rise and decay of the second-harmonic signal. The average relaxation times obtained from the SHG experiments were longer than the orientational relaxation times predicted from the dielectric relaxation data according to Williams–Landel–Ferry (WLF) behavior (42). This finding was in contrast to their measurements on contact poled (electrode poled) organic glasses (41). The difference was attributed to the corona poling process in which both molecular reorientation and charge deposition–decay processes were involved.

In addition, others at IBM also studied the orientational relaxation of a variety of guest–host and side-chain functionalized polymers by using SHG (43, 44). Both the guest–host systems and the side-chain functionalized polymers decayed nonexponentially and were fitted by using the Kohlrausch–Williams–Watts function. The temperature dependence of the relaxation time of the guest–host system was not described well by Arrhenius expression, but it did correlate well to $1/(T_g + 50 - T)$. The temperature dependence of the relaxation time of the side-chain functionalized polymers exhibited similar behavior but deviated at temperatures far below T_g. Kumar and co-workers (45) at University of Massachusetts–Lowell performed similar analysis on their SHG data from doped highly cross-linked systems. Their results differed from typical guest–host systems in that the temperature dependence of the relaxation time followed the Arrhenius expression for a certain temperature range then deviated greatly. The explanation for this behavior was that at least two different relaxation mechanisms were responsible for the SHG decay.

Time-domain dielectric spectrometry was employed to study the relaxation behavior of polystyrene (PS) containing NLO chromophores, DANS and DR1 (46). The sub-T_g β-relaxation of the dyes occurred on the same time scale as the β loss characteristic of the polymer. The variation of the β-relaxation with temperature did not follow the predicted T^{-1} behavior. Furthermore, at 20 °C below T_g, the dielectric loss was observed to rapidly increase, a result indicating that either more chromophores were participating in the relaxation process or that the angular variation over which the chromophores were relaxing was increasing. Physical aging at 20 °C below T_g in DR1-doped PS showed that the β-relaxation associated with the dye became more Cole–Cole-like with increasing aging time. To retain dopant orientation following poling for a guest–host polymer system, it is critical to anneal out the long-range T_g relaxations to longer times or higher temperatures than to diminish the sub-T_g relaxations associated with local segmental motions.

The dielectric behavior of a series of five different polymers containing the same NLO chromophore, 4'-(dialkylamino)-4-(methylsulfonyl)azobenzene, as a dopant or covalently functionalized to the side chain or main chain was studied (47). The relaxation of the chromophores was coupled to the α-relaxation of the polymer in all cases. The guest–host system showed a narrowing of the dielectric dispersion curves with increasing temperature that affected primarily the low-frequency side of the loss peak. For the side-chain and main-chain functionalized systems investigated, the time–temperature superposition principle held and the distribution functions for the relaxation times did not change with temperature. Furthermore, the fit of the HN function revealed that the parameters α and β for these systems were approximately along the contour line, a result indicating that the widths of their loss peaks did not change with temperature. Finally, an additional slow relaxation mode at frequencies below the α peaks was observed in the main chain functionalized polymer systems. This mode was attributed to the global reorientation of the end-to-end vectors of the chains. This was the first time this type of conclusion was drawn from experimental data. SHG in conjunction with thermally stimulated discharge current measurements was also performed on the same side-chain and main-chain functionalized systems (6). In the side-chain functionalized polymer system, the decay of the second-harmonic signal followed the α-relaxation peak of the thermally stimulated discharge, which occurred at the T_g measured by using differential scanning calorimetry. In the main-chain functionalized polymers, two relaxation modes were found. One mode corresponded to the rotation of the chromophores along their axis and the second-mode corresponded to the reorientation of the end-to-end vector of the polymer chain. These results were consistent with the dielectric analysis.

Dielectric relaxation was also employed to investigate how the dielectric properties change before and after thermal or chemical imidization of polyamic acids in the process of making high temperature stable NLO materials.

Wu and co-workers (*48*) at Lockheed found a loss peak near 150 °C in the dielectric loss tangent data after thermal or chemical imidization of many polyamic acid precursors, which was associated with dehydration during imide formation. An improved electrooptic (EO) thermal stability was observed in samples that were thermally imidized during poling or immediately chemically imidized following poling. This increased stability was caused by a reduction in the free volume of the polymer matrix around the aligned NLO chromophores during imidization. Dielectric relaxation experiments were later performed to determine the influence of plasticization by NLO chromophores in the fully cured polyimides, Hitachi polyimide LQ-2200 and Amoco Ultradel polyimides 4212 and 3112 (*49, 50*). The peak of the dielectric response was shifted to lower temperatures as the dopant concentration increased and indicated that the polymer host was being plasticized. The increases in the baselines and step heights of the real part of the dielectric response as dopant concentration increased were correlated with the increases in the refractive indices and poled electrooptic responses, respectively. Because the step in the real dielectric response indicates the α-transition of the polymeric material that is associated with the T_g, dielectric spectroscopy provides useful information on the poling temperatures necessary to obtain sufficient second-order NLO signal.

Recently, Torkelson and co-workers (*8, 9*) at Northwestern University examined the rotational reorientation dynamics of DR1 in poly(isobutyl methacrylate) (PIBMA), poly(ethyl methacrylate) (PEMA), and polystyrene (PS) by employing SHG and dielectric relaxation. Very good agreement was observed between the average rotational, reorientation relaxation time constants, $\langle\tau\rangle$, obtained above T_g by using SHG and dielectric relaxation, which indicated that the rotational dynamics of the NLO chromophore DR1 were directly coupled to the α-relaxation dynamics in these polymers. Even for the PEMA system, where dielectric relaxation showed distinct α- and β-relaxations, the rotational reorientation of DR1 monitored by using SHG agreed with the dielectric relaxation results associated with the α-relaxation. Furthermore, they pointed out that whereas the DR1-doped PIBMA and PEMA systems exhibited similar α-relaxation dynamics above T_g, the DR1-doped PS system showed a stronger temperature dependence and indicated that PS may be a more fragile glass-former than the methacrylate-based polymers. In all cases, $\langle\tau\rangle$ obtained by using SHG exhibited an apparent Arrhenius temperature dependence below T_g, and the dynamics were well described by the WLF equation above T_g. In addition, they studied the relative contribution of the third-order effect induced by an electric field to SHG by using a delayed-trigger approach (*51*). By assuming no randomization of chromophore orientation occurred on time scales less than 10^{-4} s, which was confirmed by dielectric measurements, it was found that the third-order effect contributed to 12% of the signal in PMMA, PEMA, and PIBMA doped with 2 wt% DR1 and to 20% of the signal in the same polymers doped with 2 wt% DANS. In these low T_g polymers, the motions observed in SHG and dielectric relaxation are not expected to be similar; these polymers are highly flexible and the backbone and side-chain motions (location of maximum polarization change) are strongly coupled. In addition, they studied the rotational reorientation of rotor probes in PIBMA and PEMA by using SHG and steady-state fluorescence using twisted intermolecular charge transfer probes, which were also NLO-active (52). They found that the average rotational reorientation relaxation times of the rotor probes determined from SHG were coupled to the α-relaxation. In contrast, the rotor motions involving internal rotations or isomerization determined from fluorescence studies were significantly decoupled from the α-relaxation and associated with sub-β-relaxation processes. This result provides additional evidence supporting the fact that different techniques are sensitive to different motions in the polymer.

The techniques of dielectric relaxation and SHG may measure different relaxations of the polymer thin film systems. However, all of these studies indicate that there are close correlations between the results obtained by using dielectric relaxation and nonlinear optics. By combining these two techniques, important information concerning polymer relaxations and mobility in the NLO polymeric materials can be obtained. Furthermore, the relationships between the relaxation modes and polymer structures and properties can be studied.

Experimental Techniques

This section describes the experimental protocols that our research group at Purdue University employs to investigate the relaxation phenomena in the homo-, doped, and side-chain functionalized polymer systems.

Dielectric Relaxation Apparatus and Experiments

Dielectric relaxation experiments were performed with GenRad 1689 RLC Digibridge, which had a frequency range between 12 Hz and 100 kHz and a tolerance of 0.005%. One volt peak to peak was used for all experiments. A convection oven (Delta Design 9023) with an accuracy of ±0.1 °C was used to control the temperature during the measurements. The Digibridge and oven were interfaced with a Macintosh Quadra 950 and remote controlled by using the LabView program.

For each dielectric experiment, the film was heated to approximately $T_g - 50$ °C in 7–10 min and maintained at this temperature for half an hour. The capacitance and $\tan\delta_\varepsilon$ were measured at 40 different frequencies between 12 Hz and 100 kHz. Values of capacitance and $\tan\delta_\varepsilon$ at each frequency were averaged over 10 measurements. From the capacitance and tanδ_ε measured, ε' and ε'' were then calculated (*16, 17, 53*). The procedure was repeated for temperatures every 10 °C up to approximately T_g + 50 °C. The dielectric values obtained from these experiments had an error of ~5%, which was mainly caused by the uncertainties in the measurements of the geometrical dimensions of the sample capacitors and film thicknesses.

SHG Apparatus and Experiments

The schematic diagram of the SHG measurement apparatus is shown in Figure 1. The laser light was generated by a Q-switched Nd–YAG (neodymium–yttrium–aluminum–garnet) laser (10 Hz, 6–8 ns pulse width) at a fundamental wavelength of 1.064 μm. The beam was *p*-polarized and split so that the sample and quartz reference could be tested simultaneously. Quartz, an inorganic crystal, crystallizes noncentrosymmetrically and is capable of SHG. The sample beam was passed through the vertically mounted polymer film at a 67° angle (measured from the normal vector of the sample

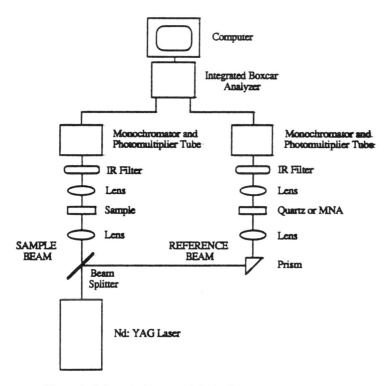

Figure 1. Schematic diagram of the SHG measurement apparatus.

surface to the incident beam). Another lens was placed after the sample to focus the beam into the monochromator. An IR filter and monochromator filtered all stray light so that only the second-harmonic light at 532 nm was detected by the photomultiplier. The second-harmonic signal was then collected by a boxcar integrator interfaced with a computer. The reference beam was collected and filtered in a similar manner. The sample optical signal was then ratioed to that of the quartz reference to minimize errors due to laser drift or changes in the ambient environment. A FORTRAN program was written to convert the experimentally measured second-harmonic intensity to $\chi^{(2)}$. The reported $\chi^{(2)}$ values were normalized to the point at which the applied voltage was turned off.

Contact and corona poling have been used to orient NLO dopants in a polymer matrix by applying an electric field across the film. Detailed discussions of these poling techniques can be found elsewhere (1–4, 8, 9, 39), so they are only briefly outlined here. When enough voltage is applied and the polymer matrix has sufficient mobility, dopants can respond to the induced torque and align in the electric field direction to produce a second-harmonic signal. When the applied field is removed, dopants in regions of sufficient mobility can rotate out of the poling-induced orientation, thus no longer contributing to the second-harmonic intensity. Corona poling (2, 4, 54, 55) creates an electric field by generating a discharge from a metallic needle tip or a wire. Because of the high field generated, the surrounding air becomes ionized and has the same polarity as the corona discharge. These ions are driven toward and deposited onto the film surface creating a field high enough to orient the dopants. In the experiments described subsequently, corona poling was the poling technique employed.

The schematic representation of a typical SHG experiment is shown in Figure 2. The polymer film was heated to $T_g + 5\,°C$ in approximately 20 min and maintained at this temperature for 1 h to insure uniform thermal history. The voltage was then applied for 25 min at this temperature and remained on until the film was cooled to the temperature at which the randomization of chromophore orientation was to be observed. The applied field was removed once the final temperature was reached. The randomization of chromophore orientation was then observed at this temperature. It is more difficult to orient chromophores when poled below T_g because of the restricted segmental mobility that hinders rotation (4). All of the films were thus poled above T_g to enhance the second-harmonic signal during poling and achieve better temporal stability of chromophore orientation following poling. A model was also developed to describe the rotational Brownian dynamics of the chromophores (10).

Material Preparation and Characterization

The amorphous, thermally stable polyarylene ether polymers synthesized by McGrath and co-workers (56) are investigated. The structures and prop-

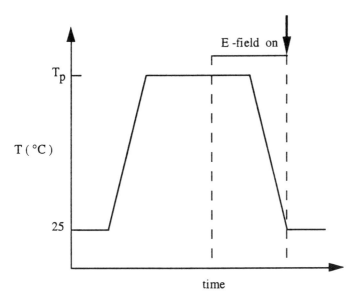

Figure 2. Schematic representation of the SHG experiments.

erties of these homo- and side-chain functionalized polymers, bis-A-poly(ar-ylene ether) phosphine oxides (PEPO), F_6-bis-A-PEPO, phenolphthalein-PEPO (PP-PEPO), phenolphthalein-poly(arylene ether) ketone (PP-PEK), phenolphthalein-poly(arylene ether) sulfone (PP-PES), PP-PEPO deriva-tized with 10% *p*-nitrophenylhydrazine (PP-PEPO-pNPH), and PP-PEPO derivatized with 10% 2,4-dinitrophenylhydrazine (PP-PEPO-DNPH), are shown in Table I. Detailed synthesis and characterization of these polymers can be found elsewhere (*56*). All of the polyarylene ethers investigated here have number-average molecular weight, $\overline{M}_n$, of 15,000 g/mol and poly-dispersity indices of approximately 2. PP-PEPO, which has an $\overline{M}_n$ of 41,000 g/mol and a polydispersity index of approximately 2, was also investigated.

Two NLO chromophores, DANS (Kodak) and 4-amino-4'-nitroazoben-zene (Disperse Orange 3, DO3) (Aldrich, 95% pure), were used in this study and their structures and properties are shown in Table II. DANS was used as received. DO3 was further purified via recrystallization with acetone. The solvent used was chloroform ($CHCl_3$).

Polymer and 10 wt% chromophore were dissolved in $CHCl_3$ to make up a solution that contained about 10 wt% solids. This solution was well mixed, filtered through a 5-μm filter, and spin-coated onto glass slides coated with indium tin oxide (ITO). The spun films were then dried to remove as much solvent as possible. They were first dried at ambient condi-tions for about 48 h, then under vacuum for about 24 h at ambient tempera-ture and heated slowly over one day to approximately T_g + 10 °C. They

Table I. Structures and Thermal and Physical Characterization Data of the Host Polymers

Polymer	T_g (°C)	TGA (°C) 5% wt. loss in air	$\overline{M}_w$ (g/mol)	$\overline{M}_n$ (g/mol)
Bis A-PEPO	205	490	30,000	15,000
F₆Bis A-PEPO	210	500	35,660	18,130
PP-PEPO	265	496	32,200	16,100
PP-PES	259	490	31,000	15,500
PP-PEK	226	490	32,000	16,000
PP-PEPO-pNPH	284	485	30,000	15,000
PP-PEPO-DNPH	278	485	30,000	15,000

NOTE: TGA is thermal gravimetric analysis.

Table II. Structures and Thermal and Physical Characterization Data of the Nonlinear Optical Chromophores

Chromophore	T_m (°C)	MW (g/mol)	$V(\text{Å}^3)^a$
4-amino-4'-nitroazobenzene (DO3)	215	242.2	179
$O_2N\text{-}\langle\bigcirc\rangle\text{-}N{=}N\text{-}\langle\bigcirc\rangle\text{-}NH_2$			
4-dimethylamino-4'-nitrostilbene (DANS)	256	268.3	206
$O_2N\text{-}\langle\bigcirc\rangle\text{-}\overset{H}{\underset{H}{C}}{=}C\text{-}\langle\bigcirc\rangle\text{-}N(CH_3)_2$			

^{a}Volume calculated using van der Waals volumes.

were then slowly cooled to 25 °C and stored in a dessicator before SHG measurements were performed.

Samples for dielectric relaxation experiments were prepared by spin-coating the undoped or doped polymers onto 50:50 ITO:glass substrates (6). After careful drying, gold electrodes were then deposited on top of the films by using an evaporator.

Film thicknesses measured via profilometry ranged between 4 to 6 μm for SHG measurements and between 12 to 20 μm for dielectric relaxation experiments. The absorbances of the undoped polymers obtained from the ultraviolet and visible spectroscopy were between 200 nm and 300 nm and those of the DO3- and DANS-doped polymeric films were between 350 nm and 550 nm. The T_g values of the undoped and doped materials were measured via differential scanning calorimetry (DSC) (10 °C/min heating rate).

Results and Discussions

In this section, the intermolecular cooperativity and segmental relaxation arising from dopant–polymer interactions, polymer backbone structures and molecular weight, and chromophore functionalization in these polymer matrices studied by using dielectric relaxation and SHG are discussed.

Effect of Dopant–Polymer Interactions

Figure 3 shows a plot of ε' versus temperature at 50.8 kHz for undoped and 4 wt% DANS-doped bis-A-PEPO. The T_g of the 4 wt% DANS-doped bis-A-PEPO was lower than that of the undoped sample. This result was expected because chromophores are plasticizers. Furthermore, the dielectric constant was larger for the doped system because the polar chromophores enabled the system to be more polarizable and contributed to the dielectric constant. Under these experimental conditions, no β-transition was observed in any of these polyarylene ethers. The normalized dielectric

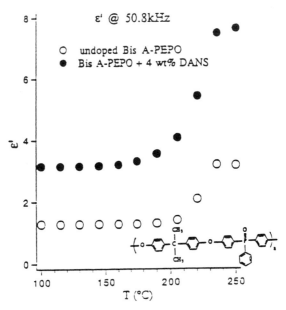

Figure 3. Values of ε′ versus T for undoped bis-A-PEPO and bis-A-PEPO doped with 4 wt% DANS measured at 50.8 kHz.

loss curve ($\varepsilon''/\varepsilon''_{max}$ vs. f/f_{max}, where f_{max} is the frequency of peak maximum) was slightly broader in the 4 wt% DANS-doped bis-A-PEPO systems than the undoped bis-A-PEPO as shown in Figure 4. This broadening was because the intermolecular coupling should become stronger in the doped system, broadening the loss curve. This result was consistent with those observed in polymers with lower T_g values and more flexible backbones such as the undoped and DR1-doped PMMA systems (*14*).

Effect of Polymer Backbone Structures

Because the polymers investigated here have very similar structures, with variations on a particular pendant group, the effect of backbone structures and steric constraints on the intermolecular cooperativity and segmental relaxation behavior was examined. The variations in the breadth of the loss curves reflect differences in the distributions of relaxation times and changes in local microenvironment and inter- and intramolecular interactions (*13, 15*). A broadening on the low frequency side of the loss curve was observed when the polymer backbone became more polar or sterically hindered and indicated that the intermolecular coupling became stronger.

Figure 5 shows the normalized dispersion curves for bis-A-PEPO, F_6-bis-A-PEPO, and PP-PEPO at approximately $T_g + 15$ °C. The breadth of

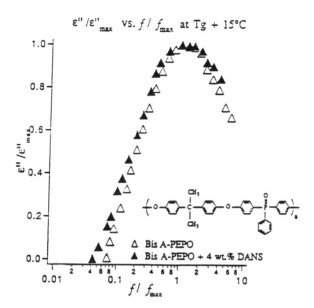

Figure 4. Values of $\varepsilon''/\varepsilon''_{max}$ versus f/f_{max} for undoped bis-A-PEPO and bis-A-PEPO doped with 4 wt% DANS measured at $T_g + 15$ °C.

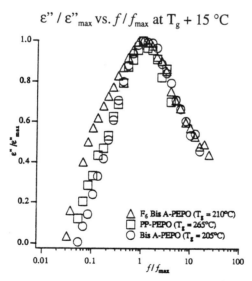

Figure 5. Values of $\varepsilon''/\varepsilon''_{max}$ versus f/f_{max} for undoped F_6-bis-A-PEPO, PP-PEPO, and bis-A-PEPO measured at $T_g + 15$ °C.

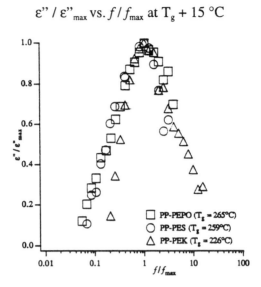

$\varepsilon'' / \varepsilon''_{max}$ vs. f / f_{max} at $T_g + 15\ °C$

Figure 6. Values of $\varepsilon''/\varepsilon''_{max}$ versus f/f_{max} for undoped PP-PEPO, PP-PES, and PP-PEK measured at $T_g + 15\ °C$.

the dispersion curves increased in the order bis-A-PEPO < PP-PEPO < F_6-bis-A-PEPO. Steric effects arising from the pendant fluorine atoms and the strong intermolecular interactions associated with their high electronegativity in F_6-bis-A-PEPO resulted in the broadest loss curve. This result showed that F_6-bis-A-PEPO exhibited the strongest intermolecular coupling. PP-PEPO, because of its bulky heterocyclic pendant lactone, resulted in a slightly broader dispersion curve than that of bis-A-PEPO. However, because the heterocyclic pendant lactone on PP-PEPO was not as polar as the fluorine atoms on F_6-bis-A-PEPO, it may not be as intermolecularly coupled as reflected by the narrower loss curve. Similar trends were also observed in the normalized loss curves for PP-PEPO, PP-PES, and PP-PEK as shown in Figure 6. The ketone functionality on PP-PEK was less bulky than the sulfone functionality on PP-PES and the phosphine oxide functionality on PP-PEPO and resulted in the narrowest loss curve. Furthermore, the sulfone functionality on PP-PES was more flexible than the phosphine oxide functionality on PP-PEPO and resulted in the slightly narrower dispersion curve. These results were consistent with those observed in the polymers with lower T_g values and more flexible backbones (15). For polymers with smoother, less polar, and more flexible and symmetrical backbones, less constraint existed on segmental relaxation from interactions with neighboring, nonbonded segments and the relaxation times showed a near-Arrhenius temperature dependence. On the other hand, those with less flexible backbones or sterically hindering pendant groups exhibited

Table III. Values of α and αβ obtained
from the Havriliak–Negami Equation Fit

Polymer	α	αβ
Bis A-PEPO	0.95	0.45
F₆Bis A-PEPO	0.81	0.38
PP-PEPO	0.88	0.51
PP-PES	0.96	0.64
PP-PEK	0.99	0.64
PP-PEPO-pNPH	0.73	0.72
PP-PEPO-DNPH	0.77	0.76

broader dielectric dispersion curves, reflecting broader distributions of relaxation times, and stronger intermolecular coupling (*15*).

The HN equation was used to fit these dielectric loss curves. It was able to fit them very well and the results of the fits are reported in Table III. Figure 7 shows the fit to the dispersion curve of the PP-PEPO polymer. For bis-A-PEPO, F₆-bis-A-PEPO, and PP-PEPO, the value of α was the smallest for F₆-bis-A-PEPO, a result indicating that it had the strongest intermolecular coupling. Similarly, for PP-PEPO, PP-PES, and PP-PEK, the value of α was the smallest for PP-PEPO, again indicating that it had the strongest intermolecular cooperativity. This attempt was the first, as far as we know, to employ the Schonhals and Schlosser model to describe the extent of intermolecular interactions arising from various steric interferences, and as shown here

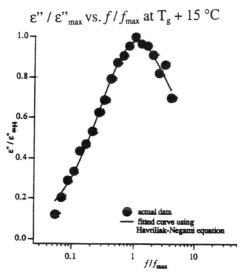

$\varepsilon'' / \varepsilon''_{max}$ vs. f / f_{max} at $T_g + 15\ °C$

Figure 7. Values of $\varepsilon''/\varepsilon''_{max}$ versus f/f_{max} for undoped PP-PEPO measured at $T_g + 15\ °C$.

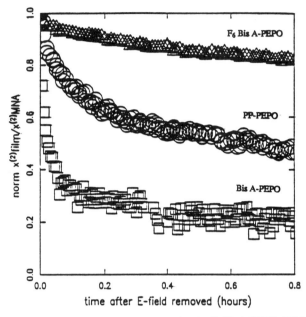

Figure 8. Temporal stability of dopant orientation in F_6-bis-A-PEPO, PP-PEPO, and bis-A-PEPO doped with 10 wt% DO3 poled at T_g + 5 °C; temperature at which chromophore disorientation was observed was approximately T_g − 140 °C.

the model can successfully correlate the values of α to the intermolecular coupling of these polymers.

Second-harmonic generation experiments were also performed on the DO3-doped bis-A-PEPO, F_6 bis-A-PEPO, and PP-PEPO systems, and the results are shown in Figure 8. The temporal stability of dopant orientation at T_g − 140 °C following poling increased in the order bis-A-PEPO < PP-PEPO < F_6-bis-A-PEPO. Because of the strongest intermolecular cooperativity of the F_6-bis-A-PEPO chain segments, chromophore motions were strongly restricted and resulted in the most stable second-harmonic signal. Because the PP-PEPO chain segments were not as intermolecularly coupled as F_6-bis-A-PEPO, chromophores could more easily rotate out of the poling induced orientation and result in the less stable second-harmonic signal. The DO3-doped bis-A-PEPO system showed the poorest temporal stability of dopant orientation because it was not as intermolecularly coupled as the F_6-bis-A-PEPO and PP-PEPO systems. This result demonstrates that by combining dielectric relaxation and SHG, information concerning the extent of intermolecular coupling and local mobility in these polymer matrices can be obtained.

It is difficult to directly compare the dispersion curves of the polymers investigated here with those that have more flexible backbones reported

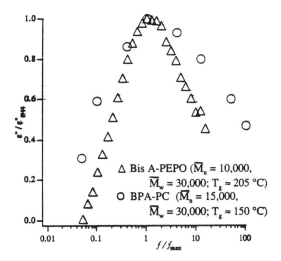

Figure 9. Values of $\varepsilon''/\varepsilon''_{max}$ versus f/f_{max} for bis-A-PEPO and BPA-PC. (Reproduced from reference 15. Copyright 1993 American Chemical Society.)

in the literature because of different molecular weight and polydispersities of the polymers. However, attempts were made to compare the polymer systems that have similar structures or polydispersities. When comparing the normalized loss curves of bis-A-PEPO ($\overline{M}_n$ = 15,000 g/mol, $\overline{M}_w$ = 30,000 g/mol) and bisphenol-A-polycarbonate (BPA-PC) ($\overline{M}_n$ = 10,000 g/mol, $\overline{M}_w$ = 30,000 g/mol) investigated by Ngai and Roland (*15*), BPA-PC exhibited a broader loss curve than that of bis-A-PEPO even though BPA-PC had less sterically hindered backbone (Figure 9). This result contradicted reports in the literature and may be explained by the different polydispersities of the bis-A-PEPO and BPA-PC polymers. For polydisperse polymers, the dispersion is inhomogeneously broadened because of the distribution of relaxation times (*57*). Another possible explanation for the discrepancy may be different sample configurations or preparations.

We also compared the normalized dispersion curves of bis-A-PEPO and poly(isobutyl methacrylate) (PIBMA) ($\overline{M}_n$ = 140,000 g/mol, $\overline{M}_w$ = 300,000 g/mol), which had a more flexible backbone and polydispersity index very close to that of bis-A-PEPO (*8*). The breadth of the loss curve of the 1 wt% DR1-doped PIBMA was very similar to that of bis-A-PEPO. One could expect that the undoped PIBMA would exhibit a slightly narrower loss curve than the 1 wt% DR1-doped PIBMA because of the weaker intermolecular coupling in the absence of dopant–polymer interactions. It is also expected that the undoped PIBMA would exhibit a slightly narrower loss curve than bis-A-PEPO. This result would agree with the literature in which polymers having less flexible backbones or sterically hindering pendant groups exhibited broader dispersion curves (*15*).

Effect of Chromophore Functionalization

The effect of chromophore functionalization on the relaxation behavior and intermolecular coupling was also investigated by using dielectric relaxation. Several advantages exist of derivatizing instead of doping the chromophores into the polymer matrices. Higher dopant concentration can be incorporated without running into the problems of phase separation or dopant aggregation when the chromophores are covalently bonded to the polymer backbone. A better temporal stability of chromophore orientation may be achieved because of the stronger coupling between the covalently bonded dopants and polymer backbone (6, 47).

Figure 10 shows the normalized dielectric loss curves of PP-PEPO, PP-PEPO-pNPH, and PP-PEPO-DNPH at a temperature of approximately T_g + 15 °C. They were broader on the low-frequency side for the two side-chain functionalized systems, PP-PEPO-pNPH and PP-PEPO-DNPH, than the undoped PP-PEPO, a result indicating stronger intermolecular coupling between the covalently bonded chromophores and polymer. PP-PEPO-DNPH exhibited a narrower loss curve than that of PP-PEPO-pNPH even though PP-PEPO-DNPH has an extra NO_2 group on each NLO-phore. This phenomenon contradicted the observations reported in the literature (15) and may be because the extra NO_2 groups could not pack as well in the PP-PEPO-DNPH matrix. This characteristic resulted in greater local mobility (private communication with Duane B. Priddy, Jr.), and thus a lower T_g (as determined by DSC) and weaker intermolecular cooperativity (as

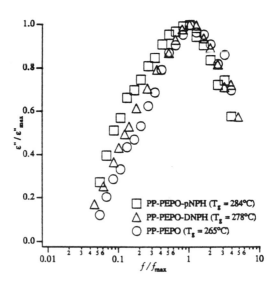

Figure 10. Values of $\varepsilon''/\varepsilon''_{max}$ versus f/f_{max} for PP-PEPO, PP-PEPO-pNPH, and PP-PEPO-DNPH measured at T_g + 15 °C.

reflected by a narrower loss curve) than PP-PEPO-pNPH. The results of the HN equation fits are shown in Table III. The values of α increased in the order PP-PEPO-pNPH < PP-PEPO-DNPH < PP-PEPO and indicated that PP-PEPO-pNPH had the strongest intermolecular coupling. The SHG measurements on these two side-chain functionalized systems will be performed, and the results will be correlated with those obtained by using dielectric relaxation and published in the near future.

Conclusions

Systematic studies on the effect of structure–property relationships on chromophore orientational dynamics and polymer relaxations in various NLO polymeric materials have been performed by several groups by using dielectric relaxation and SHG. By examining the second-order NLO properties of the doped or functionalized polymeric material as a function of time and temperature and the dielectric relaxation phenomena as a function of frequency and temperature, information concerning the local mobility and relaxation phenomena of the polymer microenvironment surrounding the NLO dopants can be obtained. In this research, we also showed that by combining dielectric relaxation and SHG, specific issues such as the effects of dopant–polymer interactions, polymer backbone structures and molecular weight, and chromophore functionalization on the segmental relaxation behavior and intermolecular cooperativity in the rigid, high T_g second-order NLO polymers can be investigated.

The dielectric loss curve was observed to be broader in the doped as compared to the undoped systems, a result indicating that the coupling between the dopants and polymer became stronger. Changes in the polymer backbone structure affected the intermolecular cooperativity and segmental relaxations. The dispersion curve became broader as the pendant groups on the polymer backbone became more polar or more sterically hindered; therefore, the segmental relaxation was indicated to be more intermolecularly coupled. Second-harmonic generation experiments also showed that chromophore disorientation following poling was retarded as the polymer chain segments became more intermolecularly coupled. For the specific polymer systems examined here, PP-PEPO with $\overline{M}_n$ of 41,000 g/mol and 16,000 g/mol, the loss curves were very similar. This result indicates that the polymers had similar distribution of relaxation times and the intermolecular cooperativity remained minimally affected in the molecular weight range examined. Finally, the dispersion curves were broader in the two side-chain functionalized PP-PEPO systems than the undoped PP-PEPO, again indicating stronger intermolecular coupling between the covalently bonded chromophores and polymer.

Acknowledgment

We gratefully acknowledge the Office of Naval Research and the Department of Education (CYSF) for supporting this work. We wish to thank Duane B. Priddy, Jr., Daniel J. Riley, and J. E. McGrath (Department of Chemistry, Virginia Polytechnic Institute and State University) for providing the polyarylene ether polymers. We also would like to thank R. P. Andres (School of Chemical Engineering, Purdue University) for the use of the evaporator.

References

1. Singer, K. D.; Sohn, J. E.; Lalama, S. J. *Appl. Phys. Lett.* **1986,** *49,* 248–250.
2. Singer, K. D.; Kuzyk, M. G.; Holland, W. R.; Sohn, J. E.; Lalama, S. J. *Appl. Phys. Lett.* **1988,** *53,* 1800–1802.
3. Hampsch, H. L.; Yang, J.; Wong, G. K.; Torkelson, J. M. *Macromolecules* **1988,** *21,* 526–528.
4. Hampsch, H. L.; Yang, J.; Wong, G. K.; Torkelson, J. M. *Macromolecules* **1990,** *23,* 3640–3647.
5. Eich, M.; Sen, A.; Looser, H.; Bjorklund, G. C.; Swalen, J. D.; Twieg, R.; Yoon, D. Y. *J. Appl. Phys.* **1989,** *66,* 2559–2567.
6. Kohler, W.; Robello, D. R.; Dao, P. T.; Willand, C. S.; Williams, D. J. *J. Chem. Phys.* **1990,** *93,* 9157–9166.
7. Ghebremichael, F.; Kuzyk, M. G.; Dirk, C. W. *Nonlinear Opt.* **1993,** *6,* 123–133.
8. Dhinojwala, A.; Wong, G. K.; Torkelson, J. M. *Macromolecules* **1993,** *26,* 5943–5953.
9. Dhinojwala, A.; Wong, G. K.; Torkelson, J. M. *J. Chem. Phys.* **1994,** *100,* 6046–6054.
10. Liu, L. Y.; Ramkrishna, D.; Lackritz, H. S. *Macromolecules* **1994,** *27,* 5987–5999.
11. Fu, C. Y. S.; Priddy, D. B., Jr.; Lyle, G. D.; McGrath, J. E.; Lackritz, H. S. *Proc. SPIE–Int. Opt. Eng.* **1994,** *2285,* 153–163.
12. Prasad, P. N.; Williams, D. J. *Introduction to Nonlinear Optical Effects in Molecules and Polymers;* John Wiley & Sons: New York, 1991.
13. Rellick, G. S.; Runt, J. *J. Polym. Sci.: Polym. Phys. Ed.* **1986,** *24,* 313–324.
14. Lei, D.; Runt, J.; Safari, A.; Newnham, R. E. *Macromolecules* **1987,** *20,* 1797–1801.
15. Ngai, K. L.; Roland, C. M. *Macromolecules* **1993,** *26,* 6824–6830.
16. McCrum, N. G.; Read, B. E.; Williams, G. *Anelastic and Dielectric Effects in Polymeric Solids;* John Wiley & Sons: London, 1967.
17. Pochan, J. M.; Pai, D. M. In *Plastics Polymer Science and Technology;* Baijal, M. D., Ed.; Wiley Interscience: New York, 1982; pp 341–394.
18. Havriliak, S.; Negami, S. *J. Polym. Sci., Polym. Symp.* **1966,** *14,* 99–117.
19. Debye, P. *Polar Molecules;* Chemical Catalogue: New York, 1929.
20. Cole, K. S.; Cole, R. H. *J. Chem. Phys.* **1941,** *9,* 341–351.
21. Davidson, D. W.; Cole, R. H. *J. Chem. Phys.* **1951,** *19,* 1484–1490.
22. Schonhals, A.; Schlosser, E. *Colloid Polym. Sci.* **1989,** *267,* 125–132.
23. Schlosser, E.; Schonhals, A. *Colloid Polym. Sci.* **1989,** *267,* 133–138.
24. Bloembergen, N. *Nonlinear Optics;* W. A. Benjamin: New York, 1965.
25. Shen, Y. R. *The Principles of Nonlinear Optics;* John Wiley & Sons: New York, 1984.

26. *Nonlinear Optical Properties of Organic and Polymeric Materials;* Williams, D. J., Ed.; ACS Symposium Series 233; American Chemical Society: Washington, DC, 1983.
27. *Nonlinear Optical Properties of Organic Molecules and Crystals;* Chemla, D. S.; Zyss, J., Eds.; Academic: Orlando, FL, 1987; Vols. 1 and 2.
28. Zyss, J. *J. Non-Cryst. Solids* **1982**, *47*, 211–226.
29. Zyss, J. *J. Mol. Eng.* **1985**, *1*, 25–45.
30. Singer, K. D.; Kuzyk, M. G.; Sohn, J. E. *J. Opt. Soc. Am. B* **1987**, *4*, 968–976.
31. Boyd, G. T.; Francis, C. V.; Trend, J. E.; Ender, D. A. *J. Opt. Soc. Am. B.* **1991**, *8*, 887–894.
32. Onsager, L. *J. Am. Chem. Soc.* **1936**, *58*, 1486–1493.
33. Meredith, G. R.; VanDusen, J. G.; Williams, D. J. *Macromolecules* **1982**, *15*, 1385–1389.
34. *Electrets;* Sessler, G. M., Ed.; Springer-Verlag: Berlin, Germany, 1980.
35. Hampsch, H. L.; Torkelson, J. M.; Bethke, S. J.; Grubb, S. G. *J. Appl. Phys.* **1990**, *67*, 1037.
36. Giacometti, J. A.; Oliveira, O. N., Jr. *IEEE Trans. Electri. Insul.* **1992**, *27*, 924.
37. Haber, K. S.; Ostrowski, M. H.; O'Sickey, M. J.; Lackritz, H. S. *Mat. Res. Soc. Symp. Proc.* **1994**, *328*, 547.
38. Ghebremichael, F.; Lackritz, H. S. *J. Appl. Phys.*, submitted.
39. Singer, K. D.; King, L. A. *J. Appl. Phys.* **1991**, *70*, 3251–3255.
40. Ren, W.; Bauer, S.; Yilmaz, S.; Wirges, W.; Gerhard-Multhaupt, R. *J. Appl. Phys.* **1994**, *75*, 7211–7219.
41. Eich, M.; Looser, H.; Yoon, D. Y.; Twieg, R.; Bjorklund, G.; Baumert, J. C. *J. Opt. Soc. Am. B* **1989**, *6*, 1590–1597.
42. Williams, M. L.; Landel, R. F.; Ferry, J. D. *J. Am. Chem. Soc.* **1955**, *77*, 3701–3714.
43. Stahelin, M.; Burland, D. M.; Ebert, M.; Miller, R. D.; Smith, B. A.; Tweig, R. J.; Volksen, W.; Walsh, C. A. *Appl. Phys. Lett.* **1992**, *61*, 1626–1628.
44. Walsh, C. A.; Burland, D. M.; Lee, V. Y.; Miller, R. D.; Smith, B. A.; Tweig, R. J.; Volksen, W. *Macromolecules* **1993**, *26*, 3720–3722.
45. Chen, J. I.; Marturunkakul, S.; Li, L.; Jeng, R. J.; Kumar, J.; Tripathy, S. K. *Macromolecules* **1993**, *26*, 7379–7381.
46. Schen, M. A.; Mopsik, F. I. *Proc. SPIE Int. Symp.* **1991**, *1560*, 315–325.
47. Kohler, W.; Robello, D. R.; Willand, C. S.; Williams, D. J. *Macromolecules* **1991**, *24*, 4589–4599.
48. Wu, J. W.; Valley, J. F.; Ermer, S.; Binkley, E. S.; Kenney, J. T.; Lytel, R. *Appl. Phys. Lett.* **1991**, 59, 2213–2215.
49. Valley, J. F.; Wu, J. W.; Ermer, S.; Stiller, M.; Binkley, E. S.; Kenney, J. T.; Lipscomb, G. F.; Lytel, R. *Appl. Phys. Lett.* **1992**, *60*, 160–162.
50. Ermer, S.; Valley, J. F.; Lytel, R.; Lipscomb, G. F.; Van Eck, T. E.; Girton, D. G. *Appl. Phys. Lett.* **1992**, *61*, 2272–2274.
51. Dhinojwala, A.; Wong, G. K.; Torkelson, J. M. *J. Opt. Soc. Am. B* **1994**, *11*, 1549–1554.
52. Hooker, J. C.; Torkelson, J. M. *Macromolecules* **1995**, *28*, 7683–7692.
53. Aklonis, J. J.; MacKnight, W. J. In *Introduction to Polymer Viscoelasticity*, 2nd ed.; John Wiley and Sons: New York, 1983; pp 189–211.
54. Hampsch, H. L.; Torkelson, J. M.; Bethke, S. J.; Grubb, S. G. *J. Appl. Phys.* **1990**, *67*, 1037–1041.
55. Giacometti, J. A.; Oliveira, O. N., Jr. *IEEE Trans. Electri. Insul.* **1992**, *27*, 924–943.
56. Priddy, D. B., Jr.; Fu, C. Y. S.; Pickering, T. L.; Lackritz, H. S.; McGrath, J. E. *Mater. Res. Soc. Symp. Proc.* **1993**, *328*, 589–594.
57. Ngai, K. L.; Rendell, R. W.; Yee, A. F. *Macromolecules* **1988**, *21*, 3396–3401.

15

Broadband Dielectric Spectroscopy on Collective and Molecular Dynamics in Ferroelectric Liquid Crystals

F. Kremer

Broadband dielectric spectroscopy (10^{-2}Hz to 10^{10}Hz) is employed to study the collective and molecular dynamics in (low molecular weight and polymeric) ferroelectric liquid crystals. Two collective loss processes are observed for frequencies smaller than 1 MHz: the Goldstone mode, which is assigned to fluctuations (respectively modulations) of the phase of the helical superstructure; and the soft-mode, which corresponds to fluctuations of the amplitude of the ferroelectric helix. In the frequency range from 10^6Hz to 10^{10}Hz one relaxation process is observed, the β-relaxation. This process is assigned to fluctuations of the chiral mesogene around its long molecular axis. At the phase transition S_A to S_C^* the relaxation peak of the β-relaxation does not split or broaden, and the temperature dependence of its relaxation rate does not show any deviation from an Arrhenius-like behavior. Its dielectric strength does not decline. These experiments prove that the β-process is a local relaxation, which is not involved in the collective dynamics taking place for frequencies below 1 MHz.

Chiral mesogenic molecules can be readily incorporated into a variety of different polymeric architectures (e.g., side-chain polymers, main-chain polymers, combined main-chain–side-group systems, functionalized co-polymers, and cross-linked systems). This enables one to tailor materials according to the current technological requirements.

Because liquid-crystalline compounds have ferroelectric properties in chiral tilted smectic phases (e.g., S_C^* phase) (1), many efforts have been made to understand and optimize their ferroelectric properties (high saturation polarization, short switching times, etc.). Besides a lot of work in the field of synthetics and in structure determination, the molecular and collective dynamics of low molecular weight ferroelectric liquid crystals (FLC) have been investigated by several techniques, such as NMR spectroscopy (2), light scattering (3), and dielectric spectroscopy (4–12), and are by now well understood. In 1984 the first ferroelectric liquid crystalline polymers were synthesized (13). These materials combine the properties of low molecular weight FLC with the typical properties of polymers (14). Further on the liquid crystalline phases were stabilized by the polymer main chain leading to much broader mesophases than in the analogous low molecular weight compounds. The combination of both, the ferroelectric and the polymeric properties, leads to a variety of new materials, such as functionalized copolymers with photochromic and chiral side groups (15) or cross-linked ferroelectric liquid crystalline rubbers (16). Materials with a tailor-made profile of properties can be realized.

In this chapter low molecular weight and polymeric ferroelectric liquid crystals are analyzed and compared with respect to their collective and molecular dynamics. It turns out that both dynamics are qualitatively comparable as long as the chiral mesogenes are sterically decoupled from the polymeric chain.

Experimental

To cover the frequency regime from 10^{-2} to 10^9 Hz, three different measurement systems (17) were combined (Figure 1):

- a frequency response analyzer Solartron Schlumberger FRA1260, which was supplemented with a high impedance preamplifier of variable gain (10^{-2} to 10^6 Hz)
- Hewlett-Packard Impedance analyzer HP4192 A (10^2 to 10^7 Hz)
- Hewlett Packard coaxial reflectometer HP4191 A (10^6 to 10^9 Hz)

The applied ac-measurement field was 1 V for all measurements if not indicated otherwise. To orient the mesogenes in bookshelf geometry (18), custom-made rubbed polyimide-coated metal electrodes (diameter, 5 mm; spacing, 20 μm) were employed. These capacitors enabled us to measure

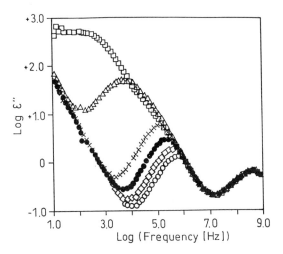

Figure 1. Goldstone mode, soft mode, and β-relaxation of Sample 1 at the phase transition S_A/S_C^*: *(○) 331.3 K, (◇) 330.4 K, (●) 329.4 K, (×) 328.5 K, (△) 327.5 K, and (□) 326.5 K. The* S_A/S_C^* *phase transition is at 327.0 K. The sample was oriented in bookshelf geometry. (Reproduced with permission from reference 6. Copyright 1994 Springer-Verlag.)*

from dc up to 1 GHz in one capacitor arrangement. The phase sequence was determined with differential scanning calorimetry (Perkin Elmer DSC-2 C, DSC), X-ray measurements (Siemens TT 500 Goniometer), and polarizing microscopy. The molecular weights were determined by analytical gel permeation chromatography against polystyrene standards. For the quantitative analysis the dielectric spectra were fitted with a generalized relaxation function according to Havriliak and Negami (*19*)

$$\varepsilon^* = \varepsilon_\infty + \frac{\varepsilon_s - \varepsilon_\infty}{[1 + (i\omega\tau)^\alpha]^\gamma} \tag{1}$$

where i is $\sqrt{-1}$, τ is the mean relaxation time, and ω is the angular frequency of the outer electric field. ε_∞ and ε_s are the real parts of the dielectric function at frequencies $\omega\tau \gg 1$ and $\omega\tau \ll 1$, respectively. The constants α and γ describe the symmetric and asymmetric broadening, respectively. At frequencies below 10^4 Hz an additional conductivity contribution was observed. It was fitted with the power law

$$\varepsilon'' = \frac{\sigma_0}{\varepsilon_0} \omega^{-1} \tag{2}$$

where σ_0 is the conductivity and ε_0 is the permittivity of the free space.

The following structures were investigated. Sample 1 (*4, 6*) has the following values: S_X, 49; S_C^*, 55; S_A, 65.

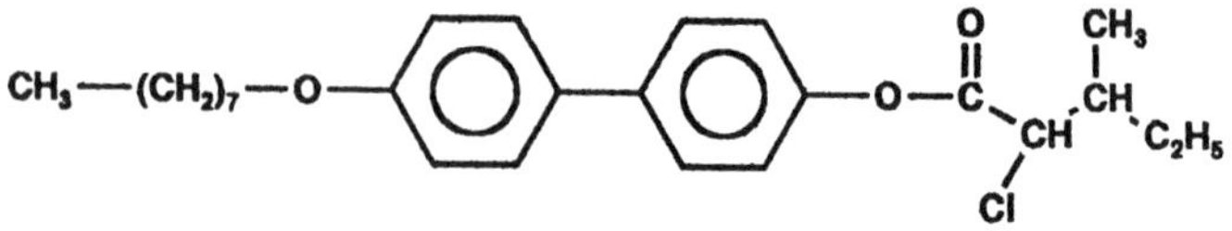

Sample 1.

Sample 2 (*20*) has the following values: c, 8; S_C^*, 63; S_A, 104.

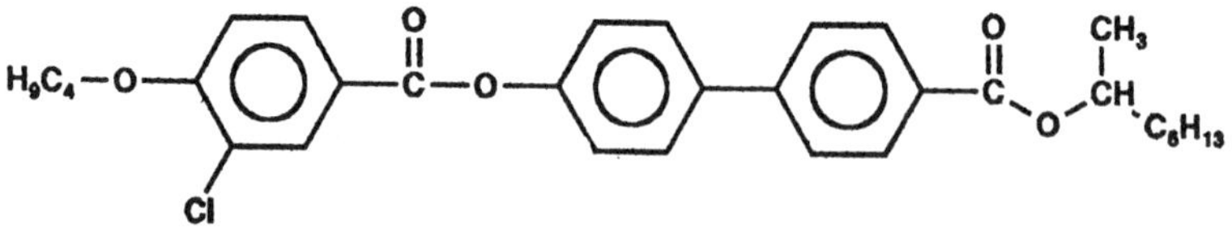

Sample 2.

Sample 3 (*21*) has the following values: c, 32; S_C^*, 68; n^+, 68.

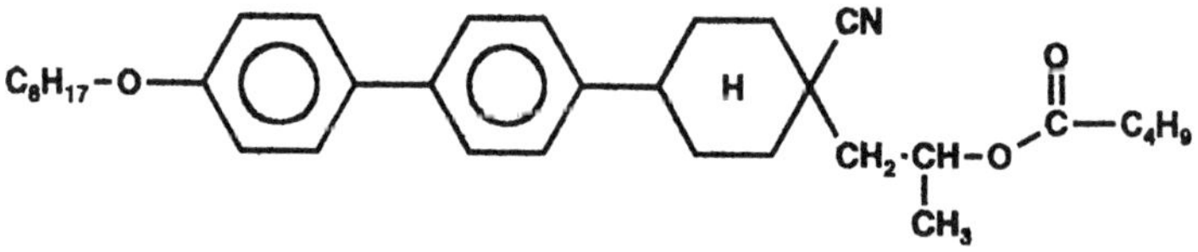

Sample 3.

For values of Sample 4 (*12*), *see* Table I.

Table I. Parameters for Structure 4

Sample	x	y	M_w	M_n	P_s (nC/cm^2)	*Phase Sequence (°C)*
4a	3.2	1	23,000	14,000	104	g, 0; S_x, 46; S_C^*, 98; S_A, 144 i
4b	1.1	1	53,000	27,000	130	g, 13; S_x, 54; S_C^*, 124; S_A, 168 i
4c	0.5		51,000	29,000	161	g, 20; S_x, 72; S_C^*, 138; S_A, 182 i
4d	0	1	28,000	13,000	211	g, 21; S_x, 57; S_C^*, 161; S_A, 183 i

NOTE: M_w and M_n are weight- and number-average molecular weights, respectively.

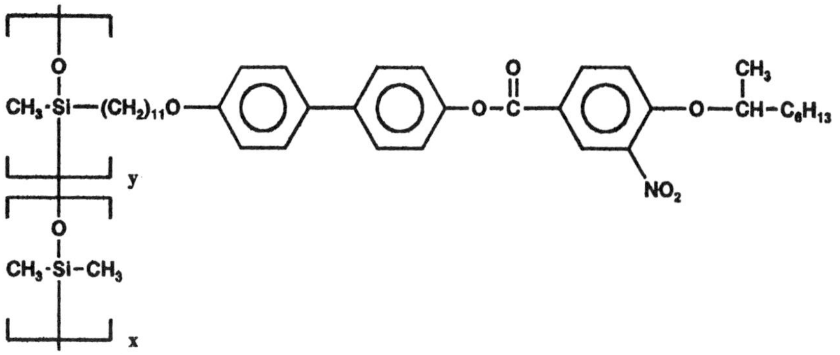

Sample 4.

Results and Discussion

Low Molecular Weight Ferroelectric Liquid Crystals

In the frequency range between 10 Hz and 10^{10} Hz, three dielectric loss processes (*11*) are observed (Figure 1):

- Goldstone mode
- soft mode
- β-relaxation

The Goldstone mode (*4–6*) is restricted to the smectic S_C^* phase only (Figure 2). On applying an additional electric dc field of increasing strength, one can continuously suppress the dielectric strength of this process (Figure 3).

The Goldstone mode is not a conventional relaxation process: It shows a strong dc field and ac field dependence of its dielectric loss, it has an extremely high dielectric strength, and it shows an asymmetry of the relaxation time distribution with the long wing on the low-frequency side (in contrast to all empirical relaxation functions such as Cole–Davidson, Havriliak–Negami, and Fuoss–Kirkwood). The molecular assignment of the Goldstone mode is the thermal fluctuation or the field-induced modulation, respectively of the phase angle of the helical superstructure, which is connected to the different polarization vectors of the smectic layers.

Analyzing the nature of the dielectrically measured Goldstone mode in more detail, one observes besides the dc-field dependence a pronounced ac-field dependence (Figure 4). By keeping the difference in strength of the ac- and dc-field fixed, all experimental data collapse into one master

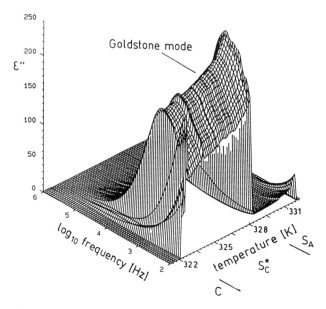

Figure 2. Dielectric loss ε″ versus frequency and temperature for Sample 1. Sample thickness, 10 μm; ac-field strength, 1 kV/cm; dc-field strength, 0 V/cm. (Reproduced with permission from reference 6. Copyright 1994 Springer-Verlag.)

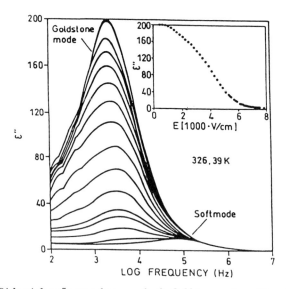

Figure 3. Dielectric loss ε″ versus frequency for the Goldstone mode and the soft mode with different external bias fields for Sample 1 ranging from 0 to 8000 V/cm: ac-field strength, 1 kV/cm; temperature, 326.39 ± 0.02 K. Inset: field dependence of the dielectric loss ε″ at a frequency of 2.51 kHz and a temperature of 326.39 ± 0.02 K: sample thickness, 10 μm. (Reproduced with permission from reference 6. Copyright 1994 Springer-Verlag.)

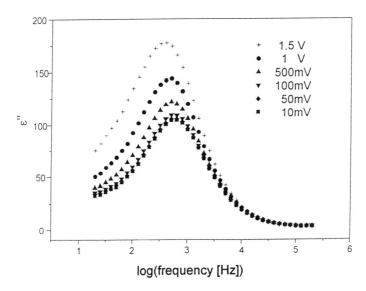

Figure 4. The Goldstone mode of Sample 2 for different strengths of the measuring ac-field: sample thickness, 4 μm.

curve (Figure 5). This result indicates that the Goldstone mode has to be described as a driven anharmonic oscillator.

By superimposing a dc-bias field to the ac field (*see* inset in Figure 3), a second loss process becomes dielectrically observable. Close to the $S_A–S_C^*$ transition its frequency of maximum loss shows a linear dependence on temperature and exhibits a critical slowing down with a minimal value at the phase transition temperature (Figure 6). For the inverse of the dielectric decrement a similar temperature dependence is observed. This loss process has, analogous to inorganic ferroelectrics, the character of a soft mode. It is assigned to fluctuations of the amplitude of the helical superstructure (*4, 6, 22, 23*).

In the microwave range ($10^6–10^9$ Hz) only one relaxation process is observed (Figure 7): the β-relaxation, which is assigned to the libration (hindered rotation) of the mesogene around its long molecular axis (*8–10a, b*). For a quantitative analysis the data were fitted with the generalized relaxation function of Havriliak and Negami (eq 1). The temperature dependence of the relaxation rate $1/\tau$ is Arrhenius-like and shows no decline (Figure 7b) at the $S_A–S_C^*$ phase transition. The dielectric strength $\Delta\varepsilon = \varepsilon_s - \varepsilon_\infty$ of the β-relaxation does not decrease at the phase transition. Instead it increases slightly (Figure 7b). This result is caused by the change of the aspect angle of the lateral dipole moments with respect to the outer electrical field at the phase transition. At the phase transition no additional broadening of the relaxation time distribution function is observed. The

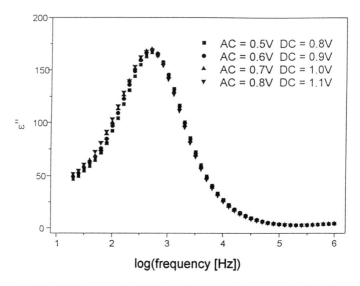

Figure 5. The Goldstone mode of Sample 2 for balancing ac and dc field: sample thickness, 4 μm.

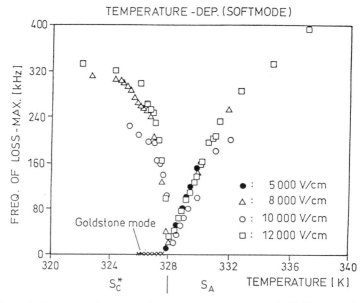

Figure 6. Frequency of maximum loss versus temperature for the Goldstone mode and the soft mode in Sample 1 at different levels of the external electric dc-bias field. The Goldstone mode (×) was measured by using an external bias field of 5000 V/cm. (Reproduced with permission from reference 6. Copyright 1994 Springer-Verlag.)

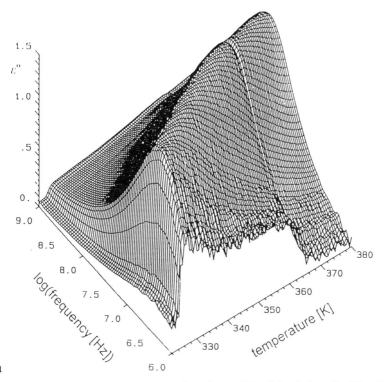

a

Figure 7a. Frequency and temperature dependence of the dielectric loss ε" of Sample 2. The sample was aligned in bookshelf geometry. Sample thickness, 10 μm; diameter of the sample capacitor, 3 mm. (Reproduced with permission from reference 10b. Copyright 1993.)

relaxation rate shows an Arrhenius-like temperature dependence, as expected for a β-relaxation (8–10). No change in the activation energy is found at the S_A–S_C^* phase transition.

The corresponding racemic mixture (Figure 8a) is fully comparable with the ferroelectric sample (Figures 7a and 7b). Also, a quantitative analysis using the Havriliak–Negami equation yields no difference (Figure 8b), neither in the temperature dependence of the relaxation rate nor in the dielectric strength.

The chiral sample deviates not only from the racemic mixture in the distribution of the lateral dipole moments, but also in the helical superstructure that is only built up in the ferroelectric phase. Superimposing a dc-bias field allows unwinding of this helical superstructure. As a result of this unwinding, the component of the lateral dipole moment interacting with the outer field changes. Hence, one expects an influence on the dielectric strength of the β-relaxation by unwinding the helical superstructure with

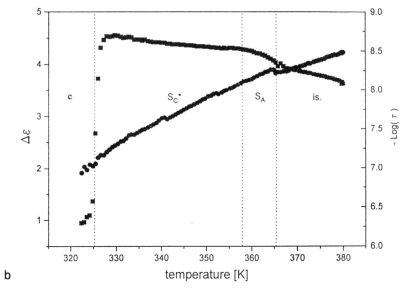

b

temperature [K]

Figure 7b. Temperature dependence of the relaxation time τ (●) and the dielectric strength Δε (■) of the β-relaxation in Sample 2 as deduced from the measurements shown in Figure 8a. (Reproduced with permission from reference 10b. Copyright 1993.)

a dc-bias field. In the nonferroelectric phases there should be no effect of an additional dc-bias field.

To check these assumptions we measured the β-relaxation with an additional superimposed variable dc-bias field of sample (*10b*). The high frequency dynamics of this sample are shown in Figure 9. As expected, it shows also one process and an Arrhenius-type temperature dependence. Superimposing a dc-bias field increases the intensity of the β-relaxation. The effect of the superimposed dc-bias field is fully reversible, which was checked by measuring with both positive and negative dc-bias fields. Furthermore, at the beginning of the measurements at each temperature we measured with 0, 20, 40, 20, and 0 V to check that the effect is not due to a change (improvement) of the alignment of the sample.

The data were fitted with the Havriliak–Negami equation. The dc-bias dependence of the dielectric strength Δε obtained by this fitting procedure is drawn in Figure 10 for various temperatures. It increases with increasing dc fields up to 15 V and then remains constant at higher fields. The increase in the dielectric strength is more pronounced at lower temperatures. This result is due to the increase of the anisotropy of the distribution of the lateral dipole moments (and the saturation polarization) with decreasing temperature. While the relaxation time remains nearly uninfluenced (Figure 11), the shape parameters α and γ show a distinct dependence on

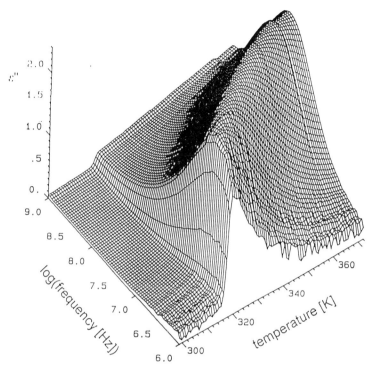

Figure 8a. Frequency and temperature dependence of the dielectric loss ε". Sample is racemic mixture of Sample 2. (Reproduced with permission from reference 10b. Copyright 1993.)

the applied dc-bias field (Figure 11). Although these are only empirical parameters they are related to short- and long-range interactions. The observed change presumably may indicate that the superimposed dc-bias field not only unwinds the helical superstructure, but also changes (improves) the arrangement of the mesogenes within each layer: that is, the anisotropy of the distribution of the lateral dipole moments.

Neither in the nonferroelectric phases nor in the S_C^* phase of a racemic mixture could a dc-bias dependence be observed. The experimental results for the β-relaxation prove that it is a libration (hindered rotational motion) around the long molecular axis of the mesogene. This libration takes place both in the ferroelectric and in the nonferroelectric phases. However, in the nonferroelectric phases (S_A, isotropic) the angular distribution of the lateral dipole moments is isotropic and becomes strongly anisotropic in the ferroelectric S_C^* phase (Figure 12). The anisotropy is induced by the chirality of the mesogene and the tilt. As expected, an Arrhenius-type temperature dependence both in the ferroelectric and in the nonferroelectric phases

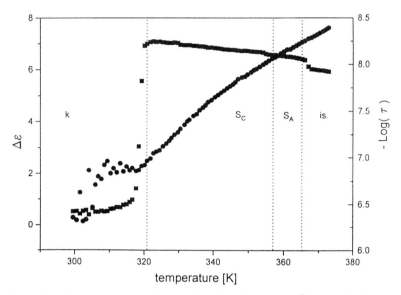

Figure 8b. Temperature dependence of the relaxation time τ (●) and the dielectric strength Δε (■) of the β-relaxation for Sample 2 deduced from the measurements shown in Figure 9a. (Reproduced with permission from reference 10b. Copyright 1993.)

and no difference between the chiral sample and its racemic mixture were found experimentally.

In summary one finds in low molecular weight FLCs three dielectric loss processes, which are quite different in their physical nature. Below 10^6 Hz two highly collective processes, the Goldstone mode and the soft mode, are observed. While the Goldstone mode can be quantitatively described as a driven oscillator, the soft mode corresponds to a fluctuation that exhibits critical behavior when approaching the phase transition $S_A/$ S_C^*. In the frequency regime between 10^6 Hz and 10^{10} Hz one dielectric loss process is observed: the β-relaxation, which is assigned to the libration of the mesogenic groups around their long molecular axis. This process is a local fluctuation that is not directly influenced by the collective rearrangements, which take place at the phase transition S_A/S_C^*. The experimental results lead to a precise picture for the origin of ferroelectricity in chiral liquid crystals.

Polymeric FLCs

In the frequency regime from 10^{-2} to 10^9 Hz three dielectric loss processes (7, 11, 12) can be observed (Figure 13). Two of them (Goldstone and

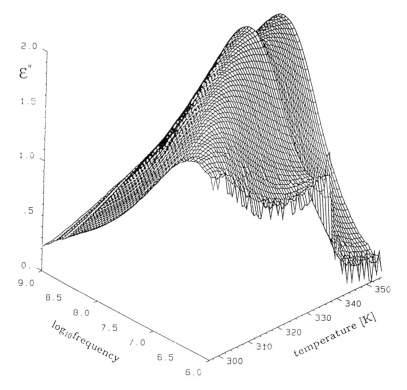

Figure 9. Frequency and temperature dependence of the β-relaxation of Sample 3. (Reproduced with permission from reference 10b. Copyright 1993.)

soft mode) occur at frequencies below 1 MHz. Their huge dielectric losses indicate their collective character. In the frequency regime above 1 MHz one dielectric relaxation is found. Its comparable low dielectric loss underlines the local character of this relaxation process, which is not restricted to the liquid crystalline phase.

Collective Dynamics of FLC Polymers

In the frequency range below 1 MHz the dielectric spectrum is dominated by one very strong relaxation (Goldstone mode), which is restricted to the S_C^* phase. It is assigned to the fluctuation of the phase of the helical super-structure. By comparing this process with the Goldstone mode in the analo-gous low molar mass compound, the relaxation is shifted nearly two decades to lower frequencies. Additionally, in FLCP the Goldstone mode shows a pronounced temperature dependence, whereas in low molar mass FLCs this process usually is only weakly temperature dependent. Superimposing a dc-bias field unwinds the helical structure, thereby continuously suppress-

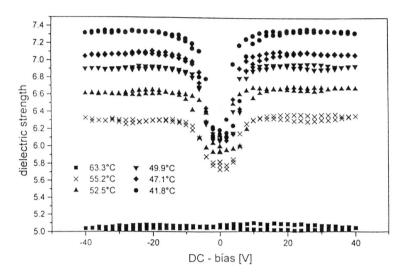

Figure 10. The dc-bias field dependence of the dielectric strength for different temperatures. The increase in $\Delta\varepsilon$ increases with decreasing temperature. (Reproduced with permission from reference 10b. Copyright 1993.)

ing the Goldstone mode with increasing bias field (Figure 14). In the unwound state a second loss process (soft mode) becomes observable. To study the soft-mode behavior more in detail the measurements were repeated with a superimposed dc-bias field (Figure 15). The soft mode, which is assigned to the fluctuation of the tilt, increases in the S_A phase, has a maximum at the phase transition S_A/S_C^*, and decreases in the S_C^* phase. Because the phase transition S_A/S_C^* is of second order, the following theoretical predictions for the soft-mode dynamics near this transition have to be fulfilled:

- the frequency position of the maximum dielectric loss should obey a Curie–Weiss law

- the slope of the plot $\omega(\Delta\varepsilon_{max})$ versus temperature in the S_C^* phase should be two times the slope in the S_A phase

- for the inverse of the dielectric strength a similar Curie–Weiss temperature dependence is predicted

In low molar mass FLCs these predictions are well fulfilled (6). To check the critical dynamics in the polymeric compounds, the experimental data were fitted by use of the generalized relaxation function according to Havriliak and Negami. The inverse of the dielectric strength shows a comparable temperature dependence (Figures 16a and 16b) as predicted,

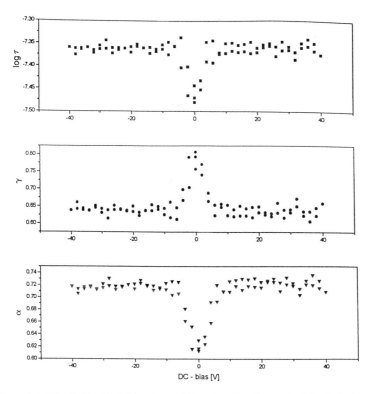

Figure 11. The dc-bias field dependence of the relaxation time and of the shape parameters α and γ. (Reproduced with permission from reference 10b. Copyright 1993.)

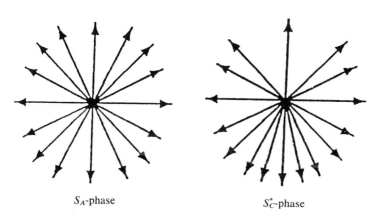

S_A-phase S_C^*-phase

Figure 12. Distribution of the lateral dipole moments in the S_A phase and S_C^* phase, respectively. (Reproduced with permission from reference 10b. Copyright 1993.)

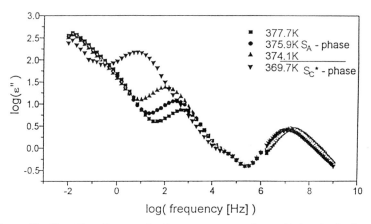

Figure 13. Plot of log(ε'') versus log(frequency) measured over 11 decades for Sample 4. The experimental error is estimated to be not larger than the size of symbols, and the slope between $10^4 Hz$ and $10^5 Hz$ is due to an artifact of the measurement system. (Reproduced with permission from reference 10b. Copyright 1993.)

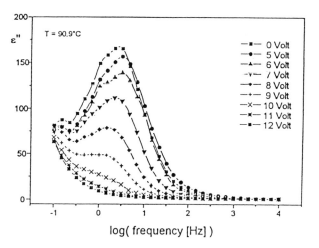

Figure 14. By superimposing a dc-bias field, the Goldstone mode can be suppressed continuously (Sample 4): sample thickness, 10 μm. (Reproduced with permission from reference 12. Copyright 1994.)

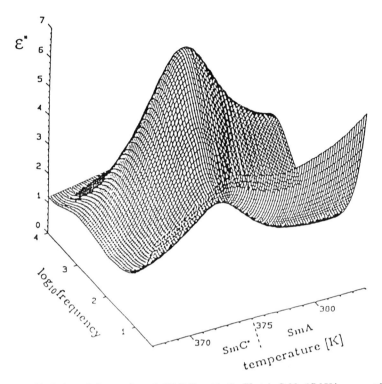

Figure 15. Soft mode in a polymeric FLC (Sample 4): Electric field: 15 kV/cm; sample thickness, 10 μm. (Reproduced with permission from reference 12. Copyright 1994.)

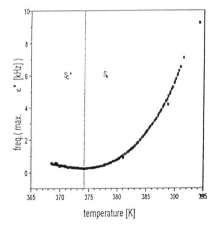

Figure 16a. Frequency of the maximum ε″ versus temperature. (Reproduced with permission from reference 12. Copyright 1994.)

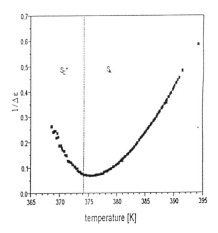

Figure 16b. Inverse dielectric strength versus temperature (Sample 4a): E, 15 kV/cm²; sample thickness, 10 μm. (Reproduced with permission from reference 12. Copyright 1994.)

similar to the low molar mass FLCs. The relaxation frequency in the S_A phase slows down when approaching the transition from high temperatures, but does not speed up when further cooling into the S_C^* phase. In contrast it remains nearly constant. This property may be due to viscosity effects from the polymer main chain or from an insufficient suppression of the Goldstone mode. Additionally, the phase transition occurs not as sharp as for low molar mass compounds, which is a general feature of polymeric FLCs.

In the frequency regime above 1 MHz, one dielectric loss process, the β-relaxation, is observed (Figure 17). It is assigned to the libration of the mesogene around its long molecular axis and has a low intensity, as expected for a local process. The measured loss curves can be well described with the Havriliak–Negami equation (eq 1). The temperature dependence of the mean relaxation time τ and the dielectric strength Δε is shown in Figure 18. The relaxation time shows no discontinuities at the phase transition S_A/S_C^*; especially, no slowing down is observed. By cooling from the isotropic phase and by entering the S_C^* phase, the dielectric strength Δε increases, because the aspect angle of the dipole moment interacting with the outer electric field changes due to the formation of the bookshelf geometry. By using the same argument one can explain the small step at the phase transition S_A/S_C^*. Since alignment in the high frequency measurements is only partially achieved, these steps are not as pronounced as for low molar mass systems.

Diluting the chiral mesogenes in the side groups results in copolymers similar to that of Sample 4. Those samples are easier to align because of

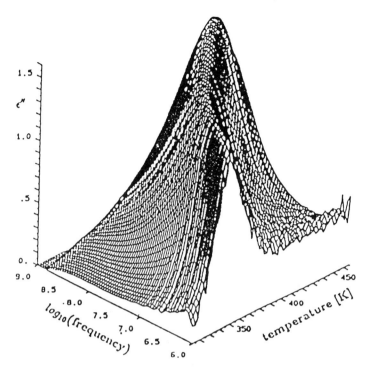

Figure 17. The β-relaxation in a polymeric FLC (Sample 4a): sample thickness, 20 μm. (Reproduced with permission from reference 12. Copyright 1994.)

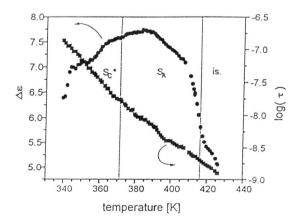

Figure 18. Temperature dependence of the dielectric strength (●) and the logarithm of the relaxation time τ (■) for the β-relaxation of Sample 4a. (Reproduced with permission from reference 12. Copyright 1994.)

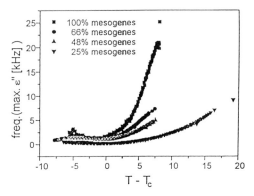

Figure 19. Critical slowing down of the soft-mode relaxation rate for the copolymer with varying content of chiral mesogenes. (Reproduced with permission from reference 12. Copyright 1994.)

the higher flexibility of the main chain. Even for the highest degree of dilution (only 25% mesogenes) a comparable phase sequence occurs. The saturation polarization shows a nearly linear dependence on the mole fraction of mesogenes.

To investigate how this influences the collective dynamics of the samples, we measured the soft-mode behavior of four different polymers. The temperature dependencies of their frequency positions are shown in Figure 19. All samples show the expected critical slowing down. By comparing the

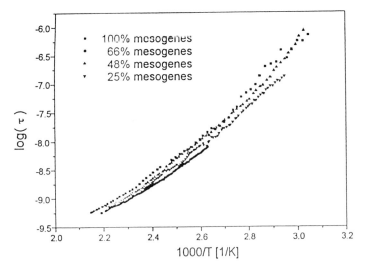

Figure 20. Activation plot of the β-relaxation for the copolymer with varying content of chiral mesogenes. (Reproduced with permission from reference 12. Copyright 1994.)

degree of dilution, the critical slowing down shows a systematic change. The more the systems are diluted the less pronounced is the critical slowing down. The dielectrically observed soft-mode relaxation agrees well with the response times obtained from measurements of the electroclinic effect.

In the high frequency regime (10^6 Hz to 10^9 Hz), where the local dynamics take place, no influence of the dilution can be observed, as one would expect (Figure 20). All the samples show the hindered rotation around the long molecular axis (β-relaxation). Because this process is very local it is not influenced by collectivity.

Conclusions

The collective and molecular dynamics (10^{-2} Hz to 10^{10} Hz) in FLCs are characterized by two collective and one molecular processes. Below 1 MHz the Goldstone and the soft modes are observed. The Goldstone mode is assigned to the thermal fluctuations or the field-induced modulation, respectively of the phase angle of the helical superstructure and can be described quantitatively in its ac- and dc-field dependence as a damped driven oscillator. The soft mode corresponds to fluctuations of the amplitude of the helical superstructure. It has a negligible ac- and dc-field dependence. In the frequency regime from 10^6 to 10^{10} Hz, one relaxation process is observed: the β-relaxation, which corresponds to the libration of the chiral mesogene around its long molecular axis. This process is not directly involved in the molecular rearrangements that take place at the phase transition.

By incorporating chiral mesogenic groups into polymeric architectures (side chain, main chain, combined side-group–main-chain systems, copolymers, networks, etc.), the electric and dielectric properties of ferroelectric liquid crystals can be combined with the viscoelastic properties of polymers. Thus, new materials can be tailored (e.g., photochromic, ferroelectric polymers, single crystal, and ferroelectric liquid crystalline rubbers) with completely new technological perspectives.

Acknowledgments

Support from the German Science Foundation within the framework of the *Sonderforschungsbereich 294 (Teilprojekt B9)* and from the *Fonds der chemischen Industrie e. V.* is gratefully acknowledged.

References

1. Meyer, R. B.; Liebert; L.; Strzlecki, L.; Keller; P. J. *J. Phys. (Paris) Lett.* **1975**, *36*, L69.

2. Yoshizawa, A.; Kikuzaki, H.; Hirai, T.; Yamane; M. *Jpn. Appl. Phys.* **1989**, *28*, 1988.
3. Musevic, J.; Blinc, R.; Zeks, B.; Filipic, C.; Copic, M.; Seppen, A.; Wyder, P.; Levanyuk, M. A. *Phys. Rev. Lett.* **1988**, *60*, 1530.
4. Levstik, A.; Carlsson, T.; Filipic, C.; Levstik, I.; Zeks, B. *Phys. Rev.* **1987**, *A35*, 3527.
5. Pavel, J.; Glogorova, M.; Bawa; S. S. *Ferroelectrics* **1987**, *76*, 221.
6. Kremer, F.; Schönfeld, A. In *Ordering in Macromolecular Systems;* Teramoto, A.; Kobayashi, M.; Novisuje, T., Eds.; Springer-Verlag: Berlin, Germany, 1994.
7. Vallerien, S. U.; Zentel, R.; Kremer, F.; Kapitza, H.; Fischer, E. W. *Makromol. Chem. Rapid Commun.* **1989**, *10*, 333.
8. Vallerien, S. U.; Kremer, F.; Geelhar, T.; Wächtler, A. *Phys. Rev.* **1990**, *A42*, 2482.
9. Kremer, F.; Vallerien, S. U.; Kapitza, H.; Zentel, R.; Fischer, E. W. *Phys. Rev.* **1990**, *A42*, 3667.
10. (a) Schönfeld, A.; Kremer, F.; Vallerien, S. U.; Poths, H.; Zentel, R. *Ferroelectrics* **1991**, *121*, 69. (b) Schönfeld, A.; Kremer; F. *Ber. Bunsenges. Phys. Chem.* **1993**, *97*, 1237.
11. Kremer, F.; Schönfeld, A.; Hofmann, R.; Zentel, R.; Poths, H. *Polym. Adv. Technol.* **1992**, *3*, 249.
12. Schönfeld, A.; Kremer, F.; Poths, H.; Zentel, R. *Mol. Cryst. Liq. Cryst.* **1994**, *254*, 49.
13. Shibaev, V. B.; Koslovsky, M. Z.; Beresnev, L. A.; Blinov, L. M.; Plate, N. A. *Polym. Bull.* **1984**, *12*, 299.
14. Mc Ardle, C. B. *Side Chain Liquid Crystal Polymers;* Blackie: Glasgow, Scotland, 1989
15. Servaty, S.; Kremer, F.; Schönfeld, A.; Zentel, R. *Zeitschrift Phys. Chem.* **1995**, *190*, 73.
16. Vallerien, S. U.; Kremer, F.; Kapitza, H.; Zentel, R.; Poths, H. *Makromol. Chem. Rapid Commun.* **1990**, *11*, 593.
17. Kremer, F.; Boese, D.; Meier, G.; Fischer, E. W. *Prog. Coll. Polym. Sci.* **1989**, *80*, 129.
18. Clark, N. A.; Lagerwall, S. T. In *Ferroelectric Liquid Crystals—Principles, Properties and Applications;* Goodby, J. W.; et al., Eds.; Gordon and Breach: Langhorne, PA, 1991.
19. Havriliak, S.; Negami, S. *J. Polym. Sci.* **1966**, *C14*, 89.
20. The sample was obtained from F. Gouda, Chelmas University, Göteborg, Sweden.
21. The sample was obtained from Merck, Darmstadt, Germany.
22. Blinc, R.; Zeks, B. *Phys. Rev.* **1978**, *A18*, 740.
23. Carlsson, T.; Zeks, B.; Filipic, C.; Levstik, A. *Phys. Rev.* **1990**, *A42*, 877.

Index

The letter "f" following a page number indicates a figure.

Copy editing: Scott Hofmann-Reardon
Production: Margaret J. Brown and Kimberly N. Lassair
Acquisition: Anne Wilson
Indexing: Colleen Stamm
Cover design: Linda M. Mattingly

Typeset by Maryland Composition Company, Inc., Glen Burnie, MD
Printed and Bound by Maple Press, York, PA

Bestsellers from ACS Books

The ACS Style Guide: A Manual for Authors and Editors
Edited by Janet S. Dodd
264 pp; clothbound ISBN 0–8412–0917–0; paperback ISBN 0–8412–0943–X

Writing the Laboratory Notebook
By Howard M. Kanare
145 pp; clothbound ISBN 0–8412–0906–5; paperback ISBN 0–8412–0933–2

Career Transitions for Chemists
By Dorothy P. Rodmann, Donald D. Bly, Frederick H. Owens, and Anne-Claire Anderson
240 pp; clothbound ISBN 0–8412–3052–8; paperback ISBN 0–8412–3038–2

Chemical Activities (student and teacher editions)
By Christie L. Borgford and Lee R. Summerlin
330 pp; spiralbound ISBN 0–8412–1417–4; teacher edition, ISBN 0–8412–1416–6

Chemical Demonstrations: A Sourcebook for Teachers, Volumes 1 and 2, Second Edition
Volume 1 by Lee R. Summerlin and James L. Ealy, Jr.
198 pp; spiralbound ISBN 0–8412–1481–6
Volume 2 by Lee R. Summerlin, Christie L. Borgford, and Julie B. Ealy
234 pp; spiralbound ISBN 0–8412–1535–9

From Caveman to Chemist
By Hugh W. Salzberg
300 pp; clothbound ISBN 0–8412–1786–6; paperback ISBN 0–8412–1787–4

The Internet: A Guide for Chemists
Edited by Steven M. Bachrach
360 pp; clothbound ISBN 0–8412–3223–7; paperback ISBN 0–8412–3224–5

Laboratory Waste Management: A Guidebook
ACS Task Force on Laboratory Waste Management
250 pp; clothbound ISBN 0–8412–2735–7; paperback ISBN 0–8412–2849–3

Reagent Chemicals, Eighth Edition
700 pp; clothbound ISBN 0–8412–2502–8

Good Laboratory Practice Standards: Applications for Field and Laboratory Studies
Edited by Willa Y. Garner, Maureen S. Barge, and James P. Ussary
571 pp; clothbound ISBN 0–8412–2192–8

For further information contact:

American Chemical Society
1155 Sixteenth Street, NW ◆ Washington, DC 20036
Telephone 800–227–9919 ◆ 202–776–8100 (outside U.S.)

The ACS Publications Catalog is available on the Internet at
http://pubs.acs.org/books

Highlights from ACS Books

Desk Reference of Functional Polymers: Syntheses and Applications
Reza Arshady, Editor
832 pages, clothbound, ISBN 0–8412–3469–8

Chemical Engineering for Chemists
Richard G. Griskey
352 pages, clothbound, ISBN 0–8412–2215–0

Controlled Drug Delivery: Challenges and Strategies
Kinam Park, Editor
720 pages, clothbound, ISBN 0–8412–3470–1

Chemistry Today and Tomorrow: The Central, Useful, and Creative Science
Ronald Breslow
144 pages, paperbound, ISBN 0–8412–3460–4

Eilhard Mitscherlich: Prince of Prussian Chemistry
Hans-Werner Schutt
Co-published with the Chemical Heritage Foundation
256 pages, clothbound, ISBN 0–8412–3345–4

Chiral Separations: Applications and Technology
Satinder Ahuja, Editor
368 pages, clothbound, ISBN 0–8412–3407–8

Molecular Diversity and Combinatorial Chemistry: Libraries and Drug Discovery
Irwin M. Chaiken and Kim D. Janda, Editors
336 pages, clothbound, ISBN 0–8412–3450–7

A Lifetime of Synergy with Theory and Experiment
Andrew Streitwieser, Jr.
320 pages, clothbound, ISBN 0–8412–1836–6

Chemical Research Faculties, An International Directory
1,300 pages, clothbound, ISBN 0–8412–3301–2

For further information contact:

American Chemical Society
Customer Service and Sales
1155 Sixteenth Street, NW
Washington, DC 20036

Telephone 800–227–9919
202–776–8100 (outside U.S.)

The ACS Publications Catalog is available on the Internet at
http://pubs.acs.org/books

1—MONTH